T0189714

Research and Development in Intelligent Systems XXXIII

Incorporating Applications and Innovations in Intelligent Systems XXIV

Max Bramer · Miltos Petridis
Editors

Research and Development in Intelligent Systems XXXIII

Incorporating Applications and Innovations in Intelligent Systems XXIV

Proceedings of AI-2016, The Thirty-Sixth SGAI International Conference on Innovative Techniques and Applications of Artificial Intelligence

Editors
Max Bramer
School of Computing
University of Portsmouth
Portsmouth
UK

Miltos Petridis
School of Computing, Engineering and Mathematics
University of Brighton
Brighton
UK

ISBN 978-3-319-47174-7 ISBN 978-3-319-47175-4 (eBook)
DOI 10.1007/978-3-319-47175-4

Library of Congress Control Number: 2016954594

Printed on acid-free paper

This Springer imprint is published by Springer Nature
The registered company is Springer International Publishing AG
The registered company address is: Gewerbestrasse 11, 6330 Cham, Switzerland

Programme Chairs' Introduction

This volume comprises the refereed papers presented at AI-2016, the Thirty-sixth SGAI International Conference on Innovative Techniques and Applications of Artificial Intelligence, held in Cambridge in December 2016 in both the technical and the application streams. The conference was organised by SGAI, the British Computer Society Specialist Group on Artificial Intelligence.

The technical papers included present new and innovative developments in the field, divided into sections on Knowledge Discovery and Data Mining, Sentiment Analysis and Recommendation, Machine Learning, AI Techniques, and Natural Language Processing. This year's Donald Michie Memorial Award for the best-refereed technical paper was won by a paper entitled "Harnessing Background Knowledge for E-learning Recommendation" by B. Mbipom, S. Craw and S. Massie (Robert Gordon University, Aberdeen, UK).

The application papers included present innovative applications of AI techniques in a number of subject domains. This year, the papers are divided into sections on legal liability, medicine and finance, telecoms and e-Learning, and genetic algorithms in action. This year's Rob Milne Memorial Award for the best-refereed application paper was won by a paper entitled "A Genetic Algorithm Based Approach for the Simultaneous Optimisation of Workforce Skill Sets and Team Allocation" by A.J. Starkey and H. Hagras (University of Essex, UK), S. Shakya and G. Owusu (British Telecom, UK).

The volume also includes the text of short papers presented as posters at the conference.

On behalf of the conference organising committee, we would like to thank all those who contributed to the organisation of this year's programme, in particular the programme committee members, the executive programme committees and our administrators Mandy Bauer and Bryony Bramer.

Max Bramer, Technical Programme Chair, AI-2016

Miltos Petridis, Application Programme Chair, AI-2016

Acknowledgements/Committees

AI-2016 Conference Committee

Prof. Max Bramer, University of Portsmouth (Conference Chair)
Prof. Max Bramer, University of Portsmouth (Technical Programme Chair)
Prof. Miltos Petridis, University of Brighton (Application Programme Chair)
Dr. Jixin Ma, University of Greenwich (Deputy Application Programme Chair)
Prof. Adrian Hopgood, University of Liege, Belgium (Workshop Organiser)
Rosemary Gilligan (Treasurer)
Dr. Nirmalie Wiratunga, Robert Gordon University, Aberdeen (Poster Session Organiser)
Andrew Lea, Primary Key Associates Ltd. (AI Open Mic and Panel Session Organiser)
Dr. Frederic Stahl, University of Reading (Publicity Organiser)
Dr. Giovanna Martinez, Nottingham Trent University and Christo Fogelberg Palantir Technologies (FAIRS 2016)
Prof. Miltos Petridis, University of Brighton and Prof. Thomas Roth-Berghofer University of West London (UK CBR Organisers)
Mandy Bauer, BCS (Conference Administrator)
Bryony Bramer, (Paper Administrator)

Technical Executive Programme Committee

Prof. Max Bramer, University of Portsmouth (Chair)
Prof. Frans Coenen, University of Liverpool
Dr. John Kingston, University of Brighton
Prof. Dan Neagu, University of Bradford
Prof. Thomas Roth-Berghofer, University of West London
Dr. Nirmalie Wiratunga, Robert Gordon University, Aberdeen

Applications Executive Programme Committee

Prof. Miltos Petridis, University of Brighton (Chair)
Mr. Richard Ellis, Helyx SIS Ltd.
Ms. Rosemary Gilligan, University of Hertfordshire
Dr. Jixin Ma, University of Greenwich (Vice-Chair)
Dr. Richard Wheeler, University of Edinburgh

Technical Programme Committee

Andreas Albrecht (Middlesex University)
Abdallah Arioua (IATE INRA France)
Raed Batbooti (University of Swansea UK (PhD Student), University of Basra (Lecturer))
Lluís Belanche (Universitat Politecnica de Catalunya, Barcelona, Catalonia, Spain)
Yaxin Bi (Ulster University, UK)
Mirko Boettcher (University of Magdeburg; Germany)
Max Bramer (University of Portsmouth)
Krysia Broda (Imperial College; University of London)
Ken Brown (University College Cork)
Charlene Cassar (De Montfort University UK)
Frans Coenen (University of Liverpool)
Ireneusz Czarnowski (Gdynia Maritime University; Poland)
Nicolas Durand (Aix-Marseille University)
Frank Eichinger (CTS EVENTIM AG & Co. KGaA, Hamburg, Germany)
Mohamed Gaber (Robert Gordon University, Aberdeen, UK)
Hossein Ghodrati Noushahr (De Montfort University, Leicester, UK)
Wael Hamdan (MIMOS Berhad., Kuala Lumpur, Malaysia)
Peter Hampton (Ulster University, UK)
Nadim Haque (Capgemini)
Chris Headleand (University of Lincoln, UK)
Arjen Hommersom (Open University, The Netherlands)
Adrian Hopgood (University of Liège, Belgium)
John Kingston (University of Brighton)
Carmen Klaussner (Trinity College Dublin Ireland)
Ivan Koychev (University of Sofia)
Thien Le (University of Reading)
Nicole Lee (University of Hong Kong)
Anne Liret (British Telecom France)
Fernando Lopes (LNEG-National Research Institute; Portugal)
Stephen Matthews (Newcastle University)
Silja Meyer-Nieberg (Universitat der Bundeswehr Munchen Germany)

Roberto Micalizio (Universita' di Torino)
Daniel Neagu (University of Bradford)
Lars Nolle (Jade University of Applied Sciences; Germany)
Joanna Isabelle Olszewska (University of Gloucestershire UK)
Dan O'Leary (University of Southern California)
Juan Jose Rodriguez (University of Burgos)
Thomas Roth-Berghofer (University of West London)
Fernando Saenz-Perez (Universidad Complutense de Madrid)
Miguel A. Salido (Universidad Politecnica de Valencia)
Rainer Schmidt (University Medicine of Rostock; Germany)
Frederic Stahl (University of Reading)
Simon Thompson (BT Innovate)
Jon Timmis (University of York)
M.R.C. van Dongen (University College Cork)
Martin Wheatman (Yagadi Ltd.)
Graham Winstanley (University of Brighton)
Nirmalie Wiratunga (Robert Gordon University)

Application Programme Committee

Hatem Ahriz (Robert Gordon University)
Tony Allen (Nottingham Trent University)
Ines Arana (Robert Gordon University)
Mercedes Arguello Casteleiro (University of Manchester)
Ken Brown (University College Cork)
Sarah Jane Delany (Dublin Institute of Technology)
Richard Ellis (Helyx SIS Ltd.)
Roger Evans (University of Brighton)
Andrew Fish (University of Brighton)
Rosemary Gilligan (University of Hertfordshire)
John Gordon (AKRI Ltd.)
Chris Hinde (Loughborough University)
Adrian Hopgood (University of Liege, Belgium)
Stelios Kapetanakis (University of Brghton)
Alice Kerly
Jixin Ma (University of Greenwich)
Lars Nolle (Jade University of Applied Sciences)
Miltos Petridis (University of Brighton)
Miguel A. Salido (Universidad Politecnica de Valencia)
Roger Tait (University of Cambridge)
Richard Wheeler (Edinburgh Scientific)

Contents

Research and Development in Intelligent Systems XXXIII

Best Technical Paper

Harnessing Background Knowledge for E-Learning Recommendation 3
Blessing Mbipom, Susan Craw and Stewart Massie

Knowledge Discovery and Data Mining

Category-Driven Association Rule Mining 21
Zina M. Ibrahim, Honghan Wu, Robbie Mallah and Richard J.B. Dobson

A Comparative Study of SAT-Based Itemsets Mining 37
Imen Ouled Dlala, Said Jabbour, Lakhdar Sais and Boutheina Ben Yaghlane

Mining Frequent Movement Patterns in Large Networks: A Parallel Approach Using Shapes 53
Mohammed Al-Zeyadi, Frans Coenen and Alexei Lisitsa

Sentiment Analysis and Recommendation

Emotion-Corpus Guided Lexicons for Sentiment Analysis on Twitter 71
Anil Bandhakavi, Nirmalie Wiratunga, Stewart Massie and P. Deepak

Context-Aware Sentiment Detection from Ratings 87
Yichao Lu, Ruihai Dong and Barry Smyth

Recommending with Higher-Order Factorization Machines 103
Julian Knoll

Machine Learning

Multitask Learning for Text Classification with Deep Neural Networks . . . 119
Hossein Ghodrati Noushahr and Samad Ahmadi

An Investigation on Online Versus Batch Learning in Predicting User Behaviour . . . 135
Nikolay Burlutskiy, Miltos Petridis, Andrew Fish, Alexey Chernov and Nour Ali

A Neural Network Test of the Expert Attractor Hypothesis: Chaos Theory Accounts for Individual Variance in Learning . . . 151
P. Chassy

AI Techniques

A Fast Algorithm to Estimate the Square Root of Probability Density Function . . . 165
Xia Hong and Junbin Gao

3Dana: Path Planning on 3D Surfaces . . . 177
Pablo Muñoz, María D. R-Moreno and Bonifacio Castaño

Natural Language Processing

Covert Implementations of the Turing Test: A More Level Playing Field? . . . 195
D.J.H. Burden, M. Savin-Baden and R. Bhakta

Context-Dependent Pattern Simplification by Extracting Context-Free Floating Qualifiers . . . 209
M.J. Wheatman

Short Papers

Experiments with High Performance Genetic Programming for Classification Problems . . . 221
Darren M. Chitty

Towards Expressive Modular Rule Induction for Numerical Attributes . . . 229
Manal Almutairi, Frederic Stahl, Mathew Jennings, Thien Le and Max Bramer

OPEN: New Path-Planning Algorithm for Real-World Complex Environment . . . 237
J.I. Olszewska and J. Toman

Encoding Medication Episodes for Adverse Drug Event Prediction 245
Honghan Wu, Zina M. Ibrahim, Ehtesham Iqbal
and Richard J.B. Dobson

Applications and Innovations in Intelligent Systems XXIV

Best Application Paper

A Genetic Algorithm Based Approach for the Simultaneous Optimisation of Workforce Skill Sets and Team Allocation 253
A.J. Starkey, H. Hagras, S. Shakya and G. Owusu

Legal Liability, Medicine and Finance

Artificial Intelligence and Legal Liability 269
J.K.C. Kingston

SELFBACK—Activity Recognition for Self-management of Low Back Pain 281
Sadiq Sani, Nirmalie Wiratunga, Stewart Massie
and Kay Cooper

Automated Sequence Tagging: Applications in Financial Hybrid Systems 295
Peter Hampton, Hui Wang, William Blackburn and Zhiwei Lin

Telecoms and E-Learning

A Method of Rule Induction for Predicting and Describing Future Alarms in a Telecommunication Network 309
Chris Wrench, Frederic Stahl, Thien Le, Giuseppe Di Fatta,
Vidhyalakshmi Karthikeyan and Detlef Nauck

Towards Keystroke Continuous Authentication Using Time Series Analytics 325
Abdullah Alshehri, Frans Coenen and Danushka Bollegala

Genetic Algorithms in Action

EEuGene: Employing Electroencephalograph Signals in the Rating Strategy of a Hardware-Based Interactive Genetic Algorithm 343
C. James-Reynolds and E. Currie

Spice Model Generation from EM Simulation Data Using Integer Coded Genetic Algorithms 355
Jens Werner and Lars Nolle

Short Papers

Dendritic Cells for Behaviour Detection in Immersive Virtual Reality Training 371
N.M.Y. Lee, H.Y.K. Lau, R.H.K. Wong, W.W.L. Tam and L.K.Y. Chan

Interactive Evolutionary Generative Art 377
L. Hernandez Mengesha and C.J. James-Reynolds

Incorporating Emotion and Personality-Based Analysis in User-Centered Modelling 383
Mohamed Mostafa, Tom Crick, Ana C. Calderon and Giles Oatley

An Industrial Application of Data Mining Techniques to Enhance the Effectiveness of On-Line Advertising 391
Maria Diapouli, Miltos Petridis, Roger Evans and Stelios Kapetanakis

Research and Development in Intelligent Systems XXXIII

Best Technical Paper

Harnessing Background Knowledge for E-Learning Recommendation

Blessing Mbipom, Susan Craw and Stewart Massie

Abstract The growing availability of good quality, learning-focused content on the Web makes it an excellent source of resources for e-learning systems. However, learners can find it hard to retrieve material well-aligned with their learning goals because of the difficulty in assembling effective keyword searches due to both an inherent lack of domain knowledge, and the unfamiliar vocabulary often employed by domain experts. We take a step towards bridging this semantic gap by introducing a novel method that automatically creates custom background knowledge in the form of a set of rich concepts related to the selected learning domain. Further, we develop a hybrid approach that allows the background knowledge to influence retrieval in the recommendation of new learning materials by leveraging the vocabulary associated with our discovered concepts in the representation process. We evaluate the effectiveness of our approach on a dataset of Machine Learning and Data Mining papers and show it to outperform the benchmark methods.

Keywords Knowledge Discovery · Recommender Systems · eLearning Systems · Text Mining

1 Introduction

There is currently a large amount of e-learning resources available to learners on the Web. However, learners have insufficient knowledge of the learning domain, and are not able to craft good queries to convey what they wish to learn. So, learners are

---

B. Mbipom · S. Craw · S. Massie (✉)
School of Computing Science and Digital Media, Robert Gordon University, Aberdeen, UK
e-mail: s.massie@rgu.ac.uk

B. Mbipom
e-mail: b.e.mbipom@rgu.ac.uk

S. Craw
e-mail: s.craw@rgu.ac.uk

M. Bramer and M. Petridis (eds.), *Research and Development in Intelligent Systems XXXIII*, DOI 10.1007/978-3-319-47175-4_1

often discouraged by the time spent in finding and assembling relevant resources to meet their learning goals [5]. E-learning recommendation offers a possible solution.

E-learning recommendation typically involves a learner query, as an input; a collection of learning resources from which to make recommendations; and selected resources recommended to the learner, as an output. Recommendation differs from an information retrieval task because with the latter, the user requires some understanding of the domain in order to ask and receive useful results, but in e-learning, learners do not know enough about the domain. Furthermore, the e-learning resources are often unstructured text, and so are not easily indexed for retrieval [11]. This challenge highlights the need to develop suitable representations for learning resources in order to facilitate their retrieval.

We propose the creation of background knowledge that can be exploited for problem-solving. In building our method, we leverage the knowledge of instructors contained in eBooks as a guide to identify the important domain topics. This knowledge is enriched with information from an encyclopedia source and the output is used to build our background knowledge. DeepQA applies a similar approach to reason on unstructured medical reports in order to improve diagnosis [9]. We demonstrate the techniques in Machine Learning and Data Mining, however the techniques we describe can be applied to other learning domains.

In this paper, we build background knowledge that can be employed in e-learning environments for creating representations that capture the important concepts within learning resources in order to support the recommendation of resources. Our method can also be employed for query expansion and refinement. This would allow learners' queries to be represented using the vocabulary of the domain with the aim of improving retrieval. Alternatively, our approach can enable learners to browse available resources through a guided view of the learning domain.

We make two contributions: firstly, the creation of background knowledge for an e-learning domain. We describe how we take advantage of the knowledge of experts contained in eBooks to build a knowledge-rich representation that is used to enhance recommendation. Secondly, we present a method of harnessing background knowledge to augment the representation of learning resources in order to improve the recommendation of resources. Our results confirm that incorporating background knowledge into the representation improves e-learning recommendation.

This paper is organised as follows: Sect. 2 presents related methods used for representing text; Sect. 3 describes how we exploit information sources to build our background knowledge; Sect. 4 discusses our methods in harnessing a knowledge-rich representation to influence e-learning recommendation; and Sect. 5 presents our evaluation. We conclude in Sect. 6 with insights to further ways of exploiting our background knowledge.

2 Related Work

Finding relevant resources to recommend to learners is a challenge because the resources are often unstructured text, and so are not appropriately indexed to support the effective retrieval of relevant materials. Developing suitable representations to improve the retrieval of resources is a challenging task in e-learning environments [8], because the resources do not have a pre-defined set of features by which they can be indexed. So, e-learning recommendation requires a representation that captures the domain-specific vocabulary contained in learning resources. Two broad approaches are often used to address the challenge of text representation: corpus-based methods such as topic models [6], and structured representations such as those that take advantage of ontologies [4].

Corpus-based methods involve the use of statistical models to identify topics from a corpus. The identified topics are often keywords [2] or phrases [7, 18]. Coenen et al. showed that using a combination of keywords and phrases was better than using only keywords [7]. Topics can be extracted from different text sources such as learning resources [20], metadata [3], and Wikipedia [14]. One drawback of the corpus-based approach is that, it is dependent on the document collection used, so the topics produced may not be representative of the domain. A good coverage of relevant topics is required when generating topics for an e-learning domain, in order to offer recommendations that meet learners' queries which can be varied.

Structured representations capture the relationships between important concepts in a domain. This often entails using an existing ontology [11, 15], or creating a new one [12]. Although ontologies are designed to have a good coverage of their domains, the output is still dependent on the view of its builders, and because of handcrafting, existing ontologies cannot easily be adapted to new domains. E-learning is dynamic because new resources are becoming available regularly, and so using fixed ontologies limits the potential to incorporate new content.

A suitable representation for e-learning resources should have a good coverage of relevant topics from the domain. So, the approach in this paper draws insight from the corpus-based methods and structured representations. We leverage on a structured corpus of teaching materials as a guide for identifying important topics within an e-learning domain. Thesc topics are a combination of keywords and phrases as recommended in [7]. The identified topics are enriched with discovered text from Wikipedia, and this extends the coverage and richness of our representation.

3 Background Knowledge Representation

Background knowledge refers to information about a domain that is useful for general understanding and problem-solving [21]. We attempt to capture background knowledge as a set of domain concepts, each representing an important topic in the domain. For example, in a learning domain, such as Machine Learning, you would

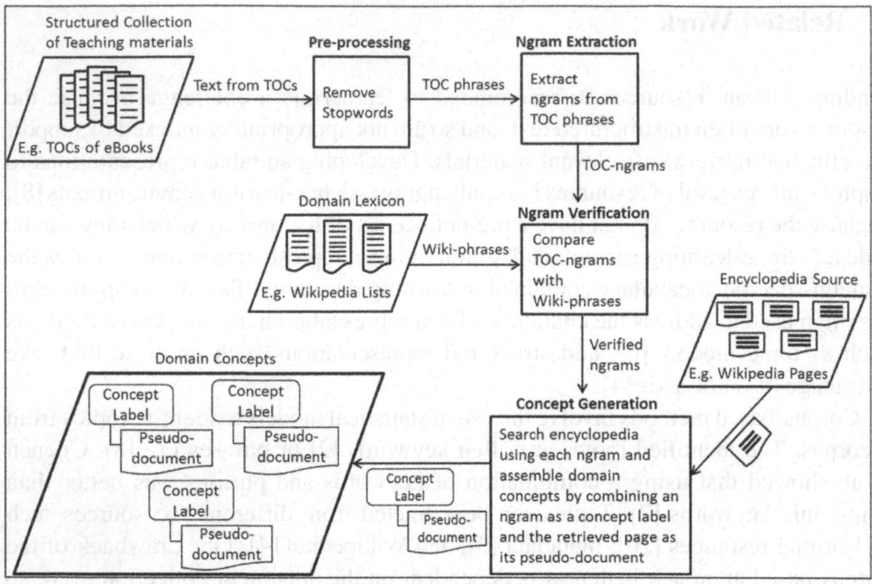

Fig. 1 An overview of the background knowledge creation process

find topics such as Classification, Clustering and Regression. Each of these topics would be represented by a concept, in the form of a concept label and a pseudo-document which describes the concept. The concepts can then be used to underpin the representation of e-learning resources.

The process involved in discovering our set of concepts is illustrated in Fig. 1. Domain knowledge sources are required as an input to the process, and we use a structured collection of teaching materials and an encyclopedia source. We automatically extract ngrams from our structured collection to provide a set of potential concept labels, and then we use a domain lexicon to validate the extracted ngrams in order to ensure that the ngrams are also being used in another information source. The encyclopedia provides candidate pages that become the concept label and discovered text for the ngrams. The output from this process is a set of concepts, each comprising a label and an associated pseudo-document. The knowledge extraction process is discussed in more detail in the following sections.

3.1 Knowledge Sources

Two knowledge sources are used as initial inputs for discovering concept labels. A structured collection of teaching materials provides a source for extracting important topics identified by teaching experts in the domain, while a domain lexicon provides a broader but more detailed coverage of the relevant topics in the domain. The lexicon is

Table 1 Summary of eBooks used

Book title and author	Cites
Machine learning; Mitchell	264
Introduction to machine learning; Alpaydin	2621
Machine learning a probabilistic perspective; Murphy	1059
Introduction to machine learning; Kodratoff	159
Gaussian processes for machine learning; Rasmussen and Williams	5365
Introduction to machine learning; Smola and Vishwanathan	38
Machine learning, neural and statistical classification; Michie, Spiegelhalter, and Taylor	2899
Introduction to machine learning; Nilsson	155
A first encounter with machine learning; Welling	7
Bayesian reasoning and machine learning; Barber	271
Foundations of machine learning; Mohri, Rostamizadeh, and Talwalkar	197
Data mining-practical machine learning tools and techniques; Witten and Frank	27098
Data mining concepts models and techniques; Gorunescu	244
Web data mining; Liu	1596
An introduction to data mining; Larose	1371
Data mining concepts and techniques; Han and Kamber	22856
Introduction to data mining; Tan, Steinbach, and Kumar	6887
Principles of data mining; Bramer	402
Introduction to data mining for the life sciences; Sullivan	15
Data mining concepts methods and applications; Yin, Kaku, Tang, and Zhu	23

used to verify that the concept labels identified from the teaching materials are directly relevant. Thereafter, an encyclopedia source, such as Wikipedia pages, is searched and provides the relevant text to form a pseudo-document for each verified concept label. The final output from this process is our set of concepts each comprising a concept label and an associated pseudo-document.

Our approach is demonstrated with learning resources from Machine Learning and Data Mining. We use eBooks as our collection of teaching materials; a summary of the books used is shown in Table 1. Two Google Scholar queries: "Introduction to data mining textbook" and "Introduction to machine learning textbook" guided the selection process, and 20 eBooks that meet all of the following 3 criteria were chosen. Firstly, the book should be about the domain. Secondly, there should be Google Scholar citations for the book. Thirdly, the book should be accessible. We use the Tables-of-Contents (TOCs) of the books as our structured knowledge source.

We use Wikipedia to create our domain lexicon because it contains articles for many learning domains [17], and the contributions of many people [19], so this provides the coverage we need in our lexicon. The lexicon is generated from 2 Wikipedia sources. First, the phrases in the *contents* and *overview* sections of the chosen domain are extracted to form a topic list. In addition, a list containing the titles of articles related to the domain is added to the topic list to assemble our lexicon. Overall, our domain lexicon consists of a set of 664 Wiki-phrases.

3.2 Generating Potential Domain Concept Labels

In the first stage of the process, the text from the TOCs is pre-processed. We remove characters such as punctuation, symbols, and numbers from the TOCs, so that only words are used for generating concept labels. After this, we remove 2 sets of stopwords. First, a standard English stopwords list,[1] which allows us to remove common words and still retain a good set of words for generating our concept labels. Our second stopwords are an additional set of words which we refer to as TOC-stopwords. It contains: structural words, such as *chapter* and *appendix*, which relate to the structure of the TOCs; roman numerals, such as *xxiv* and *xxxv*, which are used to indicate the sections in a TOC; and words, such as *introduction* and *conclusion*, which describe parts of a learning material and are generic across domains.

We do not use stemming because we found it harmful during pre-processing. When searching an encyclopedia source with the stemmed form of words, relevant results would not be returned. In addition, we intend to use the background knowledge for query refinement, so stemmed words would not be helpful.

The output from pre-processing is a set of TOC phrases. In the next stage, we apply ngram extraction to the TOC phrases to generate all 1–3 grams across the entire set of TOC phrases. The output from this process are TOC-ngrams containing a set of 2038 unigrams, 5405 bigrams and 6133 trigrams, which are used as the potential domain concept labels. Many irrelevant ngrams are generated from the TOCs because we have simply selected all 1–3 grams.

3.3 Verifying Concept Labels Using Domain Lexicon

The TOC-ngrams are first verified using a domain lexicon to confirm which of the ngrams are relevant for the domain. Our domain lexicon contains a set of 664 Wiki-phrases, each of which is pre-processed by removing non-alphanumeric characters. The 84 % of the Wiki-phrases that are 1–3 grams are used for verification. The comparison of TOC-ngrams with the domain lexicon identifies the potential domain concept labels that are actually being used to describe aspects of the chosen domain in Wikipedia. During verification, ngrams referring directly to the title of the domain, e.g. *machine learning* and *data mining*, are not included because our aim is to generate concept labels that describe the topics within the domain. In addition, we intend to build pseudo-documents describing the identified labels, and so using the title of

[1] http://snowball.tartarus.org/algorithms/english/stop.txt.

the domain would refer to the entire domain rather than specific topics. Overall, a set of 17 unigrams, 58 bigrams and 15 trigrams are verified as potential concept labels. Bigrams yield the highest number of ngrams, which indicates that bigrams are particularly useful for describing topics in this domain.

3.4 Domain Concept Generation

Our domain concepts are generated after a second verification step is applied to the ngrams returned from the previous stage. Each ngram is retained as a concept label if all of 3 criteria are met. Firstly, if a Wikipedia page describing the ngram exists. Secondly, if the text describing the ngram is not contained as part of the page describing another ngram. Thirdly, if the ngram is not a synonym of another ngram. For the third criteria, if two ngrams are synonyms, the ngram with the higher frequency is retained as a concept label while its synonym is retained as part of the extracted text. For example, 2 ngrams *cluster analysis* and *clustering* are regarded as synonyms in Wikipedia, so the text associated with them is the same. The label *clustering* is retained as the concept label because it occurs more frequently in the TOCs, and its synonym, *cluster analysis* is contained as part of the discovered text.

The concept labels are used to search Wikipedia pages in order to generate a domain concept. The search returns discovered text that forms a pseudo-document which includes the concept label. The concept label and pseudo-document pair make up a domain concept. Overall, 73 domain concepts are generated. Each pseudo-document is pre-processed using standard techniques such as removal of English stopwords and Porter stemming [13]. The terms from the pseudo-documents form the concept vocabulary that is now used to represent learning resources.

4 Representation Using Background Knowledge

Our background knowledge contains a rich representation of the learning domain and by harnessing this knowledge for representing learning resources, we expect to retrieve documents based on the domain concepts that they contain. The domain concepts are designed to be effective for e-learning, because they are assembled from the TOCs of teaching materials [1]. This section presents two approaches which have been developed by employing our background knowledge in the representation of learning resources.

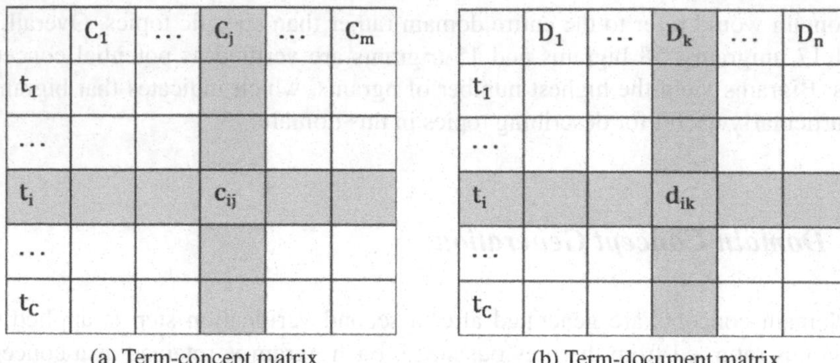

(a) Term-concept matrix (b) Term-document matrix

Fig. 2 Term matrices for concepts and documents

	D1	…	Dk	…	Dm	…	Dn
C1							
…							
Cj			yjk		yjm		
…							
Cm							

(a) Concept-document matrix representation

	D1	…	Dm	…	Dn
D1	1				
…					
Dk			zkm		
…					
Dn					1

(b) Document-document similarity

Fig. 3 Document representation and similarity using the ConceptBased approach

4.1 The ConceptBased *Approach*

Representing documents with the concept vocabulary allows retrieval to focus on the concepts contained in the documents. Figures 2 and 3 illustrate the ConceptBased method. Firstly, in Fig. 2, the concept vocabulary, $t_1 \ldots t_c$, from the pseudo-documents of concepts, $C_1 \ldots C_m$, is used to create a term-concept matrix and a term-document matrix using TF-IDF weighting [16]. In Fig. 2a, c_{ij} is the TF-IDF of term t_i in concept C_j, while Fig. 2b shows d_{ik} which is the TF-IDF of t_i in D_k.

Next, documents D_1 to D_n are represented with respect to concepts by computing the cosine similarity of the term vectors for concepts and documents. The output is the concept-document matrix shown in Fig. 3a, where y_{jk} is the cosine similarity of the vertical shaded term vectors for C_j and D_k from Fig. 2a, b respectively. Finally, the document similarity is generated by computing the cosine similarity of concept-vectors for documents. Figure 3b shows z_{km}, which is the cosine similarity of the concept-vectors for D_k and D_m from Fig. 3a.

	C_1	...	C_j	...	C_m	D_1	...	D_j	D_k	...	D_n
t_1											
...											
t_i			p_{ij}						q_{ik}		
...											
t_T											

(a) Hybrid term-document matrix representation

	D_1	...	D_k	...	D_n
D_1	1				
...					
D_j			r_{jk}		
...					
D_n					1

(b) Hybrid document similarity

Fig. 4 Representation and similarity of documents using the HYBRID approach

The CONCEPTBASED approach uses the document representation and similarity in Fig. 3. By using the CONCEPTBASED approach we expect to retrieve documents that are similar based on the concepts they contain, and this is obtained from the document-document similarity in Fig. 3b. A standard approach of representing documents would be to define the document similarity based on the term document matrix in Fig. 2b, but this exploits the concept vocabulary only. However, in our approach, we put more emphasis on the domain concepts, so we use the concept document matrix in Fig. 3a, to underpin the similarity between documents.

4.2 *The* HYBRID *Approach*

The HYBRID approach exploits the relative distribution of the vocabulary in the concept and document spaces to augment the representation of learning resources with a bigger, but focused, vocabulary. So the TF-IDF weight of a term changes depending on its relative frequency in both spaces.

First, the concepts, C_1 to C_m and the documents we wish to represent, D_1 to D_n, are merged to form a corpus. Next, a term-document matrix with TF-IDF weighting is created using all the terms, t_1 to t_T from the vocabulary of the merged corpus as shown in Fig. 4a. For example, entry q_{ik} is the TF-IDF weight of term t_i in D_k. If t_i has a lower relative frequency in the concept space compared to the document space, then the weight q_{ik} is boosted. So, distinctive terms from the concept space will get boosted. Although the overlap of terms from both spaces are useful for altering the term weights, it is valuable to keep all the terms from the document space because this gives us a richer vocabulary. The shaded term vectors for D_1 to D_n in Fig. 4a form a term-document matrix for documents whose term weights have been influenced by the presence of terms from the concept vocabulary.

Finally, the document similarity in Fig. 4b, is generated by computing the cosine similarity between the augmented term vectors for D_1 to D_n. Entry r_{jk} is the cosine similarity of the term vectors for documents, D_j and D_k from Fig. 4a. The HYBRID method exploits the vocabulary in the concept and document spaces to enhance the retrieval of documents.

5 Evaluation

Our methods are evaluated on a collection of topic-labeled learning resources by simulating an e-learning recommendation task. We use a collection from Microsoft Academic Search (MAS) [10], in which the author-defined keywords associated with each paper identifies the topics they contain. The keywords represent what relevance would mean in an e-learning domain and we exploit them for judging document relevance. The papers from MAS act as our e-learning resources, and using a query-by-example scenario, we evaluate the relevance of a retrieved document by considering the overlap of keywords with the query. This evaluation approach allows us to measure the ability of the proposed methods to identify relevant learning resources. The methods compared are:

- CONCEPTBASED represents documents using the domain concepts (Sect. 4.1).
- HYBRID augments the document representation using a contribution of term weights from the concept vocabulary (Sect. 4.2).
- BOW is a standard Information Retrieval method where documents are represented using the terms from the document space only with TF-IDF weighting.

For each of the 3 methods, the documents are first pre-processed by removing English stopwords and applying Porter stemming. Then, after representation, a similarity-based retrieval is employed using cosine similarity.

5.1 Evaluation Method

Evaluations using human evaluators are expensive, so we take advantage of the author-defined keywords for judging the relevance of a document. The keywords are used to define an overlap metric. Given a query document Q with a set of keywords K_Q, and a retrieved document R with its set of keywords K_R, the relevance of R to Q is based on the overlap of K_R with K_Q. The overlap is computed as:

$$Overlap(K_Q, K_R) = \frac{|K_Q \cap K_R|}{min\left(|K_Q|, |K_R|\right)} \tag{1}$$

We decide if a retrieval is relevant by setting an overlap threshold, and if the overlap between K_Q and K_R meets the threshold, then K_R is considered to be relevant.

Our dataset contains 217 Machine Learning and Data Mining papers, each being 2–32 pages in length. A distribution of the keywords per document is shown in Fig. 5,

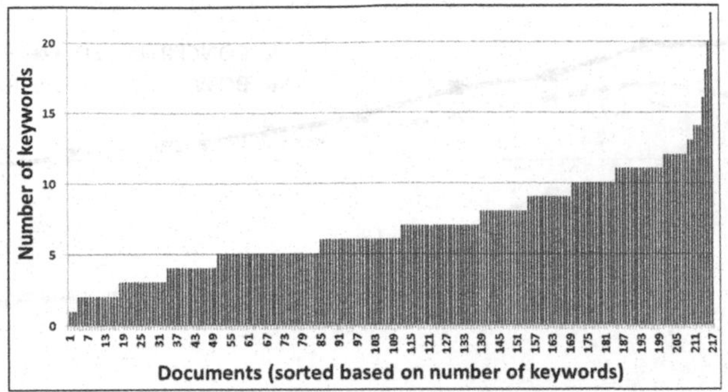

Fig. 5 Number of keywords per Microsoft document

Table 2 Overlap of document-keywords and the proportion of data

Overlap coefficient	Number of pairs	Proportion of data (%)	Overlap threshold
Zero	20251 (86 %)	10	0.14
Non-zero	3185 (14 %)	5	0.25
		1	0.5

where the documents are sorted based on the number of keywords they contain. There are 903 unique keywords, and 1497 keywords in total.

A summary of the overlap scores for all document pairs is shown in Table 2. There are 23436 entries for the 217 document pairs, and 20251 are zero, meaning that there is no overlap in 86 % of the data. So only 14 % of the data have an overlap of keywords, indicating that the distribution of keyword overlap is skewed. There are 10 % of document pairs with overlap scores that are ≥ 0.14, while 5 % are ≥ 0.25.

The higher the overlap threshold, the more demanding is the relevance test. We use 0.14 and 0.25 as thresholds, thus avoiding the extreme values that would allow either very many or few of the documents to be considered as relevant. Our interest is in the topmost documents retrieved, because we want our top recommendations to be relevant. We use precision@n to determine the proportion of relevant documents retrieved:

$$Precision@n = \frac{|retrievedDocuments \cap relevantDocuments|}{n} \quad (2)$$

where, n is the number of documents retrieved each time, *retrievedDocuments* is the set of documents retrieved, and *relevantDocuments* are those documents that are considered to be relevant i.e. have an overlap that is greater than the threshold.

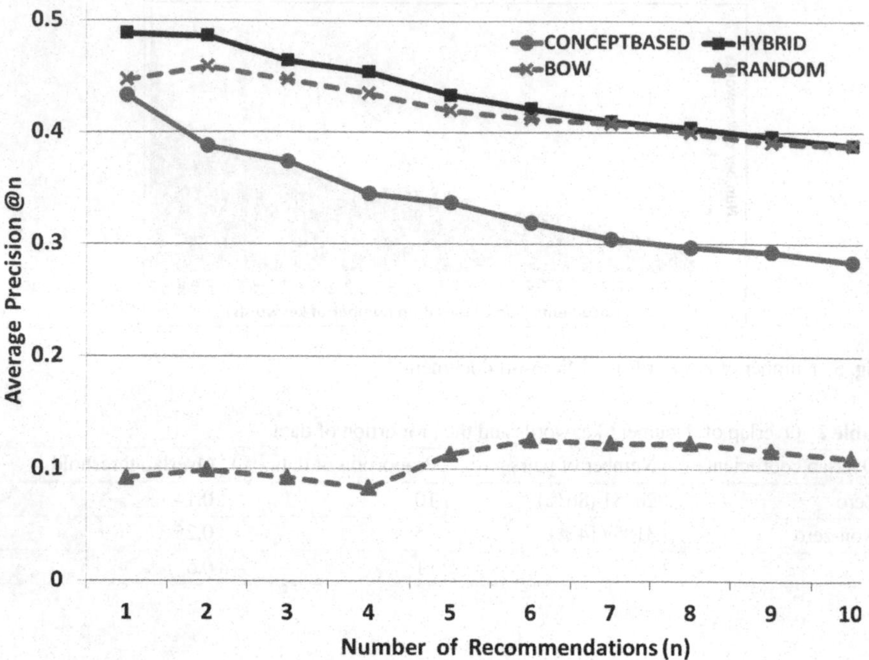

Fig. 6 Precision of the methods at an overlap threshold of 0.14

5.2 *Results and Discussion*

The methods are evaluated using a leave-one-out retrieval. In Fig. 6, the number of recommendations (n) is shown on the x-axis and the average precision@n is shown on the y-axis. RANDOM (▲) has been included to give an idea of the relationship between the threshold and the precision values. RANDOM results are consistent with the relationship between the threshold and the proportion of data in Table 2.

Overall, HYBRID (■) performs better than BOW (×) and CONCEPTBASED (•), showing that augmenting the representation of documents with a bigger, but focused vocabulary, as done in HYBRID, is a better way of harnessing our background knowledge. BOW also performs well because the document vocabulary is large, but the vocabulary used in CONCEPTBASED may be too limited. All the graphs fall as the number of recommendations, *n* increases. This is expected because the earlier retrievals are more likely to be relevant. However, the overlap of HYBRID and BOW at higher values of n may be because the documents retrieved by both methods are drawn from the same neighbourhoods.

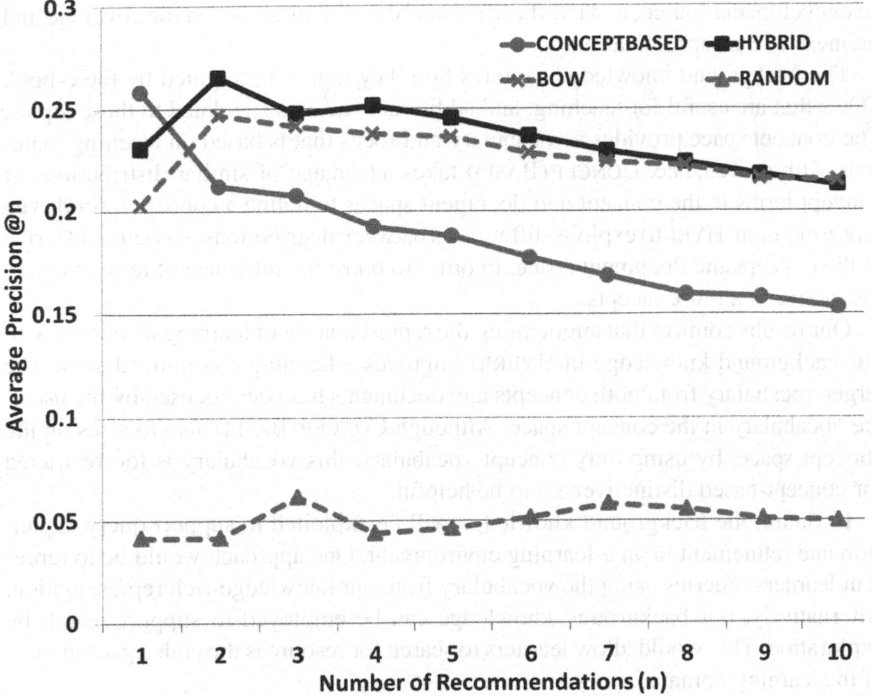

Fig. 7 Precision of the methods at an overlap threshold of 0.25

The relative performance at a threshold of 0.25 in Fig. 7, is similar to the performance at 0.14. However, at this more challenging threshold, HYBRID and BOW do not perform well on the first retrieval. This may be due to the size of the vocabulary used by both methods. Generally, the results show that the HYBRID method is able to identify relevant learning resources by highlighting the domain concepts they contain, and this is important in e-learning. The graphs show that augmenting the representation of learning resources with our background knowledge is beneficial for e-learning recommendation.

6 Conclusions

E-learning recommendation is challenging because the learning resources are often unstructured text, and so are not appropriately indexed for retrieval. One solution is the creation of a concept-aware representation that contains a good coverage of relevant topics. In this paper domain-specific background knowledge is built by exploiting a structured collection of teaching materials as a guide for identifying important concepts. We then enrich the identified concepts with discovered text from an encyclopedia source, and use these pseudo-documents to extend the coverage and richness of our representation.

The background knowledge captures both key topics highlighted by the e-book TOCs that are useful for teaching, and additional vocabulary related to these topics. The concept space provides a vocabulary and focus that is based on teaching materials with provenance. CONCEPTBASED takes advantage of similar distributions of concept terms in the concept and document spaces to define a concept term driven representation. HYBRID exploits differences between distributions of document terms in the concept and document space, in order to boost the influence of terms that are distinctive in a few concepts.

Our results confirm that augmenting the representation of learning resources with our background knowledge in HYBRID improves e-learning recommendation. The larger vocabulary from both concepts and documents has been focused by the use of the vocabulary in the concept space. Although CONCEPTBASED also focuses on the concept space, by using only concept vocabulary, this vocabulary is too restricted for concept-based distinctiveness to be helpful.

In future, the background knowledge will be exploited to support query expansion and refinement in an e-learning environment. One approach would be to represent learners' queries using the vocabulary from our knowledge-rich representation. Alternatively, our background knowledge can be employed to support search by exploration. This would allow learners to search for resources through a guided view of the learning domain.

References

1. Agrawal, R., Chakraborty, S., Gollapudi, S., Kannan, A., Kenthapadi, K.: Quality of textbooks: an empirical study. In: ACM Symposium on Computing for Development, pp. 16:1–16:1 (2012)
2. Beliga, S., Meštrović, A., Martinčić-Ipšić, S.: An overview of graph-based keyword extraction methods and approaches. J. Inf. Organ. Sci. **39**(1), 1–20 (2015)
3. Bousbahi, F., Chorfi, H.: MOOC-Rec: a case based recommender system for MOOCs. Proc. Soc. Behav. Sci. **195**, 1813–1822 (2015)
4. Boyce, S., Pahl, C.: Developing domain ontologies for course content. J. Educ. Technol. Soc. **10**(3), 275–288 (2007)
5. Chen, W., Niu, Z., Zhao, X., Li, Y.: A hybrid recommendation algorithm adapted in e-learning environments. World Wide Web **17**(2), 271–284 (2014)
6. Chen, Z., Liu, B.: Topic modeling using topics from many domains, lifelong learning and big data. In: 31st International Conference on Machine Learning, pp. 703–711 (2014)
7. Coenen, F., Leng, P., Sanderson, R., Wang, Y.J.: Statistical identification of key phrases for text classification. In: Machine Learning and Data Mining in Pattern Recognition, pp. 838–853. Springer (2007)
8. Dietze, S., Yu, H.Q., Giordano, D., Kaldoudi, E., Dovrolis, N., Taibi, D.: Linked education: interlinking educational resources and the web of data. In: 27th Annual ACM Symposium on Applied Computing, pp. 366–371 (2012)
9. Ferrucci, D., Levas, A., Bagchi, S., Gondek, D., Mueller, E.T.: Watson: beyond Jeopardy!. Artif. Intell. **199**, 93–105 (2013)
10. Hands, A.: Microsoft academic search. Tech. Serv. Q. **29**(3), 251–252 (2012)

11. Nasraoui, O., Zhuhadar, L.: Improving recall and precision of a personalized semantic search engine for e-learning. In: 4th International Conference on Digital Society, pp. 216–221. IEEE (2010)
12. Panagiotis, S., Ioannis, P., Christos, G., Achilles, K.: APLe: agents for personalized learning in distance learning. In: 7th International Conference on Computer Supported Education, pp. 37–56. Springer (2016)
13. Porter, M.F.: An algorithm for suffix stripping. Program **14**(3), 130–137 (1980)
14. Qureshi, M.A., O'Riordan, C., Pasi, G.: Exploiting Wikipedia to identify domain-specific key terms/phrases from a short-text collection. In: 5th Italian Information Retrieval Workshop, pp. 63–74 (2014)
15. Ruiz-Iniesta, A., Jimenez-Diaz, G., Gomez-Albarran, M.: A semantically enriched context-aware OER recommendation strategy and its application to a computer science OER repository. IEEE Trans. Educ. **57**(4), 255–260 (2014)
16. Salton, G., Buckley, C.: Term-weighting approaches in automatic text retrieval. Inf. Process. Manag. **24**(5), 513–523 (1988)
17. Völkel, M., Krötzsch, M., Vrandecic, D., Haller, H., Studer, R.: Semantic Wikipedia. In: 15th International Conference on World Wide Web, pp. 585–594. ACM (2006)
18. Witten, I.H., Paynter, G.W., Frank, E., Gutwin, C., Nevill-Manning, C.G.: KEA: Practical automatic keyphrase extraction. In: 4th ACM Conference on Digital libraries, pp. 254–255 (1999)
19. Yang, H.L., Lai, C.Y.: Motivations of Wikipedia content contributors. Comput. Hum. Behav. **26**(6), 1377–1383 (2010)
20. Yang, K., Chen, Z., Cai, Y., Huang, D., Leung, H.: Improved automatic keyword extraction given more semantic knowledge. In: International Conference on Database Systems for Advanced Applications, pp. 112–125. Springer (2016)
21. Zhang, X., Liu, J., Cole, M.: Task topic knowledge vs. background domain knowledge: impact of two types of knowledge on user search performance. In: Advances in Information Systems and Technologies, pp. 179–191. Springer (2013)

Knowledge Discovery and Data Mining

Category-Driven Association Rule Mining

Zina M. Ibrahim, Honghan Wu, Robbie Mallah and Richard J.B. Dobson

Abstract The quality of rules generated by ontology-driven association rule mining algorithms is constrained by the algorithm's effectiveness in exploiting the usually large ontology in the mining process. We present a framework built around superimposing a hierarchical graph structure on a given ontology to divide the rule mining problem into disjoint subproblems whose solutions can be iteratively joined to find global associations. We present a new metric for evaluating the interestingness of generated rules based on where their constructs fall within the ontology. Our metric is anti-monotonic on subsets, making it usable in an Apriori-like algorithm which we present here. The algorithm categorises the ontology into disjoint subsets utilising the hierarchical graph structure and uses the metric to find associations in each, joining the results using the guidance of anti-monotonicity. The algorithm optionally embeds built-in definitions of user-specified filters to reflect user preferences. We evaluate the resulting model using a large collection of patient health records.

Keywords Association rule mining · Ontologies · Big data

1 Introduction

Ontology-driven association rule mining seeks to enhance the process of searching for association rules with the aid of domain knowledge represented by an ontology. The body of work in this area falls within two themes: (1) using ontologies as models

---

Z.M. Ibrahim (✉) · H. Wu · R.J.B. Dobson
Department of Biostatistics and Health Informatics, King's College London, London, UK
e-mail: zina.ibrahim@kcl.ac.uk

H. Wu
e-mail: honghan.wu@kcl.ac.uk

R.J.B. Dobson
e-mail: richard.j.dobson@kcl.ac.uk

R. Mallah
The South London and Maudsley NHS Foundation Trust, London, UK
e-mail: robbie.mallah@slam.nhs.uk

M. Bramer and M. Petridis (eds.), *Research and Development in Intelligent Systems XXXIII*, DOI 10.1007/978-3-319-47175-4_2

for evaluating the usefulness of generated rules [3, 8, 9] and (2) using ontologies in a post-mining step to prune the set of generated rules to those that are interesting [10, 12, 13]. In addition to the above, the users in most application domains are usually interested in associations between specific subsets of items in the data. For example, a clinical researcher is almost never interested in associations involving all the articles that appear in her dataset but instead may ask specific queries, such as whether interesting relations exist between medication usage and adverse drug reactions, or the degree of patient conformity to testing procedures and the likelihood of relapse. As a result, many efforts have been directed towards accommodating user preferences given a domain ontology [14–16].

Regardless of the method adopted, the quality of the model is constrained by how well it utilises the (usually) large ontology in the mining process. For example, the Systematized Nomenclature of Medicine Clinical Term Top-level Ontology (SCTTO)[1] has over 100,000 entries. Exploring SCTTO to discover interesting rules from large medical records will yield many possible associations, including irrelevant ones. If SCTTO is used in an association rule mining task, complex queries will be needed to extract the relevant subsets of the ontology. Even then, it is almost inevitable that extensive manual examination is required to maximise relevance.

The above challenges give rise to the need to (1) organise ontology collections to facilitate subsetting and retrieval, (2) build association rule mining algorithms to utilise the organised ontologies to improve the mining process. Our work is motivated by these two needs and revolves around enforcing a meta-structure over an ontology graph. This meta-structure associates a category with a collection of ontology terms and/or relations, creating ontological subcommunities corresponding to categories of interest from a user's perspective. For example, categories may be defined for specific *class of diseases* or *laboratory findings* in SCTTO to investigate novel screening of patients for some disease based on laboratory test results.

This work builds an association rule mining framework which enables the formation of ontology sub-communities defined by categories. The building block of our work is the meta-ontological construct *category*, which we superimpose over domain knowledge and build a representation around. The resulting framework provides (1) translation of user preferences into constraints to be used by the algorithm, to prune domain knowledge and produce more interesting rules, (2) a new scoring metric for rule evaluation given an ontology, and (3) an algorithm that divides rule mining given an ontology into disjoint subproblems whose joint solutions provide the global output, reducing the computational burden and enhancing rule quality. We present a case study of finding associations between the occurrence of drug-related adversities and different patient attributes using hospital records. To our knowledge, meta-ontologies have not been used in conjunction with association rules mining.

[1] https://bioportal.bioontology.org/ontologies/SCTTO

2 Mining Association Rules Revisited

Given a dataset $\mathcal{D} = \{d_1, \ldots, d_N\}$ of N rows with every row containing a subset of items chosen from a set of items $\mathcal{I} = \{i_1, \ldots, i_n\}$, association rule mining finds subsets of $\mathcal{I}$ containing items which show frequent co-occurrence in $\mathcal{D}$.

An association rule is taken to be the following implication: $r : \mathcal{A} \Rightarrow \mathcal{S}$. Where $\mathcal{A}, \mathcal{S} \subseteq \mathcal{I}$ and $\mathcal{A} \cap \mathcal{S} = \emptyset$. $\mathcal{A} = \{i_1, \ldots, i_x\}$ is the set of items of the *antecedent* of an association rule and $\mathcal{S} = \{i_{x+1}, \ldots, i_y\}$ is the set of items of the *consequent* of r. The implication reads that all the rows in $\mathcal{D}$ which contain the set of items making up $\mathcal{A}$ will contain the items in $\mathcal{S}$ with some probability Pr.

Two measures establish the strength of an association rule: *support* and *confidence*. *Support* Determines how often a rule is applicable to a given data set and is measured as the probability of finding rows containing all the items in the antecedent and consequent, $Pr(\mathcal{A} \cup \mathcal{S})$, or $\frac{|\mathcal{A} \cup \mathcal{S}|}{N}$, where N is the total number of rows. *Confidence* determines how frequently items in $\mathcal{S}$ appear in rows containing $\mathcal{A}$ and is interpreted as the conditional probability $Pr(\mathcal{S}|\mathcal{A})$. Therefore, *support* is a measure of statistical significance while *confidence* is a measure of the strength of the rule. The goal of association rule mining is to find all the rules whose *support* and *confidence* exceed predetermined thresholds [17].

Support retains a useful property which states that the support of a set of items never exceeds the support of its subsets. In other words, *support* is anti-monotonic on rule subsets. More specifically, let r_1 and r_2 be two association rules where r_1 : $\mathcal{A}_1 \Rightarrow \mathcal{S}_1$ and $r_2 : \mathcal{A}_2 \Rightarrow \mathcal{S}_2$, then the following holds [1]:

$$\mathcal{A}_1 \cup \mathcal{S}_1 \subseteq \mathcal{A}_2 \cup \mathcal{S}_2 \rightarrow support(\mathcal{A}_1 \cup \mathcal{S}_1) \geq support(\mathcal{A}_2 \cup \mathcal{S}_2)$$

Althrough *confidence* does not adhere to general anti-monotonicity, the confidence of rules generated using the same itemset is anti-monotonic with respect to the size of the consequent, e.g. if $\mathcal{I}_s \subset I$ is an itemset such that $\mathcal{I}_s = \{A, B, C, D\}$, all rules generated using all elements of $\mathcal{I}_s$ will be anti-monotonic with respect to the consequents of the possible rules. e.g. $confidence(ABC \rightarrow D) \geq confidence(AB \rightarrow CD) \geq confidence(A \rightarrow BCD)$.

Unlike in *confidence*, the anti-monotonicity of *support* is agnostic to the position of the items in the rule (i.e. whether they fall within the antecedent or the consequent). Therefore, when evaluating *support*, a rule is collapsed to the unordered set of items $\mathcal{A} \cup \mathcal{S}$. This difference has been exploited by the Apriori algorithm [1] to divide the association mining process into two stages: (1) a candidate itemset generation stage aiming to reduce the search space by using *support* to extract unordered candidate itemsets that pass a predetermined frequency threshold in the dataset, and (2) a rule-generation stage which discovers rules from the frequent itemsets and uses *confidence* to return rules which pass a predetermined threshold. In both stages, anti-monotonicity prevents generating redundant itemsets and rules by iteratively generating constructs of increasing lengths and avoiding the generation of supersets that do not pass the *support* and *confidence* thresholds [1]. Our work uses these principles to build a category-aware ontology-based Apriori-like framework.

3 Category-Augmented Knowledge

The building block of this work is the meta-ontological construct *category* we use to augment an ontology. Let $\mathcal{K} = (\mathcal{O}, \mathcal{R})$ be our knowledge about some domain defined by a set of terms (classes and instances) $\mathcal{O} = \{o_1, \ldots, o_n\}$, also called the universe, and a set of relations $\mathcal{R} = \{r_1, \ldots, r_k\}$ connecting the elements of $\mathcal{O}$. Moreover, let $\mathcal{C} = c_1, \ldots, c_m$ be a non-empty set of categories describing different groups to which the elements of $\mathcal{O}$ belong, such that $m << n$. The basic idea is to superimpose $\mathcal{C}$ over $\mathcal{O}$, creating subcommunities in the ontology graph which can be processed individually. To achieve this, we first define a mapping from $\mathcal{O}$ to $\mathcal{C}$ which organises the elements of $\mathcal{O}$ into subcommunities. The intuition is that every category in $\mathcal{C}$ represents a group of interest which can be mined for associations individually or in conjunction with other groups.

We can therefore define a mapping $\mathcal{F} : \mathcal{O} \times \mathcal{C} \rightarrow \{0, 1\}$ to yield a value of 1 whenever a concept $o \in \mathcal{O}$ is associated with the category $c \in \mathcal{C}$, and 0 otherwise. $\mathcal{F}$ is exhaustive over $\mathcal{O}$, i.e. every element in the universe must belong to a category. Formally: $\forall o \in \mathcal{O}, \exists c \in \mathcal{C}$ such that $\mathcal{F}(o, c) = 1$.

A function $\sigma : \mathcal{C} \rightarrow \mathcal{O}$ can then be defined to extract the set of elements in the universe associated with a category $c \in \mathcal{C}$:

$$\sigma(c) = \mathcal{O}_c \subset \mathcal{O} : \forall o_c \in \mathcal{O}_c, \mathcal{F}(o_c, c) = 1 \tag{1}$$

Because $\mathcal{F}$ is exhaustive, the inverse of σ is also a function. $\sigma^{-1} : \mathcal{O} \rightarrow \mathcal{C}$ yields the set of categories to which an element o belongs (an element may belong to multiple categories):

$$\sigma^{-1}(o) = \mathcal{C}_o \subset \mathcal{C} \iff \forall c_o \in \mathcal{C}_o : \sigma(c_o) = o \tag{2}$$

3.1 Graphical Representation

To represent category-augmented background knowledge (ontology) graphically, we borrow the concept of a *hierarchical graph* [5], which is one whose nodes may contain other graphs and arcs can contain other arcs. The graph contained in a node is called a *subgraph* of that *parent node*. The arcs that connect two nodes belonging to the same subgraph are called *internal arcs*, while arcs connecting nodes in different subgraphs of the same hierarchical level are called *external arcs*, and the nodes that are connected in that way are called *border nodes* of their respective subgraphs. No arc is allowed to connect two nodes of different hierarchical levels.

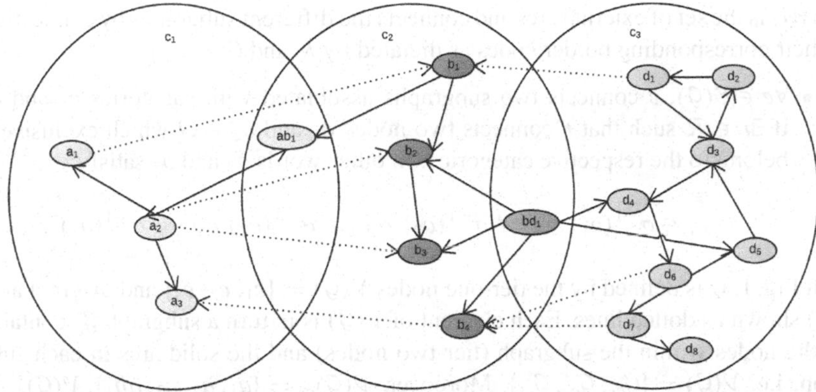

Fig. 1 A two-tier hierarchical graph

To capture the properties of a category-augmented ontology as described earlier, we define a two-tier hierarchical graph structure such as the one shown in Fig. 1. In the figure, the three subgraphs correspond to three categories c_1, c_2 and c_3. The solid arcs are internal to each subgraph while the dotted arcs are the external arcs of the graph. A formal definition of a two-tier hierarchical graph follows.

Definition 1 Let $\mathcal{K} = (\mathcal{O}, \mathcal{R})$ be some domain knowledge, and let $\mathcal{C} = \{c_1, \ldots, c_n\}$ be a set of categories such that $\mathcal{F} : \mathcal{O} \times \mathcal{C} \rightarrow \{0, 1\}$ is defined. A two-tier hierarchical graph $\mathcal{G} = (\mathcal{V}(\mathcal{G}), \mathcal{E}(\mathcal{G}))$ represents $\mathcal{K}$ with $\mathcal{C}$ superimposed such that:

1. Nodes in $\mathcal{V}(\mathcal{G})$ are subgraphs connecting subsets of the universe belonging to a single category $c \in \mathcal{C}$. We denote the elements of $\mathcal{V}(\mathcal{G})$ by **tier-one** nodes characterise them as follows:
   a. The number of tier-one nodes corresponds to the number of categories in $\mathcal{C}$, i.e. $\mathcal{V}(\mathcal{G})| = |\mathcal{C}|$
   b. The subgraphs corresponding to the nodes in $\mathcal{V}(\mathcal{G})$ comprise internal nodes which are a subset of the universe associated with the category, and arcs that are a subset of $\mathcal{E}(\mathcal{G})$ connecting the internal nodes, i.e. $\forall c \in \mathcal{C}, \exists \mathcal{G}_c \in \mathcal{V}(\mathcal{G})$ such that $\mathcal{G}_c = (\mathcal{V}(\mathcal{G}_c), \mathcal{E}(\mathcal{G}_c))$ corresponds to a subgraph of $\mathcal{G}$ given category c with $\mathcal{V}(\mathcal{G}_c)$ as nodes and $\mathcal{E}(\mathcal{G}_c)$ as its set of arcs further defined as follows:
      - The nodes in each subgraph $\mathcal{V}(\mathcal{G}_c)$ is the subset of $\mathcal{O}$ associated with c:

$$\mathcal{V}(\mathcal{G}_c) = \sigma(c)$$

        These nodes are termed the **tier-two** nodes of the graph.
      - The set of arcs $\mathcal{E}(\mathcal{G}_c)$ is mapped from a subset of the set of relations $\mathcal{R}$ which only contains the relations connecting universe elements exclusively associated with c. For any two nodes $o_1, o_2 \in \mathcal{O}$:

$$\forall e_c = (o_1, o_2) \in \mathcal{E}(\mathcal{G}_c),\ \ c \in \sigma^{-1}(o_1) \wedge c \in \sigma^{-1}(o_2)$$

2. $\mathcal{E}(\mathcal{G})$ is the set of external arcs and connects the different subgraphs by connecting their corresponding border nodes as dictated by $\mathcal{K}$ and $\mathcal{C}$:
   - $\forall e \in \mathcal{E}(\mathcal{G})$, e connects two subgraphs associated with categories c_1 and c_2 if $\exists r \in \mathcal{R}$ such that r connects two nodes o_1 and $o_2 \in \mathcal{O}$ which exclusively belong to the respective categories. In other words o_1 and o_2 satisfy:

$$c_1 \in \sigma^{-1}(o_1) \wedge c_1 \notin \sigma^{-1}(o_2) \wedge c_2 \in \sigma^{-1}(o_2) \wedge c_2 \notin \sigma^{-1}(o_1)$$

In Fig. 1, $\mathcal{G}$ is defined by the tier-one nodes $\mathcal{V}(\mathcal{G}) = \{c_1, c_2, c_3\}$ and external arcs $\mathcal{E}(\mathcal{G})$ shown as dotted lines. Each element of $\mathcal{V}(\mathcal{G})$ is in turn a subgraph $\mathcal{G}_c$ containing the nodes within the subgraph (tier-two nodes) and the solid arcs in each subgraph, i.e. $\mathcal{V}(\mathcal{G}) = \{\mathcal{G}_{c_1}, \mathcal{G}_{c_2}, \mathcal{G}_{c_3}\}$. Moreover, $\mathcal{V}(\mathcal{G})_{c_1} = \{a_1, a_2, a_3, ab_1\}$, $\mathcal{V}(\mathcal{G})_{c_2} = \{b_1, b_2, b_3, b_4, ab_1, bd_1\}$ and $\mathcal{V}(\mathcal{G})_{c3} = \{d_1, d_2, d_3, d_4, d_5, d_6, d_7, d_8, bd_1\}$. Note that ab_1 is shared between subgraphs c_1 and c_2 and bd_1 is shared between subgraphs c_2 and c_3, reflecting that an element in the universe may belong to more than one category. A similar observation can be made for the other two subgraphs.

4 Category-Augmented Rule-Association Mining Using Background Knowledge

Let $\mathcal{K} = (\mathcal{O}, \mathcal{R})$ be our ontology with $\mathcal{C}$ categories as before. Let $\mathcal{G} = (\mathcal{V}(\mathcal{G}), \mathcal{E}(\mathcal{G}))$ be a two-tier hierarchical graph representation of $\mathcal{C}$ superimposed on $\mathcal{O}$. Let $\mathcal{D} = \{d_1, \ldots, d_N\}$ be a data set of N records, where each row $d_i \in \mathcal{D}$ contains a subset of items chosen from a predefined set of items $\mathcal{I} = \{i_1, \ldots, i_n\}$. Every element of $\mathcal{I}$ corresponds to a node in $\mathcal{O}$. To represent this, we define a one-to-one and onto mapping $\mathcal{M} : \rightarrow \mathcal{O}$ which maps each item in $\mathcal{I}$ to a node in $\mathcal{O}$.

4.1 Category-Derived Constraints

The category-augmented knowledge framework introduced so far can be used to define constraints on the association rules to be discovered. We can use the constraints to determine user preferences to guide the algorithm to avoid performing an unnecessary search. Given a dataset $\mathcal{D}$, we define four types of rule constraints:

Definition 2 Let $\mathcal{K} = (\mathcal{O}, \mathcal{R})$ our domain knowledge with $\mathcal{C} = \{c_1, \ldots, c_m\}$ being the set of categories superimposed over $\mathcal{O}$ as before ($m << n$). Let $r : \mathcal{A} \Rightarrow \mathcal{S}$ be an association rule with antecedent $\mathcal{A}$ and consequent $\mathcal{S}$ where $\mathcal{A} = \{i_1, \ldots, i_x\} \subseteq \mathcal{I}$ and $\mathcal{S} = \{i_{x+1}, \ldots, i_y\} \subseteq \mathcal{I}$. Moreover, let the mapping $\mathcal{M} : \mathcal{I} \rightarrow \mathcal{O}$ hold and let $\mathcal{C}_p \subseteq \mathcal{C}$ be a subset of the categories imposed on $\mathcal{O}$.

1. r is said to adhere to a head-inclusion constraint on $\mathcal{C}_p$ if all the items in its antecedent map to concepts in $\mathcal{O}$ which are associated with a category which falls within $\mathcal{C}_p$.
$$\forall i \in \mathcal{A}(r) : \mathcal{M}(i) = o \wedge \sigma^{-1}(o) \subset \mathcal{C}_p$$
2. r is said to adhere to a head-exclusion constraint on $\mathcal{C}_p$ if none of the items in its antecedent map to concepts in $\mathcal{O}$ which are associated with a category which falls within $\mathcal{C}_p$.
$$\neg\exists i \in \mathcal{A}(r) : \mathcal{M}(i) = o \wedge \sigma^{-1}(o) \subset \mathcal{C}_p$$

Tail-inclusion and tail-exclusion constraints are similarly defined by replacing $\mathcal{A}$ with $\mathcal{S}$ in points 1 and 2 respectively.

4.2 Score Evaluation

We would like to use a scoring function that can (1) accommodate both $\mathcal{D}$ and the ontology represented by $\mathcal{G}$, and (2) retain monotonicity on the model so that the Apriori principle [1] can be used. Therefore, we formulate *interest*, a scoring metric that measures how interesting a rule is given an ontology by quantifying the goodness of the fit between the two. *interest* is based on the following two components:

1. The lengths of the paths connecting two-tier nodes that correspond to items in $\mathcal{A} \cup \mathcal{S}$. Shorter paths reflect more direct relationships and are more likely to form interesting associations. We define the distance between two tier-two nodes $d(o_i, o_j)$ as the length of the shortest undirected path connecting them. To express our preference to shorter paths, we use the ratio of between the minimum distance connecting any two tier-two nodes in the graph to $d(o_i, o_j)$. The resulting measure ζ quantifies the interestingness of the relations among $\mathcal{A} \cup \mathcal{S}$ items by the sum of their pairwise distance ratios. $\zeta : \mathcal{O} \rightarrow [0-1]$ is defined below.
$$\zeta(\mathcal{O}_k|\mathcal{G}) = \min_{o_i, o_j \in \mathcal{O}_k} d(o_i, o_j) \times \sum_{o_i, o_j \in \mathcal{O}_k} \frac{1}{d(o_i, o_j)}, \mathcal{O}_k \subseteq \mathcal{O}$$
2. The degrees of the nodes reflect their centrality within the graph, which we use as a reciprocal of interestingness. The hypothesis is that more significant relations exist among nodes which connect to fewer other nodes. For instance, in the worst case scenario where a tier-two node o connects to every other node in the graph, no information is gained from finding an association translating to (o, o_i) in the data (with o_i being any other tier-two node).

We define the degree as the number of undirected relations the node forms within the graph in question. The definition of the degree is context-specific, i.e. the degree of a node can be different depending on whether it is computed relative to the entire graph or the one induced by a given category:

$$deg(o|\mathcal{G}) = |\mathcal{E}_o|, \mathcal{E}_o \subset \mathcal{E}(\mathcal{G}), \forall e \in \mathcal{E}_o : e = (o_i, o) \vee e = (o, o_i)$$

where o_i is any other tier-two node in the graph. The reader should note that when $\mathcal{G}$ is taken as the subgraph induced by a category, then $\mathcal{E}(\mathcal{G})$ will correspond to the arcs internal to the graph, according to the definition of tier-one nodes being subgraphs of specific categories (Definition 1). Therefore the external arcs will not count towards the degree. This results in a value corresponding to the degree of the node relative to the internal structure of the graph induced by the category. Having defined the degree, we can now determine degree-based interestingness of a set of tier-two nodes given a graph as the sum of the reciprocals of their respective degrees within the graph. $\psi : \mathcal{O} \to \mathbb{R}_+$ is defined as:

$$\psi(\mathcal{O}_k|\mathcal{G}) = \sum_{o_i \in \mathcal{O}_k} \frac{1}{deg(o_i|\mathcal{G})}, \mathcal{O}_k \subseteq \mathcal{O}$$

We can now define the *interest* of a set of items given an ontology graph $\mathcal{G}$ as:

Definition 3 Let $\mathcal{G} = (\mathcal{V}(\mathcal{G}), \mathcal{E}(\mathcal{G}))$ be a two-tier hierarchical graph representation of domain knowledge $\mathcal{K} = (\mathcal{O}, \mathcal{R})$. Let $r : \mathcal{A} \Rightarrow \mathcal{S}$ be a rule with $\mathcal{A} \cup \mathcal{S}$ map to a collection of items from the universe and are represented by tier-two nodes in $\mathcal{V}(\mathcal{G})$. The *interest* of r given $\mathcal{G}$ is:

$$interest(r|\mathcal{G}) = \sqrt{\zeta(\mathcal{A} \cup \mathcal{S}) \times \psi(\mathcal{A} \cup \mathcal{S})}$$

Proposition 1 *interest(r|$\mathcal{G}$) is anti-monotonic with respect to subsets. Formally, let $r_1 : \mathcal{A}_1 \Rightarrow \mathcal{S}_1$ and $r_2 : \mathcal{A}_2 \Rightarrow \mathcal{S}_2$ be two rules, then: $\mathcal{A}_1 \cup \mathcal{S}_1 \subseteq \mathcal{A}_2 \cup \mathcal{S}_2 \to interest(\mathcal{A}_1 \cup \mathcal{S}1) \geq interest(\mathcal{A}_2 \cup \mathcal{S}_2)$*

Justification 1 *Given that $\mathcal{A}_1 \cup \mathcal{S}_1 \subseteq \mathcal{A}_2 \cup \mathcal{S}_2$ and $\zeta : \mathcal{V}(\mathcal{G}) \to [0-1]$, this implies that $\zeta(\mathcal{A}1 \cup \mathcal{S}_1) \geq \zeta(\mathcal{A}_2 \cup \mathcal{S}_2)$ because $|\mathcal{A}_1 \cup \mathcal{S}_1| \leq |\mathcal{A}_2 \cup \mathcal{S}_2|$.*

Similarly, because $|\mathcal{A}_1 \cup \mathcal{S}_1| \leq |\mathcal{A}_2 \cup \mathcal{S}_2|$, then $\psi(\mathcal{A}_1 \cup \mathcal{S}_1) \geq \psi(\mathcal{A}_2 \cup \mathcal{S}_2)$. It follows that: $interest(\mathcal{A}_1 \cup \mathcal{S}_1) \geq interest(\mathcal{A}_2 \cup \mathcal{S}_2)$.

4.3 The Algorithm

The algorithm presented here relies on a two-tier hierarchical graph and the *interest* scoring metric to mine rules from a given dataset. Category-miner is a modified Apriori algorithm and consists of the two Apriori stages: (1) a candidate itemsets generation stage, which uses our *interest* metric in addition to *support* to generate itemsets that pass the frequency test and reflect a good fit with the ontology, (2) a rule generation step, using the candidate itemsets (stage 1) to generate rules that pass the *confidence* threshold. The algorithm also considers user preferences by incorporating the category-derived constraints we defined in Sect. 4.1.

Algorithm 1 is a wrapper algorithm. It receives as input a two-tier hierarchical graph $\mathcal{G}$ representing an ontology augmented with categories $\mathcal{C}$ and a dataset $\mathcal{D}$, where $\mathcal{D}$ maps to nodes in the ontology and will be used to generate candidate itemsets. The four optional parameters hi, he, ti and te correspond to the four category-derived constraints (Sect. 4.1) and are used to specify user preferences.

Algorithm 1 Association Rule Mining Wrapper Algorithm

Input: $\mathcal{G}, \mathcal{D}, hi, he, ti, te$
Output: Set of association rules $\mathcal{R}$
Procedure:
1: $\mathcal{S} \leftarrow \emptyset$ // set of candidate itemsets, initially empty
2: $\mathcal{HS} \leftarrow (\mathcal{C} \setminus he)$; $\mathcal{TS} \leftarrow (\mathcal{C} \setminus te)$
3: $\mathcal{HS} \leftarrow \mathcal{HS} \cap hi$ **if** $hi \neq \emptyset$
4: $\mathcal{TS} \leftarrow \mathcal{TS} \cap ti$ **if** $ti \neq \emptyset$
5: **for** $c_i \in \mathcal{HS} \cup \mathcal{TS}$ **do**
6: $\mathcal{S} \leftarrow \mathcal{S} \cup$ category-miner($\mathcal{G}$,$\mathcal{D}$,c_i)
7: $\mathcal{S} \leftarrow$ expand($\mathcal{S}$, $\mathcal{G}$)
8: $\mathcal{R} \leftarrow$generate-rules($\mathcal{S}$, $\mathcal{HS}$, $\mathcal{TS}$)
9: **Return:** $\mathcal{R}$

The wrapper algorithm constrains the antecedent and consequent by user preferences (lines 2–4). The variables head set ($\mathcal{HS}$) and tail set ($\mathcal{TS}$) contain all the categories permissible in the antecedent and consequent of any generated rule respectively. $\mathcal{HS}$ and $\mathcal{TS}$ are generated by excluding the categories specified in the exclusion constraints (*he* and *te* respectively) from $\mathcal{C}$ and if inclusion sets are provided (*hi* or *ti*), they will be the only sets used in the. Providing empty *hi*, *he*, *ti* and *te* sets makes our procedure equivalent to the general (non-constrained) algorithm.

The algorithm iteratively calls category-miner (lines 5–6), which generates candidate itemsets whose constructs fall strictly within the same category, for every category c_i in $\mathcal{HS}$ and $\mathcal{TS}$. This stage is agnostic to the position of the items in the rule. Hence category-miner is called for all categories in $\mathcal{HS} \cup \mathcal{TS}$ (line 6).

In category-miner (Algorithm 2), the initial 1-item itemsets are only pruned using *support* (line 2) because *interest* evaluates relationships rather than objects (requiring at least two-item itemsets). candidate-gen (line 4) generates all k-itemset supersets of a k -1-itemsets. The results are pruned at every iteration on the data using *support* and on the ontology using *interest*. Anti-monotonicity of the two metrics guarantees correctness, ensuring that any interesting and supported k-itemsets are composed using interesting and supported $k-1$-itemsets.

Once candidate itemsets associated with all categories are generated, an informed search is performed for supersets which transcend the category boundaries using the expand procedure (Algorithm 3). The algorithm uses anti-monotonicity once again to formulate the hypothesis: since all within-category associations have been found, the rules spanning the categories can be identified by evaluating the scores of their supersets. These supersets are found by examining the external arcs $\mathcal{E}(\mathcal{G})$ and adding their connecting nodes to the existing sets if they result in associations which pass our *support* and *interest* tests. For each external arc (o_i, o_j) used for expansion search, we obtain the set of associations previously found which strictly contain node o_i (line 2) and the set of associations which were previously found to strictly contain node o_j (line 3). Supersets are found by examining pair-wise unions of the generated sets which pass the goodness test (lines 5–6).

Algorithm 2 Category-specific Mining (category-miner)

Input: $\mathcal{G}, \mathcal{D}, \mathcal{S}$
Output: Subset of associations for the category $\mathcal{S}$
Procedure:
1: $//\mathcal{S}_1$ contains interesting 1-itemset from the category subgraph
2: $\mathcal{S}_1 \leftarrow o \in \mathcal{S},\ \ support(o|\mathcal{D}) \geq t_1$
3: **for** $(k = 2; \mathcal{S}_{k-1} \neq \emptyset; k++)$ **do**
4: $\quad \lambda_k = candidate - gen(\mathcal{S}_{k-1}, \mathcal{D})$ // New Candidate set
5: $\quad \mathcal{S}_k = \{s \in \lambda_k | support(s|\mathcal{D}) > t_1; interest(s|\mathcal{G}) > t_2\}$
6: $\quad$ *remove-proper-subsets*$(\mathcal{S}_k)$
7: **Return:** $\bigcup_k \mathcal{S}_k$

Algorithm 3 Itemset Expansion Algorithm (expand)

Input: $\mathcal{S}, \mathcal{G}$
Output: Inter-category rule set
Procedure:
1: **for** $e = (o_i, o_j) \in \mathcal{E}_\mathcal{G}$ **do**
2: $\quad \mathcal{S}_{o_i} \subset \mathcal{S} : \forall s_{o_i} \in \mathcal{S}_{o_i}, o_i \in s_{o_i} \wedge o_j \notin s_{o_i}$
3: $\quad \mathcal{S}_{o_j} \subset \mathcal{S} : \forall s_{o_j} \in \mathcal{S}_{o_j}, o_j \in s_{o_j} \wedge o_i \notin s_{o_j}$
4: $\quad \mathcal{S}_{ij} = \{s_{o_i} \cup s_{o_j} | s_{o_i} \in \mathcal{S}_{o_i}, s_{o_j} \in \mathcal{S}_{o_j}\}$
5: $\quad \mathcal{S} \leftarrow \{\mathcal{S} \cup \mathcal{S}_{ij} | support(\mathcal{S}_{ij}|\mathcal{D}) > t_1; interest(\mathcal{S}_{ij}|\mathcal{G}) > t_2)\}$
6: **Return:** $\mathcal{S}$

The expand procedure also uses the anti-monotonicity to guarantee correctness. The unions are performed pairwise for every pair of itemsets connected with an external arc in the ontology, which (1) makes the procedure complete by not losing interesting associations as all pairs are considered, and (2) reduces the computational burden by only considering itemset pairs with nodes sharing an external arc (lines 2–3). The resulting itemset $\mathcal{S}$ is returned to the wrapper algorithm (line 10), which marks the end of the itemset generation step. The rule generation step (line 11 of wrapper) is not shown as it is similar to Apriori's rule generation, with additional pruning according to user preferences. *confidence* is used iteratively to generate rules from $\mathcal{S}$ and the final set of rules $\mathcal{R}$ is returned by the algorithm.

4.4 Time Complexity

Given our database $\mathcal{D}$ with N rows and n items in each row, let m be the number of categories such that $n >> m$. cateogry-miner receives as input $M = N \times \frac{n}{m}$ items on average (the number of elements in each row in the category is $\frac{n}{m}$.

In category-miner, the first iteration evaluating 1-item sets requires M comparisons. Subsequently, for each iteration k generating $|\mathcal{R}_k|$ candidate itemsets, the complexity will be $\mathcal{O}(|\mathcal{R}_{k-1}|.|\mathcal{R}_k|)$ because each iteration will compare to the itemsets generated in the previous step. The resulting complexity will be $k \times \sum_k \mathcal{O}(|\mathcal{R}_{k-1}|.|\mathcal{R}_1|)$. The worst time complexity occurs when only one category is imposed on the ontology, resulting a running time equivalent to that of the Apriori algorithm with the added computations induced by calculating *interest*. Moreover, the worst-case time complexity of the *expand* procedure is $|\mathcal{E}_{\mathcal{G}}|.|\mathcal{S}|^2$, which is only reached in the unlikely situation where every node in every subgraph shares an external arc with another node in another subgraph.

The performance gain in our algorithm is due to the use of multiple categories, which prevents the worst case of category-miner by ensuring that M is small compared to the total number of items that would have been supplied to the algorithm if no categories were enforced (which is $n \times N$), and delegating more to the expand procedure. Figure 2 shows how the number of comparisons is reduced by increasing the number of categories supplied to the algorithm using 2–7 itemsets.

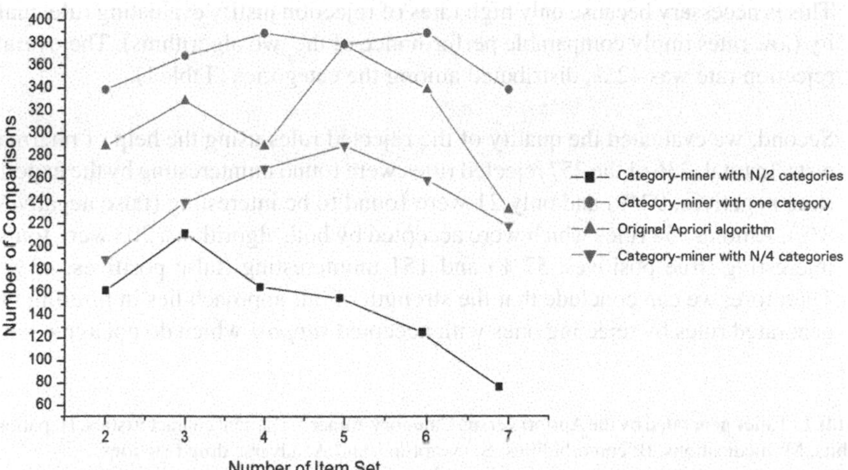

Fig. 2 Number of comparisons under different settings

5 Experimental Evaluation Using Medical Records

We now describe the process of selecting association rules from a large database of anonymised hospital medical records for 253 patients with bipolar disorder in the South London and Maudsley NHS Foundation Trust. We extracted a dataset of patient records containing $|\mathcal{I}| = 675$ items corresponding to six categories which we selected with the help of pharmacists: symptoms, medications, co-morbidities (diseases in addition to bipolar), compliance indicators, patient characteristics such as height and weight and finally occurrences of adverse drug reactions. We used UMLS (Unified Medical Language System) as a reference ontology to extract a subset containing our 675 items and superimposed a hierarchical graph on it using the six categories. At the time of writing, the task of constructing the two-tier graph is done in a semi-automated way where experts take the task of assigning class nodes to a set of predefined categories used to construct the graph. The procedure assigns instance nodes the categories of their respective class nodes. The relations are mapped into arcs with the semantics retained. We ran this procedure on the subset of UMLS we extracted for the six categories. We designed three experiments to test our framework and selected the top 30 % rules/items ranked by the scores.

1. **Evaluating the scoring metric**: To evaluate *interest*, we compared category-miner to the standard Apriori algorithm for each of the six categories individually. Since the difference between category-miner and the Apriori algorithm is the addition of the *Interest* metric, the quality of the rules accepted by the Apriori algorithm and rejected by category-miner reflects the goodness of our metric. We first compute the rate of rejecting rules generated by Apriori using *interest*. This is necessary because only high rates of rejection justify evaluating rule quality (low rates imply comparable performance of the two algorithms). The overall rejection rate was 42 %, distributed among the categories (Table 1).

   Second, we evaluated the quality of the rejected rules using the help of pharmacists. In total, 236 of the 257 rejected rules were found uninteresting by the experts (true negatives, 92 %) and only 21 were found to be interesting (false negatives, 8 %). Among 354 rules which were accepted by both algorithms, 203 were found interesting (true positives, 57 %) and 151 uninteresting (false positives, 43 %). Therefore, we can conclude that the strength of our approach lies in filtering the generated rules by rejecting ones with accepted *support* which do not agree with

Table 1 Rules generated by the Apriori versus Category-miner. C: patient characteristics, H: patient habits, M: medications, B: comorbidities, S: symptoms and A: adverse drug reactions.

Category	C	H	M	B	S	A
N. Rule (Apriori)	59	46	61	53	70	65
N. Rule (category-miner)	80	88	99	112	105	127
Rejection rate %	26.3	47.7	38.4	52.7	33.3	48.8

the ontology. It is, however, worth noting that upon manual examination, many of the rules accepted by our algorithm correspond to hierarchical and instance-of relations (e.g. drug - clozapine) which can be filtered by discarding them from the algorithm's search space. These filters are part of our ongoing work.

2. **Comparison with Other Ontology-driven Approaches** We also compare our results to knowledge-based approaches. As a model, we chose [8] as it is conceptually similar to our approach and formulates an ontology-based probabilistic measure to evaluate the generated rules. We ran both algorithms using the full set of categories (using as an ontology the full subset of UMLS extracted). The resulting set of rules includes 588 rules accepted by both approaches and 287 rules rejected by both approaches. 124 rules were rejected by our approach but not that of [8] and 111 rules were rejected by [8] and accepted by our approach. We first asked pharmacists to assess the quality of the rejected rules. The results were 85 % true negatives (and 15 % false negatives) using our algorithm versus 76 % true negatives (and 24 % false negatives using [8]). The rules jointly rejected by both approaches had 93 % true negatives (and 7 % false negatives).

   With respect to the accepted rules, those accepted by both algorithms reported 63 % true positives. The rules accepted by our approach reported 68 % true positives while the rules accepted by [8] reported 52 % true positives. Despite being significantly higher than the rate reported by [8], these results do not show conclusive improvement by our algorithm. However, once again, manual examination shows that 37 % of the false positives reported by our algorithm correspond to class hierarchy and instance-of relations which we have not accommodated in the filtration step. Therefore, we anticipate higher power than what is being reported here. The reader should note that the work in [8] accommodates the existence of is-a relations and considers them uninteresting. As a result, they do not appear in their results. In addition to these findings, the results of this experiment show a higher true positive rate than the first experiment. This shows that many of the rules found interesting by experts are in fact the inter-category rules.

3. **Evaluating the Constrained Algorithm** The last experiment evaluates the effectiveness of the constraints in producing rules which are interesting to the user. The experiment was conducted by restricting the categories permissible in the antecedents to comorbidities and medications and those permissible in the consequents to adverse drug reactions. Clinicians were asked whether the rules generated reflect known joint associations between co-morbidities, medications and the possibility of developing adverse drug reactions. 154 rules were generated by the constrained algorithm, among which 53 were true positive, and 12 were false positive. The constrained algorithm shows better rates than the general one, which can be attributed to the reduced search space but also shows the appropriateness of the definitions of constraints we presented here.

6 Summary and Conclusions

We used *categories* to superimpose a structure on ontologies and define a framework to decompose the set of rules to be generated into disjoint subsets and joined iteratively at a later step. We evaluated the framework for performance and quality of output using an extensive database of patient records.

We are currently developing methods for evaluating the relations between items to discard trivial hierarchical, instance and property relations from being used to generate itemsets and rules. The idea of decomposing the mining problem into disjoint subsets is part of a large effort to create parallel algorithms which can operate on very large and possibly distributed data sources.

Acknowledgments The authors would like to acknowledge the National Institute for Health Research (NIHR) Biomedical Research Centre and Dementia Unit at South London and Maudsley NHS Foundation Trust and Kings College London.

References

1. Agrawal, R., Imielinski, T., Swami, A.: *Mining association rules between sets of items in large databases*. ACM SIGMOID. 207-216. ACM Press (1993)
2. Baclawski, K., Schneider, T.: The open ontology repository initiative: requirements and research challenges. In: Workshop on Collaborative Management of Structured Knowledge at the ISWC (2009)
3. Cherfi, H., Napoli, A., Toussaint, Y.: Towards a text mining methodology using association rule extraction. Int. J. Soft Comput. **10**, 431–441 (2006)
4. Cornet, R., de Keizer, N.: Forty years of SNOMED: a literature review. BMC Med. Inf. Decis. Making **8**(1), S2 (2008)
5. Fernandez, J., Gonzalez, J.: Hierarchical graph search for mobile robot path planning. In: IEEE ICRA, pp. 656–661 (1998)
6. Garbacz, P., Trypuz, R.: A metaontology for applied ontology. Appl. Ontol. **8**, 1–30 (2013)
7. Herre, H., Loebe, F.: A meta-ontological architecture for foundational ontologies. On the Move to Meaningful Internet Systems, pp. 1398–1415 (2005)
8. Janetzko, D., Cherfi, H., Kennke, R., Napoli, A., Toussaint, Y.: Knowledge-based selection of association rules for text mining. In: ECAI, pp. 485–489 (2004)
9. Lieber, J., Napoli, A., Szathmary, L.: First elements on knowledge discovery guided by domain knowledge. Concept Lattices Appl. **4923**, 22–41 (2008)
10. Marinica, C., Guillet, F.: Knowledge based interactive postmining of association rules using ontologies. IEEE Trans. Knowl. Data Eng. **22**(6), 784–797 (2010)
11. Palma, P., Hartmann, J., Haase, P.: Ontology metadata vocabulary for the semantic web. Technical Report, Universidad Politcnica de Madrid, University of Karlsruhe (2008)
12. Ramesh, C., Ramana, K., Rao, K., Sastry, C.: Interactive post-mining association rules using cost complexity pruning and ontologies KDD. Int. J. Comput. Appl. **68**(20), 16–21 (2013)
13. Savasere, A., Omiecinski, E., Navathe, S.: An efficient algorithm for mining association rules in large databases. In: VLDB, pp. 432–444 (1995)
14. Singh, L., Chen, B., Haight, R., Scheuermann, P.: An algorithm for constrained association rule mining in semi-structured data. In: PAKDD, pp. 148–158 (1999)
15. Song, S., Kim, E., Kim, H., Kumar, H.: Query-based association rule mining supporting user perspective. Computing **93**, 1–25 (2011)

16. Srikant, R., Vu, Q., Agrawal, R.: Mining association rules with item constraints. In: KDD, pp. 67–73 (1997)
17. Tan, P., Steinbach, M.: *Introduction to Data Mining*. Addison-Wesley (2006)

A Comparative Study of SAT-Based Itemsets Mining

Imen Ouled Dlala, Said Jabbour, Lakhdar Sais and Boutheina Ben Yaghlane

Abstract Mining frequent itemsets from transactional datasets is a well known problem. Thus, various methods have been studied to deal with this issue. Recently, original proposals have emerged from the cross-fertilization between data mining and artificial intelligence. In these declarative approaches, the itemset mining problem is modeled either as a constraint network or a propositional formula whose models correspond to the patterns of interest. In this paper, we focus on the propositional satisfiability based itemset mining framework. Our main goal is to enhance the efficiency of SAT model enumeration algorithms. This issue is particularly crucial for the scalability and competitiveness of such declarative itemset mining approaches. In this context, we deeply analyse the effect of the different SAT solver components on the efficiency of the model enumeration problem. Our analysis includes the main components of modern SAT solvers such as restarts, activity based variable ordering heuristics and clauses learning mechanism. Through extensive experiments, we show that these classical components play an essential role in such procedure to improve the performance by pushing forward the efficiency of SAT solvers. More precisely, our experimental evaluation includes a comparative study in enumerating all the models corresponding to the closed frequent itemsets. Additionally, our experimental analysis is extended to include the Top-k itemset mining problem.

Keywords Data Mining · Frequent Itemset Mining · Propositional Satisfiability (SAT) Declarative Approaches

I. Ouled Dlala · S. Jabbour (✉) · L. Sais
CRIL CNRS UMR 8188, University of Artois, Lens, France
e-mail: jabbour@cril.fr

I. Ouled Dlala
e-mail: dlalaimen@yahoo.fr

L. Sais
e-mail: sias@cril.fr

B. Ben Yaghlane
LARODEC, High Institute of Management, University of Tunis, Tunis, Tunisia
e-mail: boutheina.yaghlane@ihec.rnu.tn

M. Bramer and M. Petridis (eds.), *Research and Development in Intelligent Systems XXXIII*, DOI 10.1007/978-3-319-47175-4_3

1 Introduction

Pattern mining constitutes a well established data mining research issue. The most important and representative problem is *frequent itemset mining* (FIM for short), which consists in finding all sets of items that have a high support in a given transaction database. Since the inception of the first algorithm of Agrawal et al. [1] for extracting frequent itemsets, this research area has attracted tremendous interest among data mining community, trying to speed up mining performance. Indeed, the existing algorithms can be classified into two main categories: The level wise and the pattern growth like approaches. The first one is based on the generate-and-test framework such that Apriori [1]. While the second category is based on the divide-and-conquer framework such as FP-Growth [11], H-Mine [22] and LCM [28] (see [26] for more details).

A new declarative data mining research trend has been proposed recently [6, 16, 23]. The main goal of this new framework is to overcome the lack of declarativity of traditional data mining approaches. Indeed, in classical data mining techniques, a slight modification in the problem specification leads to an implementation from scratch, while such changes can be taken into account by a simple addition/removal of constraints in declarative data mining. In addition to the nice declarative and flexibility properties, this new framework allows us to benefit from generic and efficient CP/SAT solving techniques. In the same vein, several contributions have been proposed in order to address other data mining task using propositional satisfiability (SAT) (e.g. [10, 14, 16]).

Today, propositional satisfiability has gained a considerable audience with the advent of a new generation of solvers able to solve large instances encoding real-world problems. In addition to the traditional applications of SAT to hardware and software formal verification, this impressive progress led to increasing use of SAT technology to solve new real-world applications such as data mining, planning, bioinformatics and cryptography. In the majority of these applications, we are mainly interested in the decision problem and some of its optimisation variants (e.g. Max-SAT). Compared to other issues in SAT, the SAT model enumeration problem has received much less attention. Most of the recent proposed model enumeration approaches are built on the top of SAT solvers. Usually, these implementations are based on the use of additional clauses, called blocking clauses, to avoid producing repeated models [5, 18, 20, 21]. Several improvements have been proposed to this blocking clause based model enumerator (e.g. [18, 21]). In particular, the Morgado and Marques-Silva [21] proposed several optimizations obtained through clause learning and simplification of blocked clauses. However, these kind of approaches are clearly impractical in SAT based data mining. Indeed, the size of the output (number of models) even using condensed representations, might be exponential in the worst case. In addition to the clauses learnt from each conflict, adding a blocked clause each time a model is found leads to an exponential number of additional clauses. This approach is heavy and impractical. In [9], the authors elaborate an interesting approach for enumerating answer sets of a logic program (ASP), centered around

First-UIP learning scheme and backjumping. In [13], the authors propose an approach based on a combination of a DPLL-like procedure with CDCL-based SAT solvers in order to mainly avoid the space complexity limitation mentioned above.

In this paper, we focus on the SAT encoding of the problem of frequent itemset mining introduced in [16]. In such encoding, there is a one-to-one mapping between the models of the propositional formula and the set of interesting patterns of the transaction database. In our experimental analysis, we consider two itemset mining problems: the problem of enumerating the whole set of frequent closed itemsets and the problem of mining the Top-k frequent closed itemsets of length greater than a given lower bound min, where k is the desired number of frequent closed itemsets to be mined, and min is the minimal length of each itemset [16]. The work presented in this paper is mainly motivated by this interesting SAT application to data mining and by the lack of efficient model enumerator.

Our aim is to study through an extensive empirical evaluation, the effects on model enumeration of the main components of CDCL based SAT solvers including restarts, branching heuristics and clauses learning.

2 Background

Let us first introduce the propositional satisfiability problem (SAT) and some necessary notations. We consider the conjunctive normal form (CNF) representation for the propositional formulas. A *CNF formula* Φ is a conjunction ($\wedge$) of clauses, where a *clause* is a disjunction ($\vee$) of literals. A *literal* is a positive (p) or negated ($\neg p$) propositional variable. The two literals p and $\neg p$ are called *complementary*. A CNF formula can also be seen as a set of clauses, and a clause as a set of literals. Let us mention that any propositional formula can be translated to CNF using linear Tseitin's encoding [27]. We denote by $Var(\Phi)$ the set of propositional variables occurring in Φ.

A *Boolean interpretation* $\mathscr{B}$ of a propositional formula Φ is a function which associates a value $\mathscr{B}(p) \in \{0, 1\}$ (0 corresponds to $false$ and 1 to $true$) to the propositional variables $p \in Var(\Phi)$. It is extended to CNF formulas as usual. A *model* of a formula Φ is a Boolean interpretation $\mathscr{B}$ that satisfies the formula, i.e., $\mathscr{B}(\Phi) = 1$. We note $\mathscr{M}(\Phi)$ the set of models of Φ. *SAT problem* consists in deciding if a given formula admits a model or not.

Let us informally describe the most important components of modern SAT solvers. They are based on a reincarnation of the historical Davis, Putnam, Logemann and Loveland procedure, commonly called DPLL [7]. It performs a backtrack search; selecting at each level of the search tree, a decision variable which is set to a Boolean value. This assignment is followed by an inference step that deduces and propagates some forced unit literal assignments. This is recorded in the implication graph, a central data-structure, which encodes the decision literals together with there implications. This branching process is repeated until finding a model or a conflict. In the first case, the formula is answered satisfiable, and the model is reported, whereas

Algorithm 1: CDCL Based Enumeration Solver

```
   Input: 𝓕: a CNF formula
   Output: 𝓜: all the models of 𝓕
1  𝓘 = ∅, Δ = ∅, Γ = ∅, 𝓜 = ∅, dl = 0;
2  while (true) do
3      γ = unitPropagation(𝓕,𝓘);
4      if (γ!=null) then
5          β = conflictAnalysis(𝓕,𝓘,γ);
6          btl = computeBackjumpLevel(β,𝓘);
7          if (btl == 0) then return 𝓜;
8          Δ = Δ ∪ {β};
9          if (restart()) then btl = 0;
10         backjumpUntil(btl);
11         dl = btl;
12     else
13         if (𝓘 ⊨ 𝓕) then
14             𝓜 = 𝓜 ∪ {𝓘};
15             γ = generateBlockedClause(𝓘);
16             Γ = Γ ∪ {γ};
17             backjumpUntil(0);
18             dl = 0;
19         else
20             if (timeToReduce()) then reduceDB(Δ);
21             ℓ = selectDecisionVariable(𝓕);
22             dl = dl + 1;
23             𝓘 = 𝓘 ∪ {selectPhase(ℓ)};
24         end
25     end
26 end
```

in the second case, a conflict clause (called learnt clause) is generated by resolution following a bottom-up traversal of the implication graph [19, 29]. The learning or conflict analysis process stops when a conflict clause containing only one literal from the current decision level is generated. Such a conflict clause asserts that the unique literal with the current level (called asserting literal) is implied at a previous level, called assertion level, identified as the maximum level of the other literals of the clause. The solver backtracks to the assertion level and assigns that asserting literal to *true*. When an empty conflict clause is generated, the literal is implied at level 0, and the original formula can be reported unsatisfiable. In addition to this basic scheme, modern SAT solvers use other components such as activity based heuristics and restart policies. An extensive overview can be found in [4].

Algorithm 1 depicts the general scheme of a CDCL SAT solver slightly extended for model enumeration (lines 16–21). Indeed, when a model is found (line 16), it is added to the current set of models (line 18). Then a blocked clause γ (the negation of the model $\mathscr{I}$) is generated (line 19) and added to the blocked clauses database (line 20). Such addition, prevents the solver from enumerating the same models. Finally, the solver backtrack to the root of the search tree to continue the search for the remaining models. When the search space is covered exhaustively (backtrack to level 0), the whole set of models is returned (line 11).

3 Frequent Itemset Mining

Given a finite set of distinct items $I = \{a_1, a_2, \ldots, a_n\}$. A transactional database $\mathcal{D}=\{T_1, T_2, \ldots, T_m\}$ is a set of transactions, where each transaction $T_i \in \mathcal{D}, (1 \leq i \leq m)$ is a subset of I and has an unique identifier Tid_i. An itemset $X = \{a_1, a_2, \ldots, a_l\}$ is a set of l distinct items, where $a_j \in I$, $1 \leq j \leq l$, and l is the length of X. A l-itemset is an itemset of length l. An itemset X is said to be contained in a transaction T_i if $X \subseteq T_i$.

Definition 1 The support of an itemset X is the number of transactions containing X in $\mathcal{D}$ and denoted as $SC(X)$. The frequency of X is defined as the ratio of $SC(X)$ to $|\mathcal{D}|$.

Let $\mathcal{D}$ be a transaction database over I and λ a minimum support threshold. The *frequent itemset mining problem* consists in computing the following set:

$$FIM(\mathcal{D}, \lambda) = \{X \subseteq I \mid SC(X) \geq \lambda\}.$$

Mining of the complete set of frequent itemsets may lead to a huge number of itemsets. In order to face this issue, this problem can be limited on the extraction of closed itemsets. Thus, enumerating all closed itemsets allows us to reduce the size of the output.

Definition 2 (*Closed Frequent Itemset*) Let $\mathcal{D}$ be a transaction database (over I). An itemset X is a closed itemset if there exists no itemset X' such that 1) $X \subseteq X'$ and 2) $\forall T \in \mathcal{D}, X \in T \rightarrow X' \in T$

Extracting all the elements of $FIM(\mathcal{D}, \lambda)$ can be obtained from the closed itemsets by computing their subsets. We denote by $CFIM(\mathcal{D}, \lambda)$ the subset of all closed itemsets in $FIM(\mathcal{D}, \lambda)$.

For instance, consider the transaction database described in Table 1. The set of closed frequent itemsets with the minimal support threshold equal to 2 are: $CFIM(\mathcal{D}, 2) = \{A, D, G, AB, AC, AF, ABC, ACD\}$.

However, when mining databases, data mining algorithms still encounter some performance bottlenecks like the huge size of output. Consequently, for practical data mining, it is important to reduce the size of the output, by exploiting the structure of the patterns. Most of the works on itemset and sequential mining require the specification of a minimum support threshold λ. This constraint allows the user to control at least to some extent the size of the output by mining only patterns covering at least λ transactions (locations). However, in practice, it is difficult for users to provide an appropriate threshold. As pointed out in [8], a too small threshold may lead to the generation of a huge number of patterns, whereas a too high value of

Table 1 A transaction database $\mathscr{D}$

Tid	Itemset
1	A, B, C, D
2	A, B, E, F
3	A, B, C
4	A, C, D, F
5	G
6	D
7	D, G

the threshold may result in no answer. Indeed, in [8], based on a complete ranking between itemsets, the authors propose to mine the n most interesting itemsets of arbitrary length. In [12], the proposed task consists in mining Top-k frequent closed itemsets of length greater than a given lower bound min, where k is the desired number of frequent closed itemsets to be mined, and min is the minimal length of each itemset. The authors demonstrate that setting the minimal length of the itemsets to be mined is much easier than setting the usual frequency threshold. This new framework can be seen as a way to mine the k preferred patterns according to some specific constraints or measures.

Definition 3 A closed itemset X is a top-k frequent closed itemset of minimal length min if there exist no more than $(k-1)$ closed itemsets of length at least min whose support is higher than that of X.

4 Encoding of Frequent Itemset Mining Using Constraint

Propositional logic is known to be a powerful knowledge representation language. Many constraints can be expressed easily using such language. Constraint programming community has tackled recently the problem of mining itemsets by encoding such a problem through a constraint network/propositional formula. Such encodings allow to have a one to one mapping between the solutions of the constraint network/propositional formula and the itemsets to be mined. The use of constraint programming or propositional satisfiability to model data mining problems, allows to benefit from the nice declarative features of the modeling languages and their associated efficient generic solving techniques.

In [16], the authors formulate such set of constraints by introducing Boolean variables p_a (resp. q_i) to represent each item $a \in I$ (resp. each transaction T_i).

$$\bigwedge_{i=1}^{m}(\neg q_i \leftrightarrow \bigvee_{a \in I \setminus T_i} p_a) \tag{1}$$

$$\sum_{i=1}^{m} q_i \geq \lambda \tag{2}$$

$$\bigwedge_{a \in I}((\bigvee_{a \notin T_i} q_i) \vee p_a) \tag{3}$$

The formula (1) allows to model the transaction database and then to catch the itemsets when an itemset appear in a transaction T_i ($q_i = 1$) if and only iff the variables not involved in T_i are set to false. The formula ($\neg q_i \leftrightarrow \bigvee_{a \in I \setminus T_i} p_a$) can be translated to the following CNF formula:

$$\bigwedge_{a \in I \setminus T_i} (\neg q_i \vee \neg p_a) \wedge (q_i \vee \bigvee_{a \in I \setminus T_i} p_a)$$

Formula (2) allows us to consider the item sets having a support greater than or equal to λ. This encoding is defined as a 0/1 linear inequality, usually called cardinality constraint. Because of the presence of such constraint in several applications, many efficient CNF encodings have been proposed over the years. Mostly, such encodings try to derive the best compact representation while preserving the efficiency of constraint propagation (e.g. [2, 3, 15, 24, 25]). The cardinality networks is the best known encoding of the constraint $\sum_{i=1}^{m} q_i \geq \lambda$ [2], it leads to a CNF formula with $\mathcal{O}(m \times log_2^2(\lambda))$ clauses and $\mathcal{O}(m \times log_2^2(\lambda))$ variables.

Formula (3) capture the closure property. Intuitively, if the itemset is involved in all transactions containing an item a then a must be added to the candidate itemset. In other words, when in all the transactions where a does not appear, the candidate itemset is not included, we deduce that the candidate itemset appears only in transactions containing the item a. Consequently, to be closed, the item a must be added to the final itemset.

Our proposed framework is declarative, as any new requirement can be easily considered by simply adding new constraints. The basic task of enumerating itemsets of size at most max, can be expressed by adding the following constraint:

$$\sum_{a \in I} p_a \leq max$$

Example 1 Let us consider the transaction database of Table 1. The Problem encoding the enumeration of frequent closed itemsets with a threshold 4 can be written as:

$$\begin{array}{ll}
\{\neg q_1 \leftrightarrow (p_E \vee p_F \vee p_G), & (q_5 \vee q_6 \vee q_7 \vee p_A),\\
\neg q_2 \leftrightarrow (p_C \vee p_D \vee p_G), & (q_4 \vee q_5 \vee q_6 \vee q_7 \vee p_B),\\
\neg q_3 \leftrightarrow (p_D \vee p_E \vee p_F \vee p_G), & (q_2 \vee q_5 \vee q_6 \vee q_7 \vee p_C),\\
\neg q_4 \leftrightarrow (p_B \vee p_E \vee p_G), & (q_2 \vee q_3 \vee q_5 \vee p_D),\\
\neg q_5 \leftrightarrow (p_A \vee p_B \vee p_C \vee p_D \vee p_E \vee p_F), & (q_2 \vee q_3 \vee q_4 \vee q_5 \vee q_6 \vee q_7 \vee p_E),\\
\neg q_6 \leftrightarrow (p_A \vee p_B \vee p_C \vee p_E \vee p_F \vee p_G), & (q_1 \vee q_3 \vee q_4 \vee q_5 \vee q_6 \vee q_7 \vee p_F),\\
\neg q_5 \leftrightarrow (p_A \vee p_B \vee p_C \vee p_E \vee p_F), & (q_1 \vee q_2 \vee q_3 \vee q_4 \vee q_6 \vee p_G),\\
 & q_1 + q_2 + q_3 + q_4 + q_5 + q_6 + q_7 \geq 4\}
\end{array}$$

In [16], the authors propose a new algorithm for enumerating top-k sat models that represent the set of closed itemsets. An application is also developed to enhance this task. The main idea behind their approach is to add constraints dynamically in order to take off the models that does not belong to the top-k. Formally, the SAT solver starts by finding k models $\{X_1, \ldots, X_k\}$, a lower bound η which is equal to $min\{|SC(X_1)|, \ldots, |SC(X_k)|\}$ is computed and a new constraint $\sum_{i=1}^{m} q_i \geq \eta$ is added to cut models that are not top-k. When a new model is found, we compare its support to the supports of $\{X_1, \ldots, X_k\}$ and the lower bound η is updated accordingly.

5 Enumerating all Models of CNF Formulae

A naive way to extend modern SAT solvers for the problem of enumerating all models of a CNF formula consists in adding a blocking clause (line 19 in Algorithm 1) to prevent the search to return the same model again. This approach is used in the majority of the model enumeration methods in the literature. The main limitation of this approach concerns the space complexity, since the number of blocking clauses may be exponential in the worst case. Indeed, in addition to the clauses learned at each conflict by the CDCL-based SAT solver, the number of added blocking clauses is very important on problems with a huge number of models. This explains why it is necessary to design methods avoiding the need to keep all blocking clauses. It is particularly the case for encodings of data mining tasks where the number of interesting patterns is often significant, even when using condensed representations such as closed patterns.

Our main aim is to experimentally study the effects of each component of modern SAT solvers on the efficiency of the model enumeration, in the case of the encoding described in Sect. 4. As the number of frequent closed itemsets is usually great, the enumerated models for the considered encoding is huge too. Consequently, it is not suitable to store the found models using blocking clauses during the enumeration process.

We proceed by removing incrementally some components of modern SAT solvers in order to evaluate their effects on the efficiency of model enumeration. The first removed component is the restart policy. Indeed, we inhibit the restart in order to allow solvers to avoid the use of blocking clauses. Thus, our procedure performs a simple backtracking at each found model. The second removed component is that of clause learning, which leads to a DPLL-like procedure. Considering a DPLL-like procedure, we pursue our analysis by considering the branching heuristics. Indeed, our goal is to find the heuristics suitable to the considered SAT encoding. To this end, we consider three branching heuristics. We first study the performance of the well-known VSIDS (Variable State Independent, Decaying Sum) branching heuristic. In this case, at each conflict an analysis is only performed to weight variables (no learnt clause is added). The second considered branching heuristic is based on the maximum number of occurrences of the variables in the formula. The third one consists in selecting the variables randomly.

6 Experimental Validation

We carried out an experimental evaluation to analyze the effects of adding blocking clauses, adding learned clauses and branching heuristics. To this end, we implemented a DPLL-like procedure, denoted `DPLL-Enum`, without adding blocking and learned clauses. We also implemented a procedure on the top of the state-of-the-art CDCL SAT solver MiniSAT 2.2, denoted `CDCL-Enum`. In this procedure, each time a model is found, we add a blocking clause (no-good) and perform a restart. We considered a variety of datasets taken from the `FIMI`[1] and `CP4IM`[2] repositories. All the experiments were done on Intel Xeon quad-core machines with 32GB of RAM running at 2.66 GHz. For each instance, we used a timeout of 15 min of CPU time.

In our experiments, we compare the performances of `CDCL-Enum` to three variants of `DPLL-Enum`, with different branching heuristics, in enumerating all the models corresponding to the closed frequent itemsets. The considered variants of `DPLL-Enum` are the following:

- `DPLL-Enum+VSIDS`: `DPLL-Enum` with the VSIDS branching heuristic;
- `DPLL-Enum+JW`: `DPLL-Enum` with a branching heuristic based on the maximum number of occurrences of the variables [17];
- `DPLL-Enum+RAND`: `DPLL-Enum` with a random variable selection.

6.1 *Experiments on Frequent Closed Itemset Mining*

Our comparison is depicted by the the cactus plots of Fig. 1. Each dot (x, y) represents an instance with a fixed minimal support threshold λ (Quorum). Each cactus plot represents an instance and the evolution of CPU time needed to enumerate all models with the different algorithms while varying the quorum. For each instance, we tested different values of λ. The x-axis (respectively y-axis) represents the CPU time (in seconds) needed for the enumeration of all closed frequent itemsets.

---

[1]FIMI: http://fimi.ua.ac.be/data/.

[2]CP4IM: http://dtai.cs.kuleuven.be/CP4IM/datasets/.

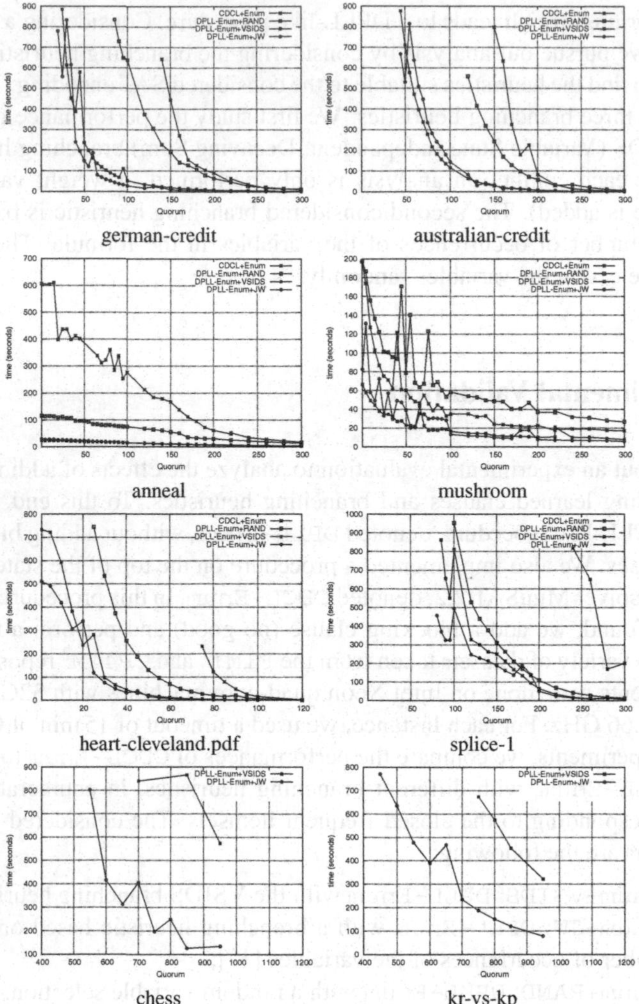

Fig. 1 Frequent closed itemsets: CDCL versus DPLL-like enumeration

Notice that the DPLL-like procedures outperform `CDCL-Enum` on the majority of instances. It is reasonable because the number of enumerated models is very large when the number of closed frequent itemsets is also very high. Hence, we can conclude that DPLL based proposal is more suitable for SAT-based items mining tasks. Furthermore, `DPLL-Enum+RAND` is clearly less efficient than `DPLL-Enum+VSIDS` and `DPLL-Enum+JW`, which shows that the branching heuristic plays a key role in model enumeration algorithms. Moreover, our experiments show that `DPLL-Enum+JW` is better than `DPLL-Enum+VSIDS`, even if `DPLL-Enum+VSIDS` compete with the procedure `DPLL-Enum+JW` on datasets such as `anneal` and mushroom. Indeed, DPLL-Enum+JW clearly outperforms DPLL-Enum+VSIDS on several datasets, such as chess, kr-vs-kp and splice-1. Note that for these two datasets, the solvers DPLL-Enum+RAND and CDCL-Enum are not able to enumerate completely the set of all models of all considered quorums.

As a summary, our experimental evaluation suggests that a DPLL-like procedure is a more suitable approach when the number of models of a propositional formula is significant. It also suggests that the branching heuristic is a key point in such a procedure to improve the performance.

6.2 Experiments on Top-*k*

In this section, we continue our investigation by exploring the behavior of SAT-Based itemset mining in the context of Top-*k* computation. Indeed, the main point in the introduction of Top-*k* is that it allows a faster computation of a subset of itemsets. The latter are considered as the most interesting ones modulo a desired preference relation over itemsets.

We perform two kind of experiments. We launch the solver DPLL-Enum+JW by assigning each variable to its truth value true DPLL-Enum+JW-Pos (respectively false DPLL-Enum+JW-Neg). In fact, our goal is to study the impact of the literal polarity on the performance of SAT-Based Top-*k* itemsets enumeration. Table 2 shows the behavior of Top-*k* Based SAT enumeration with both experiments. Each couple (x, y) represents the time needed by DPLL-Enum+JW-Neg and DPLL-Enum+JW-Pos respectively. Unlike the problem of enumerating all frequent closed itemsets, mining Top-*k* itemsets is more faster when variables are assigned to the false polarity. Indeed, DPLL-Enum+JW-Neg outperforms considerably the DPLL-Enum+JW-Pos. For instance, with splice-1 data, the procedure DPLL-Enum+JW-Neg needs some seconds to enumerate all Top-*k* itemsets while DPLL-Enum+JW-Neg needs several minutes of CPU time.

To understand the difference between the two solvers, in Table 3 we reports for each instance and for each considered k, the Top-*k* models. The couple (x, y) represents the number of models including Top-*k* and intermediary non Top-*k* models, founded by DPLL-Enum+JW-Neg (x) respectively DPLL-Enum+JW-Pos (y). We observed that with DPLL-Enum+JW-Neg, to compute the Top-*k* models, we enumerate a factor of four of intermediate non Top-*k* models. In contrast, DPLL-Enum+JW-Pos can enumerate a factor of one hundred of intermediary non Top-*k* models. This explains the time difference between the two models enumerators.

Table 2 Top-*k* (CPU time): `DPLL-Enum+JW-Pos` versus `DPLL-Enum+JW-Neg`

Instance/k	1000	5000	10000	50000	75000	100000
Primary-tumor	(0.01, 0.01)	(0.01, 0.03)	(0.03, 0.05)	(0.15, 0.14)	(0.18, 0.17)	(0.28, 0.16)
Hepatitis	(0.02, 0.15)	(0.07, 0.06)	(0.14, 0.37)	(0.51, 1.42)	(0.84, 1.71)	(0.84, 1.99)
Lymph	(0.01, 0.07)	(0.05, 0.14)	(0.08, 0.20)	(0.32, 0.38)	(0.38, 0.41)	(0.43, 0.38)
Soybean	(0.03, 0.12)	(0.12, 0.23)	(0.20, 0.30)	(0.38, 0.29)	(0.38, 0.29)	(0.38, 0.29)
Tic-tac-toe	(0.14, 0.22)	(0.39, 0.47)	(0.57, 0.56)	(0.76, 0.59)	(0.76, 0.59)	(0.78, 0.59)
Zoo-1	(0.007)	(0.01, 0.01)	(0.01, 0.01)	(0.01, 0.01)	(0.01, 0.01)	(0.01, 0.01)
Vote	(0.15, 0.24)	(0.44, 0.45)	(0.66, 0.6)	(1.41, 1.62)	(1.60, 1.43)	(1.5, 1.77)
Anneal	(0.04, 0.23)	(0.11, 0.58)	(0.18, 0.91)	(0.67, 2.7)	(0.91, 3.42)	(1.18, 4.02)
Heart-cleveland	(0.05, 0.38)	(0.16, 1.17)	(0.29, 1.8)	(1.2, 5.79)	(1.73, 7.67)	(2.32, 9.2)
Chess	(0.09, 1.33)	(0.23, 3.6)	(0.4, 5.57)	(1.95, 17.44)	(2.96, 22.7)	(3.87, 30.01)
German-credit	(0.17, 1.23)	(0.55, 3.02)	(0.88, 4.62)	(3.32, 12.74)	(5.48, 17.10)	(7.01, 20.47)
Austrial-credit	(0.09, 1.06)	(0.32, 2.97)	(0.62, 4.92)	(2.54, 14.54)	(3.62, 19.5)	(4.74, 23.44)
kr-vs-kp	(0.15, 1.49)	(0.37, 3.95)	(0.66, 6.22)	(3.02, 19.20)	(3.7, 25.59)	(4.48, 31.29)
Mushroom	(2.3, 10.29)	(5.8, 26.60)	(6.89, 37.46)	(33.09, 66.98)	(53.89, 66.70)	(65.78, 70.04)
Audiology	(0.06, 2.81)	(0.26, 9.35)	(0.52, 16.01)	(2.32, 56.10)	(3.20, 80.36)	(3.8, 102.05)
Connect	(5.6, 82.03)	(10.59, 189.30)	(15.13, –)	(55.61, –)	(83.09, –)	(102.32, –)

Finally, we compared, our method with a LCM algorithm proposed by Takeaki Uno et al. in [28], one of the best specialized algorithm, using the Top-*k* option. In Table 4, we provide the CPU time in seconds needed by LCM (Top-*k* option) and `DPLL-Enum+JW-Neg` respectively (separated by the slash symbol). As expected, on all the datasets and all tested values of k, LCM is able to enumerate the Top-k frequent closed itemsets in less than 10 s. It is important to note, that `DPLL-Enum+JW-Neg` is able to compute the Top-k frequent closed itemsets on most of the tested instances in less than 100 s except for splice-1. As a summary, comparatively, to previously published results, the gap with specialized approaches is significantly reduced.

Table 3 Number of top-k models and total number of enumerated models

Instance/k	100	1000	5000	10000	50000	75000	100000
Austrial-credit	103 (261, 7621)	1013 (2717, 49156)	5070 (19001, 186720)	10072 (39142, 330230)	50389 (215691, 1237416)	76975 (330280, 1734846)	100569 (435093, 2203668)
Heart-cleveland	103 (324, 6370)	1021 (3909, 37016)	5172 (21855, 132402)	10507 (43641, 232279)	51841 (219591, 899188)	78926 (332703, 1255339)	100678 (448248, 1581012)
Primary-tumor	101 (188, 3416)	1005 (2030, 14957)	5007 (10935, 40160)	10242 (20873, 55570)	50039 (71173, 86478)	77336 (85003, 87226)	87231 (87231, 87231)
Anneal	100 (268, 6742)	1004 (2942, 32673)	5029 (12650, 105220)	10148 (24745, 170889)	50403 (115708, 526281)	75524 (161776, 679300)	100996 (203796, 809313)
Hepatitis	121 (335, 4413)	1025 (3397, 27884)	5409 (20211, 97875)	11062 (42950, 165469)	51823 (191396, 526890)	83047 (274096, 697102)	104836 (346937, 847568)
Zoo-1	103 (282, 1040)	1027 (1991, 3415)	4568 (4568, 4568)	4568 (4568, 4568)	4568 (4568, 4568)	4568 (4568, 4568)	4568 (4568, 4568)
Lymph	100 (246, 3351)	1004 (2415, 14784)	5226 (12583, 41770)	10356 (23549, 62619)	51862 (89372, 134554)	79185 (115912, 147945)	108720 (135816, 152831)
Audiology	221 (929, 120649)	1583 (10745, 737803)	8615 (52307, 2753878)	38202 (107407, 4836316)	144016 (496001, 18633877)	144016 (716473, 26289896)	144016 (964565, 33419043)
Soybean	100 (207, 1680)	1029 (2043, 8052)	5057 (9509, 20653)	10041 (17014, 27459)	31760 (31760, 31760)	31760 (31760, 31760)	31760 (31760, 31760)
Tic-tac-toe	103 (341, 1210)	1011 (3300, 6549)	5411 (13413, 19029)	10717 (22568, 28441)	42712 (42712, 42712)	42712 (42712, 42712)	42712 (42712, 42712)
Chess	100 (152, 14817)	1005 (2716, 79601)	5019 (17283, 262391)	10018 (35613, 443580)	50174 (182140, 1539786)	75251 (285942, 2126693)	100051 (377538, 2673396)
Vote	105 (416, 2322)	1062 (4738, 11144)	5147 (21932, 32793)	10324 (39959, 47500)	50855 (126138, 133713)	75016 (159608, 164694)	101717 (182252, 183691)
German-credit	100 (302, 7486)	1013 (3889, 44790)	5031 (21814, 163615)	10126 (46044, 286502)	50460 (253203, 1042938)	75145 (379607, 1439813)	101001 (504992, 1807313)
kr-vs-kp	100 (150, 16485)	1005 (3589, 91877)	5019 (19929, 310933)	10018 (43726, 526486)	50174 (222536, 1775261)	75251 (331336)	75251 (436830,)
Mushroom	100 (202, 2337)	1000 (2316, 13420)	5016 (11632, 45457)	10127 (22014, 74974)	50914 (94294, 177773)	75364 (133102, 201894)	108114 (160138, 213301)

Table 4 CPU time (seconds) to compute top-k itemsets: LCM (Top-k option) versus DPLL-Neg

Instance/k	100	1000	5000	10000	50000	75000	100000
Chess	0.04/0.04	0.04/0.09	0.05/0.23	0.07/0.4	0.25/1.95	0.31/2.96	0.44/3.87
Heart-cleveland	0.01/0.01	0.03/0.05	0.05/0.16	0.13/0.29	0.53/1.2	0.8/1.73	1.04/2.32
Primary-tumor	0.01/0.03	0.01/0.01	0.01/0.03	0.03/0.05	0.15/0.14	0.18/0.17	0.28/0.16
Austrial-credit	0.01/0.02	0.04/0.09	0.15/0.32	0.23/0.62	0.98/2.54	1.25/3.62	1.62/4.74
Anneal	0.07/2.57	0.02/0.04	0.04/0.11	0.07/0.18	0.28/0.67	0.37/0.91	0.55/1.18
Hepatitis	0.001/001	0.01/002	0.06/007	0.10/0.14	0.43/0.51	0.61/0.84	1.2/0.84
Lymph	0.001/0.005	0.01/0.01	0.03/0.05	0.08/0.08	0.38/0.3	0.59/0.41	0.8/0.48
Vote	0.01/0.03	0.03/0.15	0.1/0.44	0.16/0.66	0.47/1.41	0.59/1.60	0.71/1.77
Soybean	0.001/0.01	0.03/0.03	0.03/0.12	0.12/0.20	0.12/0.38	0.12/0.38	0.12/0.38
German-credit	0.01/0.07	0.04/0.17	0.11/0.55	0.17/1.10	0.76/3.94	1.1/5.48	1.43/7.01
kr-vs-kp	0.01/0.07	0.04/0.15	0.11/0.37	0.12/0.66	0.25/3.02	0.43/4.3	0.61/5.67
Tic-tac-toe	0.001/0.03	0.01/0.14	0.05/0.39	0.06/0.57	0.17/0.77	0.17/0.76	0.17/0.78
Zoo-1	0.001/0.003	0.01/0.007	0.01/0.01	0.01/0.01	0.01/0.01	0.01/0.01	0.01/0.01
Audiology	0.01/0.07	0.01/0.06	0.15/0.26	0.43/0.52	1.71/2.32	2.10/3.20	2.69/4.23
Mushroom	0.01/0.73	0.08/2.3	0.17/5.8	0.26/9.1	0.74/43.48	1.01/59.47	1.03/72.99
Connect	0.81/3.81	1.18/5.6	1.55/10.59	2.64/15.13	2.21/55.61	2.56/83.09	2.8/102.32
Splice-1	0.07/2.57	0.32/6.07	0.93/24.77	1.75/70.00	6.28/332.67	8.32/451.09	10.27/557.92

7 Conclusion

In this paper, we conducted a deeper analysis of SAT based itemset mining approaches, by designing an efficient enumerator of the models of Boolean formula. Our goal is to measure the impact of the classical components of CDCL-based SAT solvers on the efficiency of model enumeration in the context of propositional formulas encoding itemset mining problems. Our results suggest that on formula with a huge number of models, SAT solvers must be adapted to efficiently enumerate all the models. We showed that the DPLL solver augmented with the classical Jeroslow-Wang heuristic achieve better performance. Concerning the enumeration of top-k itemsets, we demonstrated that the assignment literal polarity might have a great impact on the search efficiency.

As a future work, we plan to pursue our investigation in order to find the best heuristics for enumerating models of propositional formulas encoding data mining problem. For example, it would be interesting to integrate some background knowledge in such heuristic design. Finally, clause learning, an important component for the efficiency of SAT solvers, admits several limitation in the context of model enumeration. An important issue, is to study how such important mechanism can be efficiently integrated to dynamically learn relevant structural knowledge from formula encoding data mining problems.

References

1. Agrawal, R., Imieliński, T., Swami, A.: Mining association rules between sets of items in large databases. In: Proceedings of the 1993 ACM SIGMOD International Conference on Management of Data. SIGMOD '93, pp. 207–216. ACM, New York, NY, USA (1993)
2. Asín, R., Nieuwenhuis, R., Oliveras, A., Rodríguez-Carbonell, E.: Cardinality networks: a theoretical and empirical study. Constraints **16**(2), 195–221 (2011)
3. Bailleux, O., Boufkhad, Y.: Efficient CNF encoding of boolean cardinality constraints. In: CP, pp. 108–122 (2003)
4. Biere, A., Heule, M.J.H., van Maaren, H., Walsh, T. (eds.): Handbook of Satisfiability, Frontiers in AI and Applications, vol. 185. IOS Press (2009)
5. Chauhan, P., Clarke, E.M., Kroening, D.: Using sat based image computation for reachability analysis. Tech. rep., Technical Report CMU-CS-03-151 (2003)
6. Coquery, E., Jabbour, S., Saïs, L., Salhi, Y.: A SAT-based approach for discovering frequent, closed and maximal patterns in a sequence. In: Proceedings of the 20th European Conference on Artificial Intelligence (ECAI'12), pp. 258–263 (2012)
7. Davis, M., Logemann, G., Loveland, D.W.: A machine program for theorem-proving. Commun. ACM **5**(7), 394–397 (1962)
8. Fu, A.W.C., Kwong, R.W.W., Tang, J.: Mining n-most interesting itemsets. In: Proceedings of the 12th International Symposium on Methodologies for Intelligent Systems (ISMIS 2000), Lecture Notes in Computer Science, pp. 59–67. Springer (2000)
9. Gebser, M., Kaufmann, B., Neumann, A., Schaub, T.: Conflict-driven answer set enumeration. In: Baral, C., Brewka, G., Schlipf, J. (eds.) Logic Programming and Nonmonotonic Reasoning. Lecture notes in computer science, vol. 4483, pp. 136–148. Springer, Berlin (2007)
10. Guns, T., Nijssen, S., Raedt, L.D.: Itemset mining: a constraint programming perspective. Artif. Intell. **175**(12–13), 1951–1983 (2011)
11. Han, J., Pei, J., Yin, Y.: Mining frequent patterns without candidate generation. SIGMOD Rec. **29**, 1–12 (2000)
12. Han, J., Wang, J., Lu, Y., Tzvetkov, P.: Mining top-k frequent closed patterns without minimum support. In: Proceedings of the 2002 IEEE International Conference on Data Mining (ICDM 2002), 9–12 December 2002, Maebashi City, Japan, pp. 211–218 (2002)
13. Jabbour, S., Lonlac, J., Sais, L., Salhi, Y.: Extending modern SAT solvers for models enumeration. In: Proceedings of the 15th IEEE International Conference on Information Reuse and Integration, IRI 2014, Redwood City, CA, USA, August 13–15, 2014, pp. 803–810 (2014)
14. Jabbour, S., Sais, L., Salhi, Y.: Boolean satisfiability for sequence mining. In: 22nd ACM International Conference on Information and Knowledge Management (CIKM'13), pp. 649–658. ACM (2013)
15. Jabbour, S., Sais, L., Salhi, Y.: A pigeon-hole based encoding of cardinality constraints. TPLP **13**(4-5-Online-Supplement) (2013)
16. Jabbour, S., Sais, L., Salhi, Y.: The top-k frequent closed itemset mining using top-k sat problem. In: European Conference on Machine Learning and Knowledge Discovery in Databases (ECML/PKDD'03), pp. 403–418 (2013)
17. Jeroslow, R.G., Wang, J.: Solving propositional satisfiability problems. Ann. Math. Artif. Intell. **1**, 167–187 (1990)
18. Jin, H., Han, H., Somenzi, F.: Efficient conflict analysis for finding all satisfying assignments of a boolean circuit. In: In TACAS'05, LNCS 3440, pp. 287–300. Springer (2005)
19. Marques-Silva, J.P., Sakallah, K.A.: GRASP—A new search algorithm for satisfiability. In: Proceedings of IEEE/ACM CAD, pp. 220–227 (1996)
20. McMillan, K.L.: Applying sat methods in unbounded symbolic model checking. In: Proceedings of the 14th International Conference on Computer Aided Verification (CAV'02), pp. 250–264 (2002)
21. Morgado, A.R., Marques-Silva, J.A.P.: Good Learning and Implicit Model Enumeration. In: International Conference on Tools with Artificial Intelligence (ICTAI'2005), pp. 131–136. IEEE (2005)

22. Pei, J., Han, J., Lu, H., Nishio, S., Tang, S., Yang, D.: H-mine: hyper-structure mining of frequent patterns in large databases. In: Proceedings IEEE International Conference on Data Mining, 2001. ICDM 2001, pp. 441–448 (2001)
23. Raedt, L.D., Guns, T., Nijssen, S.: Constraint programming for itemset mining. In: ACM SIGKDD, pp. 204–212 (2008)
24. Silva, J.P.M., Lynce, I.: Towards robust cnf encodings of cardinality constraints. In: CP, pp. 483–497 (2007)
25. Sinz, C.: Towards an optimal cnf encoding of boolean cardinality constraints. In: CP'05, pp. 827–831 (2005)
26. Tiwari, A., Gupta, R., Agrawal, D.: A survey on frequent pattern mining: current status and challenging issues. Inf. Technol. J **9**, 1278–1293 (2010)
27. Tseitin, G.: On the complexity of derivations in the propositional calculus. In: H. Slesenko (ed.) Structures in Constructives Mathematics and Mathematical Logic, Part II, pp. 115–125 (1968)
28. Uno, T., Kiyomi, M., Arimura, H.: LCM ver. 2: Efficient mining algorithms for frequent/closed/maximal itemsets. In: FIMI '04, Proceedings of the IEEE ICDM Workshop on Frequent Itemset Mining Implementations, Brighton, UK, November 1, 2004 (2004)
29. Zhang, L., Madigan, C.F., Moskewicz, M.W., Malik, S.: Efficient conflict driven learning in Boolean satisfiability solver. In: IEEE/ACM CAD'2001, pp. 279–285 (2001)

Mining Frequent Movement Patterns in Large Networks: A Parallel Approach Using Shapes

Mohammed Al-Zeyadi, Frans Coenen and Alexei Lisitsa

Abstract This paper presents the Shape based Movement Pattern (ShaMP) algorithm, an algorithm for extracting Movement Patterns (MPs) from network data that can later be used (say) for prediction purposes. The principal advantage offered by the ShaMP algorithm is that it lends itself to parallelisation so that very large networks can be processed. The concept of MPs is fully defined together with the realisation of the ShaMP algorithm. The algorithm is evaluated by comparing its operation with a benchmark Apriori based approach, the Apriori based Movement Pattern (AMP) algorithm, using large social networks generated from the Cattle tracking Systems (CTS) in operation in Great Britain (GB) and artificial networks.

Keywords Knowledge Discovery and Data Mining · Distributed AI Algorithms · Systems and Applications

1 Introduction

Networks of all kinds feature with respect to many application domains; social networks [1], computer networks [2], Peer-to-Peer Networks [3], road traffic networks [4] and so on. A network G is typically defined in terms of a tuple of the form $\langle V, E \rangle$, where V is a set of vertices and E is a set of edges [5]. The vertices can represent individuals (as in the case of social networks), inanimate entities (as in the case of computer networks) or locations (as in the case of distribution and road traffic networks). The edges then indicate connections between vertices (virtual or actual). These edges might be indicative of some relationship, such as a friend relationship,

M. Al-Zeyadi (✉) · F. Coenen · A. Lisitsa
Department of Computer Science, University of Liverpool, Ashton Building,
Ashton Street, Liverpool L69 3BX, UK
e-mail: m.g.a.al-zeyadi@liv.ac.uk

F. Coenen
e-mail: Coenen@liv.ac.uk

A. Lisitsa
e-mail: lisitsa@liv.ac.uk

M. Bramer and M. Petridis (eds.), *Research and Development in Intelligent Systems XXXIII*, DOI 10.1007/978-3-319-47175-4_4

as in the case of social networks; or be indicative of a "hard" connection as in the case of a wired computer network or a road traffic network. However in this paper we conceive of edges as indicating traffic flow. For example: the number of messages sent from one individual to another in a social network, the volume of data exchanges between two computers in a compute network, the quantity of goods sent in a distribution network or the amount of traffic flow from one location to another in a road traffic flow network. As such our edges are directed (not necessarily the case in all net works). To distinguish such networks from other networks we will use the term *movement network*; a network $G(V, E)$ where the edges are directed and indicate some kind of "traffic flow" (for example messages, data, goods or vehicles) moving from one vertex to another.

An example movement network is given in Fig. 1 where $V = \{\phi_1, \phi_2, \ldots, \phi_5\}$ and $E = \{\varepsilon_1, \varepsilon_2, \ldots, \varepsilon_9\}$. Note that some of the vertices featured in the network double up as both "to" and "from" vertices. Thus the set of from vertices and the set of to vertices are not disjoint. Note also that some vertices in the figure are connected by more than one edge. This is because we do not simply wish to consider traffic flow in a binary context (whether it exists or does not exist) but in terms of the nature of the traffic. Thus where vertices in the figure are connected by more than one edge this indicates more than one kind of traffic flow. For example in a distribution movement network two edges connecting one vertex to a another might indicate the dispatch of two different commodities. As such edges have a set of attributes associated with them A_E. Similarly the vertices will also have a set of attributes associated with them, A_V. The nature of these attribute sets will depend on the nature of the application domain of interest, however each attribute will have two or more values associated with them. Where necessary we indicate a particular value *j* belonging to attribute *i* using the notation v_{i_j}.

Given a movement network of the form shown in Fig. 1 the nature of the traffic exchange between the vertices in the network can be described in terms of a *Movement Pattern* (MP); a three part pattern describing some traffic exchange comprising:

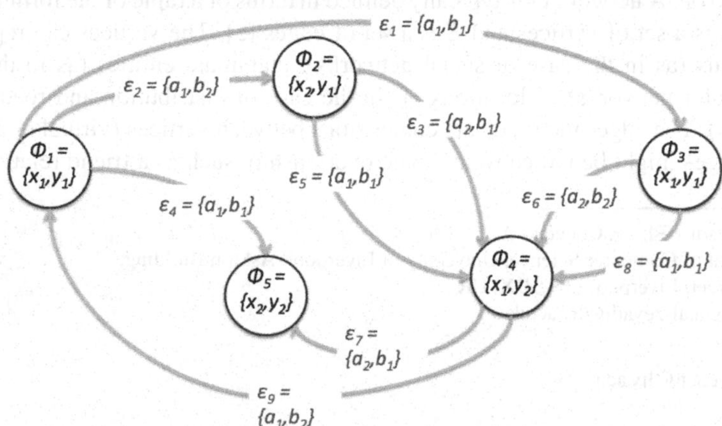

Fig. 1 Example movement network ($V = \{\phi_1, \phi_2, \ldots, \phi_5\}$ and $E = \{\varepsilon_1, \varepsilon_2, \ldots, \varepsilon_9\}$)

(i) a from vertex (F), (ii) the nature of the traffic (E) and (iii) a to vertex (T). Thus "sender", "details of movement" and "receiver". A movement pattern can thus be described as a tuple of the form $\langle F, E, T \rangle$, where F and T subscribe to the attribute set A_V, and E to the attribute set A_E. In the figure $A_V = \{x, y\}$ and $A_E = \{a, b\}$; each attribute x, y, a and b has some value set associated with it, in the figure the value sets are $\{x_1, x_2\}$, $\{y_1, y_2\}$, $\{a_1, a_2\}$ and $\{b_1, b_2\}$ respectively (value sets do not all have to be of the same length). Given a movement network G we wish to find the set of movement patterns M in G so that they can be used to predict movement in a previously unseen unconnected graph $G'(V')$ so as to generate a predicted connected graph $G'(V', E')$. More specifically we wish to be able to extract the set M from very large networks comprising more than one million vertices. To this end the Shape Movement Pattern (ShaMP) algorithm is proposed and evaluated in this paper. The algorithm leverages knowledge of the restrictions imposed by the nature of MPs to obtain efficiency advantages not available to more traditional pattern mining algorithms that might otherwise be adopted such as Frequent Itemset Mining (FIM) algorithms. A further advantage of the ShaMP algorithm, as will be demonstrated, is that it readily lends itself to parallelisation [6]. The algorithm is fully described and its operation compared to a benchmark Apriori style approach using both artificial movement networks and real life networks extracted from the Cattle Tracking System (CTS) database in operation in Great Britain.

The rest of this paper is organised as follows. Section 2 presents a review of previous and related work on Movement Patterns in large networks and how parallel processing can improve the efficiency of such algorithms. Section 3 provides a formal definition of the movement pattern concept. Section 4 then describes the proposed ShaMP algorithm and the Apriori Movement Pattern (AMP) benchmark algorithm used for comparison proposes in the following evaluation section, Sect. 5. Section 6 summarises the work, itemises the main findings and indicates some future research directions.

2 Literature Review

In the era of big data the prevalence of networks of all kinds has growing dramatically. Coinciding with this growth is a corresponding desire to analyse (mine) such networks, typically with a view to some social and/or economic gain. A common application is the extraction of customer buying patterns [7, 8] to support business decision making. Network mining is more focused but akin to graph mining [9, 10]. Network mining can take many forms; but the idea presented in this paper is the extraction of Movement Patterns (MPs) from movement networks. The concept of Movement Pattern Mining (MPM) as conceived of in this paper, to the best knowledge of the authors, has not been previously addressed in the literature. However, patten mining in general has been extensively studied. This section thus presents some background to the work presented in the remainder of this paper. The section commences with a brief review of the pattern mining context with respect to large graph networks, then continues with consideration of current work on frequent movement patterns and the usage of parallel algorithms to find such patterns in large data networks.

A central research theme within the domain of data mining has been the discovery of patterns in data. The earliest examples are the Frequent Pattern Mining (FPM) algorithms proposed in the early 1990s [11]. The main objective being to discover sets of attribute-value pairings that occur frequently. The most well established approach to FPM is the Apriori approach presented in [11]. A frequently quoted disadvantage of FPM is the significant computation time required to generate large numbers of patterns (many of which may not be relevant). The MPM concept presented in this paper shares some similarities with the concept of frequent pattern mining and hence a good benchmark algorithm would be an Apriori based algorithm such as the AMP algorithm presented later in this paper. The distinction between movement patterns and traditional frequent patterns is that movement patterns are more prescriptive, as will become apparent from the following section; they have three parts "sender, details of movement and receiver"; none of these parts can be empty. Note also that the movement patterns of interest with respect to this paper are network movement patterns and not the patterns associated with the video surveillance of individuals, animals or road traffic; a domain where the term "movement pattern" is also used. To the best knowledge of the authors there has been very little (no?) work on movement patterns as conceptualised in this paper.

The MPM concept also has some overlap with link prediction found in social network analysis where we wish to predict whether two vertices, representing individuals in a social network, will be linked at some time in the future. MPM also has some similarity the problem of inferring missing links in incomplete networks. The distinction is that link prediction and missing link resolution is typically conducted according to graph structure, dynamic in the case of link prediction [12] and static in the case of missing link resolution [13].

As noted above, the main challenge of finding patterns in network data is the size of the networks to be considered (in terms of vertices and edges). Parallel or distributed computing is seen as a key technology for addressing this challenge. One of the main advantages offered by the proposed ShaMP algorithm is that it readily lends itself to parallelisation (unlike traditional FPM algorithms which require significant interchange of information). The design of parallel algorithms involves a number of challenges in addition to those associated with the design of serial algorithms [6]. In the case of MPM the main challenge is the distribution of the data and tasks across the available processes.

In the context of the proposed ShaMP algorithm a Distributed Memory Systems (DMS) was used; a memory model in which each process has a dedicated memory space. Using a distributed memory approach only the input data and task information is shared. There are several programming models that may be used in the context of DMS. However, by far the most widely used for large-scale parallel applications is the *Message Passing Interface* (MPI) [14]; a library interface specification that supports distributed-memory parallelisation, where data is divided among a set of "processes" and each process runs in its own memory address space. Processes typically identify each other by "ranks" in the range $\{0, 1, \ldots, p-1\}$, where p is the number of processes [15]. MPI is a very powerful and versatile standard used for parallel programming of all kinds. APIs that adopt the MPI standard, at their simplest, typically provide send and receive functionality, however most such APIs also provide a wide variety of additional functions. For example, there may be functions for various "collective" communications, such as "broadcast" or "reduction" [15]. MPI based APIs exist in a range of programming languages such as C, C++, Fortran and Java; well-tested and efficient implementations include: MPICH [16], MPICH-G2 [17] and G-JavaMPI [18]. However, in the context of the research presented in this paper MPJ Express [19] was adopted; this is an implementation of MPI using the Java programming language. Interestingly, MPJ has two ways of configuring the same code; Multicore configuration and Cluster configuration. In this paper we have adopted both Multicore configuration (by using a single multicore machine) and Cluster configuration using a Linux cluster.

3 Formalism

From the introduction to this paper we are interested in mining Movement Patterns (MPs) from movement networks. A MP comprises a tuple of the form $\langle F, E, T \rangle$ where F, E and T are sets of attribute values. The minimum number of attribute values in each part must be at least one. The maximum number of values depends on the size of the attribute sets to which F, E and T subscribe, an MP can only feature a maximum of one value per subscribed attribute. More specifically the attribute value set F represents a "From" vertex (sender), T a "To" vertex (receiver), and E an "Edge" connecting the two vertices describing the nature of the traffic ("details of movement"). The attribute set to which F and T subscribe is given by $A_V = \{\phi_1, \phi_2, \ldots\}$, whilst the attribute set for E is given by $A_E = \{\varepsilon_1, \varepsilon_2, \ldots\}$. Note that F and T subscribe to the same attribute set because they are both movement network vertices, and every vertex (at least potentially) can be a "from" or a "to" vertex in the context of MPM (as illustrated in Fig. 1).

The movement networks from which we wish to extract MPs (the training data) can also be conceived of as comprising $\langle F, E, T \rangle$ tuples. In fact an entire network G can be represented as a tabular data set D where each row comprises a $\langle F, E, T \rangle$ tuple defined as described above. The movement network presented in Fig. 1 can thus be presented in tabular form as shown in Table 1 (for ease of understanding the rows are ordered according to the edge identifiers used in the figure). We refer to such data as FET data (because of the nature of the tuples that the rows describe). Thus MPM can be simplistically thought of as searching a training set D and extracting the set M of all frequently occurring MPs, that can then be used to build a model of D. This model can then be used to predict traffic in some previously unseen network D' (G') comprised solely of vertices (no known edges, the edges are what we wish

Table 1 Movement network from Fig. 1 presented in tabular form

$\langle\{x_1,y_1\},\{a_1,b_1\},\{x_1,y_1\}\rangle$
$\langle\{x_1,y_1\},\{a_1,b_2\},\{x_2,y_1\}\rangle$
$\langle\{x_2,y_1\},\{a_2,b_1\},\{x_1,y_2\}\rangle$
$\langle\{x_1,y_1\},\{a_1,b_1\},\{x_2,y_2\}\rangle$
$\langle\{x_2,y_1\},\{a_1,b_1\},\{x_1,y_2\}\rangle$
$\langle\{x_1,y_1\},\{a_2,b_2\},\{x_1,y_2\}\rangle$
$\langle\{x_1,y_1\},\{a_1,b_1\},\{x_1,y_1\}\rangle$
$\langle\{x_1,y_2\},\{a_2,b_1\},\{x_2,y_2\}\rangle$
$\langle\{x_1,y_2\},\{a_1,b_2\},\{x_1,y_1\}\rangle$

Table 2 The MPs (the set M) extracted from the movement network given in Fig. 1 using $\sigma = 30\,\%$

$\langle\{x_1\},\{a_1\},\{x_1\}\rangle$	#3
$\langle\{y_1\},\{a_1,b_1\},\{x_1\}\rangle$	#3
$\langle\{y_1\},\{a_1,b_1\},\{y_2\}\rangle$	#3
$\langle\{y_1\},\{a_1\},\{x_1\}\rangle$	#3
$\langle\{y_1\},\{a_1\},\{y_2\}\rangle$	#3
$\langle\{y_1\},\{b_1\},\{x_1\}\rangle$	#4
$\langle\{y_1\},\{b_1\},\{x_1,y_2\}\rangle$	#3
$\langle\{y_1\},\{b_1\},\{y_2\}\rangle$	#4

to predict). An MP is said to be frequent, as in the case of traditional FPM [20], if its occurrence count in D is in excess of some threshold σ expressed as a proportion of the total number of FETs in D. With reference to the movement network given in Fig. 1, and assuming $\sigma = 30\,\%$, the set of MPS, M, will be as listed in Table 2 (the numbers indicated by the # are the associated occurrence counts).

4 Movement Pattern Mining

In this section the proposed ShaMP algorithm is presented together with the Apriori Movement Pattern (AMP) algorithm used for comparison purposes later in this paper (Sect. 5). The AMP algorithm is a variation of the traditional Apriori frequent itemset mining algorithm [7] redesigned so as to address the frequent MP mining problem. We refer to this benchmark algorithm as the Apriori Movement Pattern (AMP) algorithm. The major limitation of the Apriori strategy used by the AMP algorithm is the large number of candidate patterns that are generated during the process, which limits the size of the networks that can be considered. Also, when using the AMP algorithm, with respect to a given network G (dataset D), the run time is inversely proportional with the support threshold σ used. The ShaMP algorithm, however, uses knowledge of the nature of the FET "shapes" to be considered (the fact that they have three parts and that each part must feature at least one attribute-value) of which these are only a limited number. Broadly, the ShaMP algorithm operates by generating a set of MP shapes to which records are fitted (there is no candidate generation). The algorithm offers two advantages: (i) the nature of the value of σ has very little (no?) effect on the algorithm's run time and (ii) individual shapes can be considered in isolation hence the algorithm is well suited to parallelisation (not the case with respect to the AMP algorithm which operates level by level in a top down manner and, when parallelised, features significant message passing after each level). Consequently, as will be demonstrated in Sect. 5, the ShaMP algorithm is able to process much larger networks than in the case of the AMP algorithm. The ShaMP algorithm is considered in Sect. 4.1 below, while the AMP algorithm is considered in Sect. 4.2.

4.1 The Shape Based Movement Pattern (ShaMP) Algorithm

From the foregoing the ShaMP algorithm, as the name suggests, is founded on the concept of "shapes". A shape in this context is a MP template with a particular configuration of attributes taken from A_L and A_E (note, shapes do not specify particular attribute values combinations). The total number of shapes that can exist in a FET dataset D can be calculated using Eq. 1, where: (i) $|A_L|$ is the size of the attribute set A_L and (ii) $|A_E|$ is the size of the attribute set A_E. Recall that attributes for F and T are drawn from the same domain. Thus if $|A_L| = 2$ and $|A_E| = 2$, as in the case of the movement network given in Fig. 1, there will be $(2^2 - 1) \times (2^2 - 1) \times (2^2 - 1) = 2 \times 2 \times 2 = 16$ different shapes. If we increase $|A_E|$ to 5 there will be $(2^2 - 1) \times (2^5 - 1) \times (2^2 - 1) = 3 \times 31 \times 3 = 279$ different shapes. In other words the efficiency of the ShaMP algorithm is directly related to the number of different shapes that need to be considered.

$$(2^{|A_V|} - 1) \times (2^{|A_E|} - 1) \times (2^{|A_V|} - 1) \quad (1)$$

The pseudo code for the ShaMP Algorithm is given in Algorithm 1. The input is a network G, represented in terms of a FET dataset D, and a desired support threshold σ. The output is set of frequently occurring MPs M together with their support value represented in tuples of the form $\langle MP_i, count_i \rangle$. The algorithm commences (line 5) by generating the available set of shapes, *ShapeSet*, and setting M to the empty set (line 6). We then loop through the *ShapeSet* (lines 7–15), and for each shape loop through D comparing each record $r_j \in D$ with the current shape $shape_i$. A record r_j matches a $shape_i$ if the attributes featured in the shape also feature in r_j. Where a match is found the relevant attribute values in r_i form a MP. If the identified MP is already contained in M we simply update the associated support value, otherwise we add the MP to M with a support value of 1. Once all shapes have been processed we loop through M (lines 15–16) and remove all MPs whose support count is less than σ.

```
   Input:
1  D = Collection of FETs, {r1, r2, ...} describing a network G
2  σ = Support threshold
   Output:
3  M = Set of frequently occurring MPs {⟨MP_1, count_1⟩, langleMP_2, count_2⟩, ...}
4  Start:
5  ShapeSet = the set of possible shapes {shape_1, shape_2, ...}
6  M = ∅
7  forall shape_i ∈ ShapeSet do
8      forall r_i ∈ D do
9          if r_i matches shape_i then
10         MP_k = MP extracted from r_i
11         if MP_k in M then increment support
12         else M = M ⋃ ⟨MP_k, 1⟩
13     end
14 end
15 forall MP_i ∈ M do
16 if count for MP_i < σ then remove MP_i from M
17 end
```

Algorithm 1: Shape-Based Movement Pattern Algorithm

4.2 The Apriori Based Movement Pattern (AMP) Algorithm

The AMP algorithm is presented in this section. The significance is that it is used for evaluation purposes later in this paper. As noted above, the AMP algorithm is founded on the the Apriori frequent itemset mining algorithm presented in [11] (see also [10]). The pseudo code for the AMP algorithm is presented in Algorithm 2. At face value it operates in a very similar manner to Apriori in that it adopts a candidate generation, occurrence count and prune cycle. However, the distinction is that we are dealing with three part MPs ($\langle F, E, T\rangle$) and that none of these parts should be empty. Thus, where the variable k in the traditional Apriori algorithm refers to itemset size (starting with $k = 1$), the variable k in the AMP algorithm refers to levels. We start with the $k =$ level 1 MPs which feature one attribute value in each part ($\langle F, E, T\rangle$). At level two, $k =$ level 2, we add an additional attribute-value to one of the parts; and so on until no more candidates can be generated. Candidate generation and occurrence counting is therefore not straight-forward, it is not simply a matter of uniformly "growing" K-itemsets to give $K + 1$-itemsets and passing through D looking for records where each K-itemset is a subset. Candidate set growth is thus a complex procedure (not included in Algorithm 2) and requires knowledge of the values for

A_L and A_E. It should also be noted that when conducting occurrence counting only certain columns in D can correspond to each element in a candidate FET.

```
   Input:
1    D = Binary valued input data set ; σ = Support threshold
   Output:
2    M = Empty set of frequently occurring MPs {⟨MP_1, count_1⟩, langleMP_2, count_2⟩, ...}
3  Start:
4  M = ∅, k = level 1
5  C_k = Level k candidate movement patterns
6  while C_k ≠ ∅ do
7      S = {0, 0, ...} Set of length |C_k| to hold occurrence counts (one-to-one C_k
         correspondence)
8      G = ∅ Empty set to hold frequently occurring level K movement patterns
9      forall r_i ∈ D do
10         forall c_j ∈ C_k do
11             if c_j ⊂ r_i then s_j = s_j + 1 (s_j ∈ S)
12         end
13     end
14     forall c_j ∈ C_k do
15         if s_j ≥ σ then
16             G = G ∪ c_j
17             M = M ∪ c_j
18     end
19     k = k + 1
20     C_k = Level k candidate movement patterns drawn from G
21 end
```

Algorithm 2: Apriori based Movement Pattern (AMP) Algorithm

Returning to Algorithm 2, as in the case of the ShaMP algorithm, the input is a FET dataset D, and a desired support threshold σ. The output is a set of frequently occurring MPs M together with their support values. The algorithm commences (line 3) by setting M to the empty set. The algorithm then proceeds level by level starting with the level one ($k = 1$) candidate sets (line 5), and continues until no more candidates can be generated. On each iteration a set S and a set G is defined to hold occurrence counts and the eventual size K MPs (if any). We then loop through the records in D (line 9), and for each record $r_i \in D$ loop through C_k (line 10). For each each $c_j \in C_k$ if c_j is a subset of r_i we increment the relevant support count in S (line 12). Once all the records in D have been process we loop through C_k again, compare each associated s_j value with σ and if it is larger add it to the sets G and M. We then increment k (line 20), generate a new candidate set C_k from G and repeat. The main weaknesses of the algorithm, as noted above, are the large number of candidates that may be generated (the maximum number equates to $2^{|V_{A_L}|+|V_{E_I}|+|V_{A_L}|} - 1$). Note that, in common with more general Apriori algorithms, the number of relevant candidates to be considered becomes larger when low values for σ are used (as opposed to the ShaMP algorithm where the σ value dose not have a significant effect).

5 Experiments and Evaluation

This section reports on the experiments conducted to analyse the operation of the proposed ShaMP algorithm and compare this operation with the benchmark AMP algorithm. The evaluation was conducted using two categories of network data: (i) networks extracted from the CTS database introduced in Sect. 1 and (ii) artificial networks where the parameters can be easily controlled. The objectives of the evaluation were as follows:

1. To determine whether the nature of the σ threshold used would in anyway adversely affect the ShaMP algorithm (unlike as anticipated in the case of the AMP algorithm).
2. To determine the effect on the ShaMP algorithm of the size of the network under consideration, in terms of the number of FETs, and in comparison with the AMP algorithm.
3. To determine the effect on the ShaMP algorithm on the number shapes to be considered, and if there is a point where the number of shapes is so large that it is more efficient to use an algorithm such as the AMP algorithm.
4. The effect of increasing the number of processes available with respect to the ShaMP algorithm (the first three sets of experiments were conducted using a single processor).

The remainder of this section is divided into five sections as fallows. Section 5.1 gives an overview of the CTS and artificial networks used with respect to the reported evaluation. The remaining four section, Sects. 5.2–5.5, report respectively on the experimental results obtained with respect to the above four objectives.

5.1 Data Sets

The CTS database was introduced in the introduction to this paper. The database was used to generate a collection of time stamped networks where for each network the vertices represent locations (cattle holding areas) and the edges represent occurrences of cattle movement between vertexes. The database was preprocessed so that each record represented a group of animals moved of the same type, bread and gender, from a given "from location" to a given "to location" on the same day. The set A_L comprised: (i) holding area type and (ii) county name. While the set A_E comprised: (i) number of cattle moved, (ii) breed, (iii) gender, (iv) whether the animals moved are beef animals or not, and (v) whether the animals moved are dairy animals or not. Thus $A_V = \{holdingAreaType, county\}$, and $A_E = \{numCattle, bread, gender, beef, dairy\}$. Note that the attributes *beef* and *dairy* are boolean attributes, the attribute *numCattle* is a numeric attribute, while the remaining attributes are categorical (discrete) attributes. Both the ShaMP and AMP algorithms were designed to operate with binary valued data (as in the case of traditional frequent item set mining). The values for the *numCattle* attribute were thus ranged into five sub ranges: $n \leq 10$, $11 \leq n \leq 20$, $21 \leq n \leq 30$, $31 \leq n \leq 40$ and $n > 40$. Thus each record represents a FET. The end result of the normalisation/discretisation exercise was an attribute value set comprising 391 individual attribute values.

We then used this FET dataset to define movement networks where vertices represent locations and edges traffic (animals moved), to which the proposed algorithms were applied to extract MPs. In total four networks were generated covering the years 2003, 2004, 2005 and 2006. The number of vertices in each network was about 43, 000, while the number of edges was about 270,000. Using Eq. 1 the number of shapes that will need to be considered by the ShaMP algorithm will be:

$$|ShapeSet| = 2^2 - 1 \times 2^5 - 1 \times 2^2 - 1 = 3 \times 15 \times 3 = 279$$

The maximum number of candidate MPs that the AMP algorithm may need to consider on iteration one will be $102 \times 188 \times 102 = 1{,}955{,}952$, and this is likely to increase exponentially on the following iterations (thus a significant difference).

The purpose of also using artificially generated networks was that the parameters could be controlled, namely the number of FETs and the number of attributes in the sets A_L and A_E. Further detail concerning individual, or groups, of artificial movement networks will be given where relevant in the following sections where the evaluation results obtained with respect to individual experiments are reported and discussed.

5.2 Support Threshold

In terms of Frequent Itemset Mining (FIM) the lower the σ value the more frequent itemsets that will be found, low σ values are thus seen as desirable because by finding many frequent itemsets there is less chance of missing anything significant. The same is true for MPM. In terms of FIM it is well established that efficiency decreases as the number of potential frequent itemsets increases (as the value of σ decreases). It was anticipated that this would also be true with respect to the AMP algorithm. How this would effect the ShaMP algorithm was unclear, although it was conjectured that changing σ values would have little effect. A sequence of experiments was thus conducted using a range of σ values from 5.0 to 0.5 decreasing in steps of 0.5, and using the four CTS movement networks described above. The results are presented in Fig. 2 where, for each graph, the x-axis gives the σ values and the y-axis runtime in seconds. Similar results are recorded in all cases. As expected, in the case of the AMP algorithm, run time increases exponentially as σ decreases. Also, as conjectured, from the figures it can be seen that *sigma* has little effect on the ShaMP algorithm. It is interesting to note that it is not till σ drops below 1.0 that usage of the the ShaMP algorithm becomes more advantageous than usage of the AMP algorithm. This last is significant because we wish to identify as many relevant MPs as possible and to do this we would need to use low σ values ($\sigma \leq 1.0$).

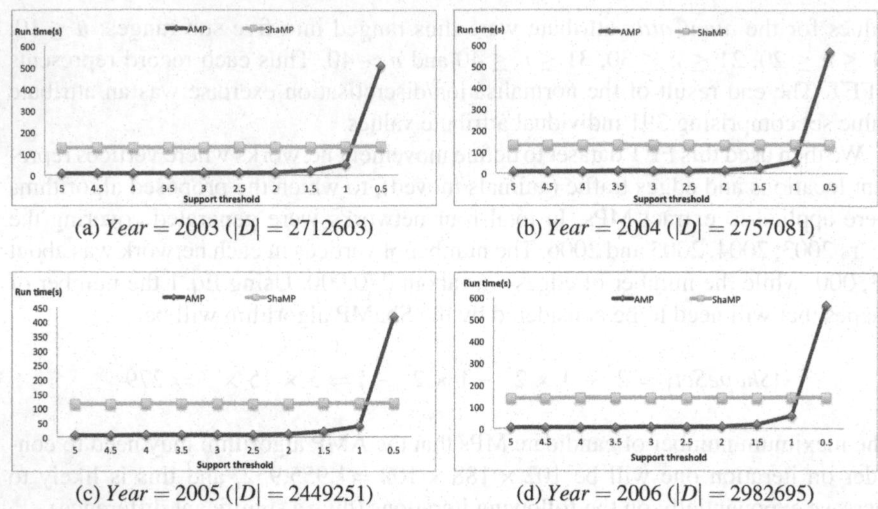

(a) $Year = 2003$ ($|D| = 2712603$) (b) $Year = 2004$ ($|D| = 2757081$)

(c) $Year = 2005$ ($|D| = 2449251$) (d) $Year = 2006$ ($|D| = 2982695$)

Fig. 2 Runtime (secs.) comparison between ShaMP and AMP using CTS network datasets from 2003–2006 with σ values

5.3 Number of FETs

The correlation between the algorithms and number of FETs was tested using five artificial datasets of increasing numbers of FETs 100,000, 200,000, 300,000, 500,000 and 1,000,000. The algorithms were applied to each of these data sets twice, once using $\sigma = 1.0$ and once using $\sigma = 0.5$. Note that for the artificial data the size of the attribute value sets used were $|V_{A_L}| = 102$ and $|V_{A_E}| = 188$. The results are presented in Fig. 3. From the figures it can clearly be seen that, as was to be expected, the run time increases as the number of records considered increases. However, the ShaMP algorithm is clearly more efficient than the AMP algorithm. It can also be seen that the run time with respect to the AMP algorithm increases dramatically as the number of FETs increases, while the increase with respect to the ShaMP algorithm is less

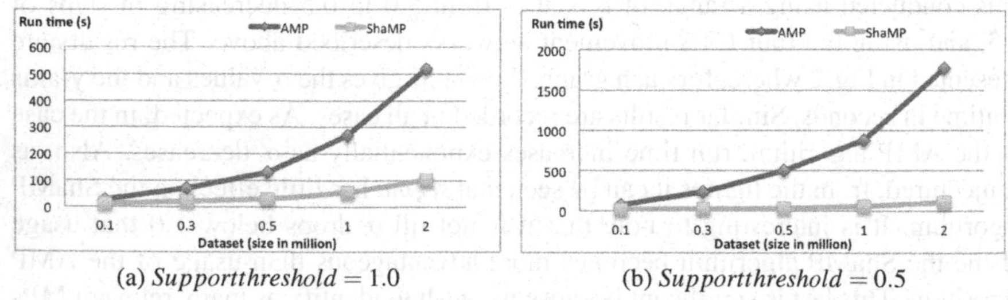

(a) $Supportthreshold = 1.0$ (b) $Supportthreshold = 0.5$

Fig. 3 Runtime (secs.) comparison between ShaMP and AMP using artificial networks

dramatic. If we increase the number of FETs beyond 1,000,000 the AMP algorithm is not able to cope. However the ShaMP algorithm, can process 10,000,000 FETs using a single processor with $\sigma = 0.5$, in 480 s of runtime.

5.4 Number of Shapes

The experiments discussed in the foregoing two sections indicated the advantages offered by the ShaMP algorithm. However as $|A_L|$ and/or $|A_E|$ increases the number of shapes will also increase. There will be a point where the number of shapes to be considered is such that it is no longer effective to use the ShaMP algorithm and it would be more beneficial to revert to the AMP algorithm. To determine where this point might be a sequence of artificial data sets were generated using $|A_V| = 2$ but increasing $|A_E|$ from 500 to 5000 in steps of 500. Each attribute had two possible attribute-values so as to allow the AMP algorithm to perform to its best advantage. (For the experiments $\sigma = 2$ was used). The results are shown in Fig. 4. As expected, the run time for both algorithms increased as the number of Shapes (attribute values) increased. It is interesting to note that there is "cross-over" when the number of shapes reaches about 2600; in practice it is difficult to image movement networks where the traffic has to be described in terms of 2600 attributes or more (Fig. 5).

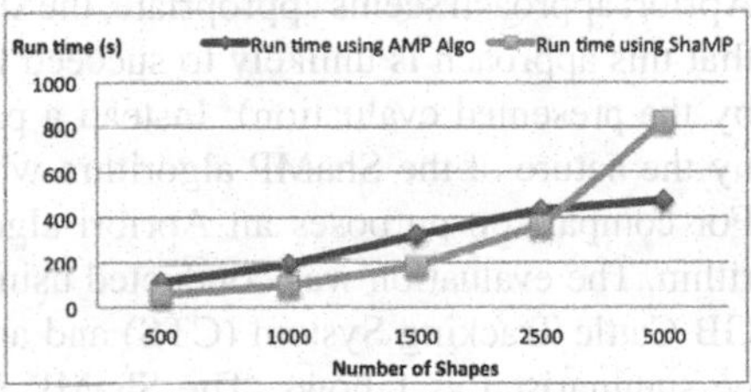

Fig. 4 Runtime plotted against increasing numbers of shapes ($\sigma = 2$)

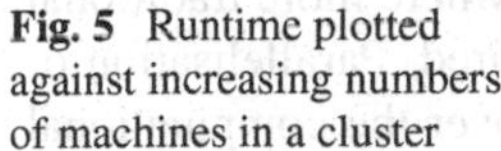

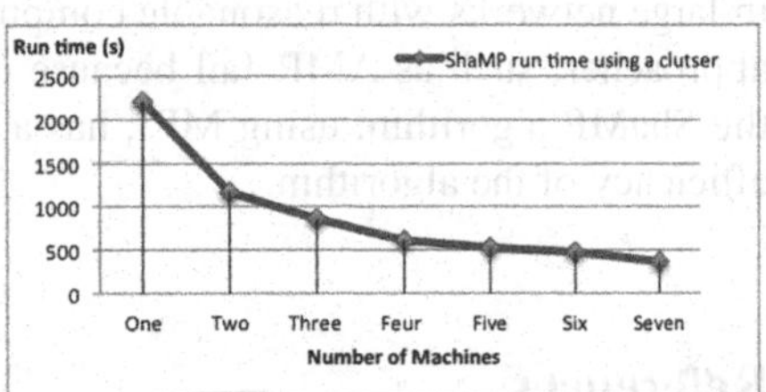

Fig. 5 Runtime plotted against increasing numbers of machines in a cluster

5.5 *Distributed ShaMP*

In the context of the experiments conducted to determine the effect of distributing the ShaMP algorithm the desired parallelisation was achieved using MPJ Express [19]. A cluster of Linux machines was used, of increasing size from 1 to 7, each with 8 cores (thus a maximum of $7 \times 8 = 56$ cores). Shapes were distributed evenly across machines. For the experiments an artificial data set comprising 4600 different Shapes were used and $\sigma = 2$. Note that the AMP algorithm was unable to process FET datasets of this size. The results are shown in Fig. 4. As predicted, as the number of machines increases the run time decreased significantly, there is no significant "distribution overhead" as a result of increasing the parallelisation. A good result.

6 Conclusion and Future Work

In this paper, the authors have proposed the ShaMP algorithm for identifying Movement Pattens (MPs) in network data. The MPs were defined in terms of three parts: From (F), Edge (E) and To (T). The acronym FET was coined to describe such patterns. The MP concept has similarities with traditional frequent itemsets except that the attributes that can appear in a particular part is limited, consequently the search space can be considerably reduced. A particular challenge was the size of the networks that we wish to analyse. Thus, although in the first instance a traditional Apriori approach seems appropriate, the size of the networks to be considered means that this approach is unlikely to succeed for any realistic application (as illustrated by the presented evaluation). Instead a parallel approach was proposed facilitated by the nature of the ShaMP algorithm which readily lends itself to parallelisation. For comparison purposes an Apriori algorithm was also defined, the AMP algorithm. The evaluation was conducted using movement networks extracted from the GB Cattle Tracking System (CTS) and artificial networks. The main findings may be summarised as fallows. The ShaMP Algorithm can successfully identify MPs in large networks with reasonable computational efficiency; where more traditional approaches, such as AMP, fail because of the resource required. Parallelisation of the ShaMP algorithm, using MPJ, has a significant impact over the computational efficiency of the algorithm.

References

1. Matsumura, N., Goldberg, D.E., Llorà, X.: Mining directed social network from message board. In: Special Interest Tracks and Posters of the 14th International Conference on World Wide Web, pp. 1092–1093. ACM (2005)
2. Chandrasekaran, B.: Survey of Network Traffic Models, vol. 567. Washington University, St. Louis CSE (2009)

3. Datta, S., Bhaduri, K., Giannella, C., Wolff, R., Kargupta, H.: Distributed data mining in peer-to-peer networks. IEEE Internet Comput. **10**(4), 18–26 (2006)
4. Gonzalez, H., Han, J., Li, X., Myslinska, M., Sondag, J.P.: Adaptive fastest path computation on a road network: a traffic mining approach. In: Proceedings of the 33rd International Conference on Very Large Data Bases, VLDB Endowment, pp. 794–805 (2007)
5. Galloway, J., Simoff, S.J.: Network data mining: methods and techniques for discovering deep linkage between attributes. In: Proceedings of the 3rd Asia-Pacific Conference on Conceptual Modelling, vol. 53, pp. 21–32. Australian Computer Society, Inc. (2006)
6. Grama, A.: Introduction to Parallel Computing. Pearson Education (2003)
7. Raorane, A., Kulkarni, R.: Data mining techniques: a source for consumer behavior analysis (2011). arXiv:1109.1202
8. Gudmundsson, J., Laube, P., Wolle, T.: Movement patterns in spatio-temporal data. In: Encyclopedia of GIS, pp. 726–732. Springer (2008)
9. Campbell, W.M., Dagli, C.K., Weinstein, C.J.: Social network analysis with content and graphs. Lincoln Lab. J. **20**(1) (2013)
10. Han, J., Kamber, M., Pei, J.: Data Mining: Concepts and Techniques. Elsevier (2011)
11. Agrawal, R., Srikant, R., et al.: Fast algorithms for mining association rules. In: Proceedings of 20th International Conference on Very Large Data Bases, VLDB, vol. 1215, pp. 487–499 (1994)
12. Bliss, C.A., Frank, M.R., Danforth, C.M., Dodds, P.S.: An evolutionary algorithm approach to link prediction in dynamic social networks. J. Comput. Sci. **5**(5), 750–764 (2014)
13. Kim, M., Leskovec, J.: The network completion problem: Inferring missing nodes and edges in networks. In: SDM, vol. 11, pp. 47–58. SIAM (2011)
14. Forum, M.P.I.: Mpi: a message passing interface standard: version 2.2; message passing interface forum, September 4, 2009
15. Brawer, S.: Introduction to Parallel Programming. Academic Press (2014)
16. Gropp, W., Lusk, E., Doss, N., Skjellum, A.: A high-performance, portable implementation of the mpi message passing interface standard. Parallel Comput. **22**(6), 789–828 (1996)
17. Karonis, N.T., Toonen, B., Foster, I.: Mpich-g2: a grid-enabled implementation of the message passing interface. J. Parallel Distrib. Comput. **63**(5), 551–563 (2003)
18. Chen, L., Wang, C., Lau, F.C.: A grid middleware for distributed java computing with mpi binding and process migration supports. J. Comput. Sci. Technol. **18**(4), 505–514 (2003)
19. Baker, M., Carpenter, B., Shaft, A.: Mpj express: towards thread safe java hpc. In: 2006 IEEE International Conference on Cluster Computing, pp. 1–10. IEEE (2006)
20. Aggarwal, C.C.: Applications of frequent pattern mining. In: Frequent Pattern Mining, pp. 443–467. Springer (2014)

Sentiment Analysis and Recommendation

Emotion-Corpus Guided Lexicons for Sentiment Analysis on Twitter

Anil Bandhakavi, Nirmalie Wiratunga, Stewart Massie and P. Deepak

Abstract Conceptual frameworks for emotion to sentiment mapping have been proposed in Psychology research. In this paper we study this mapping from a computational modelling perspective with a view to establish the role of an emotion-rich corpus for lexicon-based sentiment analysis. We propose two different methods which harness an emotion-labelled corpus of tweets to learn word-level numerical quantification of sentiment strengths over a positive to negative spectrum. The proposed methods model the emotion corpus using a generative unigram mixture model (UMM), combined with the emotion-sentiment mapping proposed in Psychology (Cambria et al. 28th AAAI Conference on Artificial Intelligence, pp. 1515–1521, 2014) [1] for automated generation of sentiment lexicons. Sentiment analysis experiments on benchmark Twitter data sets confirm the quality of our proposed lexicons. Further a comparative analysis with standard sentiment lexicons suggest that the proposed lexicons lead to a significantly better performance in both sentiment classification and sentiment intensity prediction tasks.

Keywords Emotion · Sentiment · Domain-specific Lexicons · Expectation Maximization · Emotion Theories · Twitter

1 Introduction

Sentiment analysis concerns the computational study of natural language text (e.g. words, sentences and documents) in order to identify and effectively quantify its

A. Bandhakavi (✉) · N. Wiratunga (✉) · S. Massie (✉)
School of Computing, Robert Gordon University, Aberdeen, UK
e-mail: a.s.bandhakavi@rgu.ac.uk

N. Wiratunga
e-mail: n.wiratunga@rgu.ac.uk

S. Massie
e-mail: s.massie@rgu.ac.uk

P. Deepak (✉)
Queen's University, Belfast, UK
e-mail: deepaksp@acm.org

M. Bramer and M. Petridis (eds.), *Research and Development in Intelligent Systems XXXIII*, DOI 10.1007/978-3-319-47175-4_5

polarity (i.e. positive or negative) [2]. Sentiment lexicons are the most popular resources used for sentiment analysis, since they capture the polarity of a large collection of words. These lexicons are either hand-crafted (e.g. opinion lexicon [3], General Inquirer [4] and MPQA subjectivity lexicon [5]) or generated (e.g. SentiWordNet [6] and SenticNet [1]) using linguistic resources such as WordNet [7] and ConceptNet [8]. However, on social media (e.g. Twitter), text contains special symbols resulting in non-standard spellings, punctuations and capitalization; sequence of repeating characters and emoticons for which the aforementioned lexicons have limited or no coverage.

As a result domain-specific sentiment lexicons were developed to capture the informal and creative expressions used on social media to convey sentiment [9, 10]. The extraction of such lexicons is possible with limited effort, due to the abundance of weakly-labelled sentiment data on social media, obtained using emoticons [11, 12]. However, sentiment on social media is not limited to conveying positivity and negativity. Socio-linguistics suggest that on social media, people express a wide range of emotions such as *anger, fear, joy, sadness* etc. [13]. Following the trends in lexicon based sentiment analysis, research in the textual emotion detection also developed lexicons that can not only capture the emotional orientation of words [14, 15], but also quantify their emotional intensity [16, 17].

Though research in psychology defines sentiment and emotion differently [18], it also provides a relationship between them [19]. Further research in emotion classification [20, 21] demonstrated the usefulness of sentiment features extracted using a lexicon for document representation. Similarly emoticons used as features to represent documents improved sentiment classification [10, 12]. However, the exploration of emotion knowledge for sentiment analysis is limited to emoticons [12, 22, 23], leaving a host of creative expressions such as emotional hashtags (e.g. #loveisbliss), elongated words (e.g. haaaappyy!!!) and their concatenated variants unexplored. An emotion-corpus crawled on Twitter using seed words for different emotions as in [21, 24] can potentially serve as a knowledge resource for sentiment analysis. Adopting such corpora for sentiment analysis, e.g. sentiment lexicon extraction is particularly interesting, given the challenges involved in developing effective models which can cope with the lexical variations on social media.

Therefore, in this work we explore the role of a Twitter emotion corpus for extracting a sentiment lexicon, which can be used to analyse the sentiment of tweets. We do a qualitative comparison between standard sentiment lexicons and the proposed sentiment lexicons. Our contributions in this paper are as follows:

1. We propose two different methods to generate sentiment lexicons from a corpus of emotion-labelled tweets by combining our prior work on domain-specific emotion lexicon generation [25, 26], with the emotion-sentiment mapping presented in Psychology (see Fig. 1) [19]; and
2. We comparatively evaluate the quality of the proposed sentiment lexicons, and the standard sentiment lexicons found in literature through different sentiment analysis tasks: *sentiment intensity prediction and sentiment classification* on benchmark Twitter data sets.

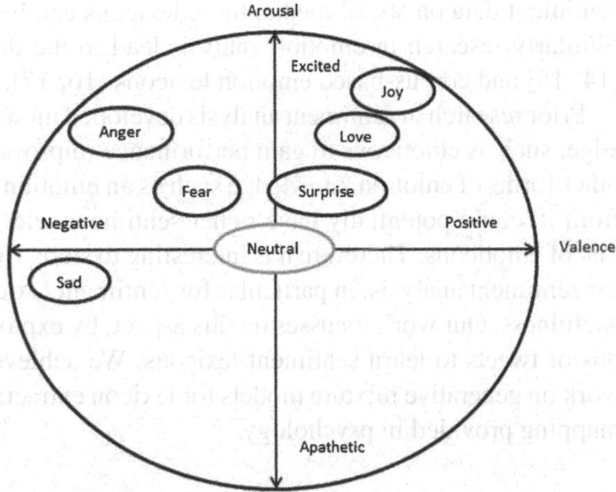

Fig. 1 Parrot's emotions in the valence-arousal plane of the dimensional model

In the rest of the paper we review related literature in Sect. 2. In Sects. 3 and 4 we formulate the methods to extract sentiment lexicons from an emotion corpus of tweets. In Sect. 5 we describe our experimental set up and analyse the results. Section 6 presents our conclusions.

2 Related Work

In this section we review the literature concerning sentiment lexicons, followed by a review of different emotion theories and their relationship with sentiments proposed in Psychology.

2.1 Lexicons for Sentiment Analysis

Broadly sentiment lexicons are of two types: *hand-crafted and automatic*. Hand-crafted lexicons such as opinion lexicon [3], General Inquirer [4] and MPQA subjectivity lexicon [5] have human assigned sentiment scores. On the other hand automatic lexicons are of two types: *corpus-based and resource-based*. Lexicons such as SentiWordNet [6] and SenticNet [1] are resource-based, since they are extracted using linguistic resources such as WordNet [7] and ConceptNet [8].

A common limitation of resource-based and hand-crafted lexicons is that, they have static vocabulary, making them limitedly effective to mine sentiment on social media, which is inherently dynamic. Corpus-based lexicons such as in [9, 10], gauge the corpus level variations in sentiment using statistical models, and are found to be very effective on social media. Further with the abundance of weakly-labelled sentiment data on social media, these lexicons can be updated with very low costs. Similarly research in emotion analysis lead to the development of resource-based [14, 15] and corpus-based emotion lexicons [16, 17].

Prior research in sentiment analysis developed models that exploit emotion knowledge, such as emoticons to gain performance improvements [12, 22, 23]. However, other forms of emotion knowledge such as an emotion corpus and the lexicons learnt from it, could potentially have richer sentiment-relevant information, compared to that of emoticons. Therefore it is interesting to study role of such emotion knowledge for sentiment analysis, in particular for sentiment lexicon generation and validate its usefulness. Our work focusses on this aspect, by exploiting an emotion-labelled corpus of tweets to learn sentiment lexicons. We achieve this by combining our prior work on generative mixture models for lexicon extraction and the emotion-sentiment mapping provided in psychology.

2.2 *Emotion Theories*

Research in psychology proposed many emotion theories, wherein each theory organizes a set of emotions into some structural form (e.g. taxonomy). In the following sections we detail the most popular emotion theories studied in psychology.

2.2.1 Ekman Emotion Theory

Paul Ekman, an American psychologist focused on identifying the most basic set of emotions that can be expressed distinctly in the form of a facial expression. The emotions identified as basic by Ekman are *anger, fear, joy, sadness, surprise and disgust* [27].

2.2.2 Plutchik's Emotion Theory

Unlike the Ekman emotion model Plutchik's emotion model defines eight basic emotions such as *anger, anticipation, disgust, joy, fear, sadness and surprise* [28]. These basic emotions are arranged as bipolar pairs namely: *joy-sadness, trust-disgust, fear-anger, surprise-anticipation*.

2.2.3 Parrot's Emotion Theory

Parrot organised emotions in a three level hierarchical structure [29]. The levels represent primary, secondary and tertiary emotions respectively. Parrot identified emotions such as *love, joy, surprise, anger, sadness and fear*, as the primary emotions. Though Ekman and Plutchik emotion models are popular, research in Twitter emotion detection [21, 30] focussed on emotions that largely overlap with that of Parrot, given their popular expressiveness on social media. We use the Parrot emotion-labelled twitter corpus [21] in this study for generating sentiment lexicons.

2.2.4 Emotion-Sentiment Relationship in Psychology

One of the popular approaches for emotion modelling in Psychology is the dimensional approach, wherein each emotion is considered as a point in the continuous multidimensional space where each aspect or characteristic of an emotion is represented as a dimension. Affect variability is captured by two dimensions namely valence and arousal [31]. Valence (pleasure - displeasure) depicts the degree of positivity or negativity of an emotion. Arousal (activation- deactivation) depicts the excitement or the strength of an emotion. The dimensional approach depicting parrot's primary emotions in the valence arousal 2D space is shown in Fig. 1 [32].

3 Emotion-Aware Models for Sentiment Analysis

In this section we formulate two different methods which utilize a corpus of emotion-labelled documents for sentiment analysis of text. The first method learns an emotion lexicon and further transforms it into a sentiment lexicon using the emotion-sentiment mapping (refer Sect. 2.2) proposed in Psychology. The second method on the other hand learns the sentiment labels for the documents in the emotion corpus using the emotion-sentiment mapping, followed by a sentiment lexicon extraction. The two proposed methods are illustrated visually in Fig. 2a, b.

3.1 Emotion Corpus-EmoSentilex

A simple way to utilize a corpus of emotion-labelled documents, X_E for sentiment analysis is to first learn an emotion lexicon, and further transform it into a sentiment lexicon. An emotion lexicon *Emolex* in our case is a $|V| \times (k+1)$ matrix, where $Emolex(i, j)$ is the emotional valence of the ith word in vocabulary V to the jth emotion in E (set of emotions) and $Emolex(i, k+1)$ corresponds to its neutral valence (refer Sect. 4). Further using the emotion-sentiment mapping proposed in Psychology we transform the emotion lexicon *Emolex* into a sentiment lexicon *EmoSentilex*, which is a $|V| \times 1$ matrix as follows:

$$EmoSentilex(i) = Log\left(\frac{\sum_{m \in E^+} Emolex(i, m)}{\sum_{n \in E^-} Emolex(i, n)}\right) \quad (1)$$

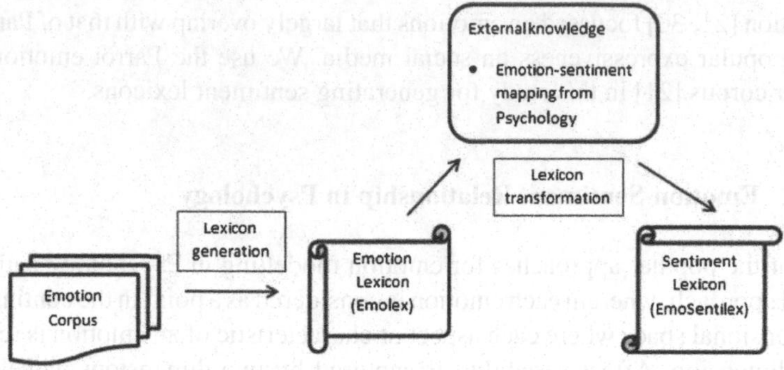

(a) Emotion Corpus-EmoSentilex

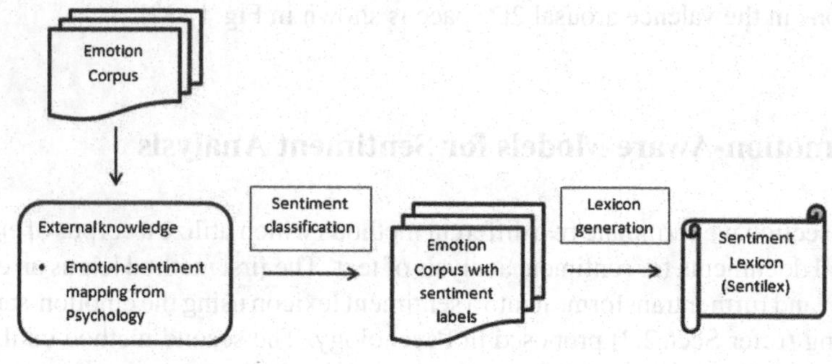

(b) Emotion Corpus-Sentilex

Fig. 2 Emotion-aware models for sentiment analysis

where $E^+ \subset E$ and $E^- \subset E$ are the set of positive and negative emotions according to the emotion-sentiment mapping. Note that the log scoring assigns a positive value for words having stronger associations with emotions such as *Joy, Surprise and Love* and negative values for words having stronger associations with emotions such as *Anger, Sadness and Fear*. Therefore we expect that sentiment knowledge for words is implicitly captured in an emotion lexicon, which can be easily extracted using this simple transformation.

Using the above method, any automatically generated emotion lexicon can be converted into a sentiment lexicon. This is very useful on Twitter, since data (tweets) corresponding to the lexicons is not always available. Further it can also avoid the additional overheads involved in re-crawling the original data using the Twitter API. However, the above method does not model the document-sentiment relationships to learn the lexicon, which is important to quantify word-sentiment associations. Therefore we introduce an alternate method which overcomes this limitation while utilizing an emotion corpus for sentiment lexicon generation.

3.2 Emotion Corpus-Sentilex

An alternate way to utilize the emotion corpus, X_E for sentiment analysis is to transform it into a sentiment corpus, X_S by learning the sentiment label for each document $d \in X_E$. This is done by using the emotion-sentiment mapping as follows:

$$Sentiment(d) = \begin{cases} positive \text{ if emotion(d)} \in E^+ \\ negative \text{ if emotion(d)} \in E^- \end{cases} \tag{2}$$

The sentiment lexicon *Sentilex* learnt from the corpus X_S is a $|V| \times 3$ matrix, where $Sentilex(i, 1)$, $Sentilex(i, 2)$ and $Sentilex(i, 3)$ are the positive, negative and neutral valences corresponding to the ith word in vocabulary V. Observe that unlike the method which learns *EmoSentilex*, by aggregating word-level emotion scores into sentiment scores, this method learns the sentiment-class knowledge corresponding to the documents, before learning a word-sentiment lexicon. We expect this additional layer of supervision, to benefit performance, following the findings of earlier research in supervised and unsupervised sentiment analysis. In the following section we briefly explain our proposed method to generate *Sentilex* and *Emolex*. Further details about our proposed method can be found in [25, 26]

4 Mixture Model for Lexicon Generation

In this section we describe our proposed unigram mixture model (UMM) applied to the task of emotion lexicon (*Emolex*) generation. Sentiment lexicon (*Sentilex*) generation is a special case of emotion lexicon generation, where the k emotion classes are reduced to positive and negative classes. Therefore we continue the presentation for the general case, i.e. *Emolex* generation.

We model real-world emotion data to be a mixture of emotion bearing words and emotion-neutral (background) words. For example consider the tweet *going to Paris this Saturday #elated #joyous*, which explicitly connotes emotion *joy*. However, the word *Saturday* is evidently not indicative of *joy*. Further *Paris* could be associated with emotions such as *love*. Therefore we propose a generative model which assumes a mixture of two unigram language models to account for such word mixtures in documents. More formally our generative model is as follows to describe the generation of documents connoting emotion e_t:

$$\begin{aligned} P(D_{e_t}, Z|\theta_{e_t}) = \prod_{i=1}^{|D_{e_t}|} \prod_{w \in d_i} [(1 - Z_w)\lambda_{e_t} P(w|\theta_{e_t}) \\ +(Z_w)(1 - \lambda_{e_t}) P(w|N)]^{c(w,d_i)} \end{aligned} \tag{3}$$

where θ_{e_t} is the emotion language model and N is the background language model. λ_{e_t} is the mixture parameter and Z_w is a binary hidden variable which indicates the language model that generated the word w.

The estimation of parameters θ_{e_t} and Z can be done using expectation maximization (EM), which iteratively maximizes the complete data (D_{e_t}, Z) by alternating between E-step and M-step. The E and M steps in our case are as follows:

E-step:

$$P(Z_w = 0|D_{e_t}, \theta_{e_t}^{(n)}) = \frac{\lambda_{e_t} P(w|\theta_{e_t}^{(n)})}{\lambda_{e_t} P(w|\theta_{e_t}^{(n)}) + (1 - \lambda_{e_t}) P(w|N)} \tag{4}$$

M-step:

$$P(w|\theta_{\theta_{e_t}}^{(n+1)}) = \frac{\sum_{i=1}^{|D_{e_t}|} P(Z_w = 0|D_{e_t}, \theta_{e_t}^{(n)}) c(w, d_i)}{\sum_{w \in V} \sum_{i=1}^{|D_{e_t}|} P(Z_w = 0|D_{e_t}, \theta_{e_t}^{(n)}) c(w, d_i)} \tag{5}$$

where n indicates the EM iteration number. The EM iterations are terminated when an optimal estimate for the emotion language model θ_{e_t} is obtained. EM is used to estimate the parameters of the k mixture models corresponding to the emotions in E. The emotion lexicon *Emolex* is learnt by using the k emotion language models and the background model N as follows:

$$Emolex(w_i, \theta_{e_j}) = \frac{P(w_i|\theta_{e_j}^{(n)})}{\sum_{t=1}^{k} [P(w_i|\theta_{e_t}^{(n)})] + P(w_i|N)} \tag{6}$$

$$Emolex(w_i, N) = \frac{P(w_i|N)}{\sum_{t=1}^{k} [P(w_i|\theta_{e_t}^{(n)})] + P(w_i|N)} \tag{7}$$

where k is the number of emotions in the corpus, and *Emolex* is a $|V| \times (k + 1)$ matrix.

5 Evaluation

Our evaluation is a comparative study, of the performance of the standard sentiment lexicons, and the proposed emotion corpus based sentiment lexicons through a variety of sentiment analysis tasks on benchmark Twitter data sets. Significance is reported using a paired one-tailed t-test using 95 % confidence (i.e. with p-value $\leq$0.05). Observe that in all our experimental results, the best performing methods are highlighted in bold.

5.1 Evaluation Tasks

Our evaluation includes the following sentiment analysis tasks.

1. *Sentiment intensity prediction*: Given a collection of words/phrases extracted from sentiment bearing tweets, the objective is to predict a sentiment intensity score for each word/phrase and arrange them in decreasing order of intensity. The predictions are validated against a ranking given by humans. Formally, given a phrase P, the sentiment intensity score for the phrase is calculated as follows:

$$SentimentIntensity(P) = \sum_{w \in P} Log\left(\frac{Lex(w,+)}{Lex(w,-)}\right) \times count(w,P) \quad (8)$$

where w is a word in the phrase P, $count(w,P)$ is the number of times w appears in P. $Lex(w,+)$, $Lex(w,-)$ are the positive and negative valences for the word w in a lexicon. Some lexicons offer the sentiment intensity scores (e.g. SenticNet, S140 lexicon), in which case we use them directly. The aforementioned computation applies to the UMM based lexicons like *Sentilex* and S140-UMM lexicon.

2. *Sentiment classification*: Given a collection of documents (tweets), the objective is to classify them into positive and negative classes. The predictions are validated against human judgements. Formally, given a document d, the sentiment class is predicted using a lexicon as follows:

$$d[+] = \sum_{w \in d} Lex(w,+) \times count(w,d) \quad (9)$$

where $d[+]$ is the positive intensity of d. Similarly $d[-]$ indicates the negative intensity of d. Finally the sentiment class of d is determined as follows:

$$Sentiment(d) = \begin{cases} positive \text{ if } d[+] > d[-] \\ negative \text{ if } d[-] > d[+] \end{cases} \quad (10)$$

5.2 Datasets

We use four benchmark data sets in our evaluation. Note that the emotion corpus is used in two different ways to learn sentiment lexicons (refer Sects. 3 and 4). Further the S140 training data is used to learn a sentiment lexicon using the proposed method (refer Sect. 4). The remaining data sets are used for evaluation. We expect our evaluation to test the transferability of each of the lexicons, given that the training and test data are not always from the same corpus, albeit from similar genre.

5.2.1 Emotion Dataset

A collection of 0.28 million emotional tweets crawled from Twitter streaming API[1] using emotion hashtags provided in [21]. The emotion labels in the data set correspond to Parrot's [29] primary emotions and were obtained through distant-supervision.[2] Parrot's emotion theory identifies an equal number of positive and negative emotions. Therefore we expect the sentiment lexicons learnt on this corpus to be able to mine both positive and negative sentiment in the test corpora.

5.2.2 S140 Dataset

A collection of 1.6 million (0.8 million positive and 0.8 million negative) sentiment bearing tweets harnessed by Go et.al [11] using the Twitter API. Further the data set also contains a collection of 359 (182 positive and 177 negative) manually annotated tweets. We generate a sentiment lexicon using the proposed method in Sect. 4 on the 1.6 million tweets and compare it with the S140 lexicon [10].

5.2.3 SemEval-2013 Dataset

A collection of 3430 (2587 positive and 843 negative) tweets hand-labelled for sentiment using Amazon Mechanical Turk [33]. Note that unlike the S140 test data, there is high skewness in the class distributions. Therefore it would be a greater challenge to transfer the lexicons learnt on the emotion corpus and also those learnt on the S140 training corpus to sentiment classification.

5.2.4 SemEval-2015 Dataset

A collection of 1315 words/phrases hand-labelled for sentiment intensity scores [34]. A higher score indicates greater positivity. Further the words/phrases are arranged in decreasing order of positivity. We used this data set to validate the performance of different lexicons in ranking words/phrases for sentiment.

5.3 Baselines and Metrics

The following different models are used in our comparative study:

1. Resource-based sentiment lexicons SentiWordNet and SenticNet;
2. Corpus-based sentiment lexicons S140 lexicon [10] and NRCHashtag lexicon [10];

---

[1]https://dev.twitter.com/streaming/public.

[2]http://www.gabormelli.com/RKB/Distant-Supervision-Learning-Algorithm.

3. Corpus-based sentiment lexicon (S140-UMM lexicon) learnt using the proposed method on S140 corpus (refer Sect. 4); and
4. Corpus-based sentiment lexicons (*EmoSentilex* and *Sentilex*) learnt on the emotion corpus (refer Sect. 5.2.1) using the proposed method (refer Sects. 3 and 4)

Performance evaluation is done using using Spearman's rank correlation coefficient and F-score for sentiment ranking and sentiment classification respectively. F-score is chosen for the classification task since it measures the performance of an algorithm in terms of both precision and recall.

5.4 *Results and Analysis*

In this section we analyse the sentiment ranking results and the sentiment classification results obtained using the different lexicons.

5.4.1 Sentiment Ranking

Table 1 summarizes the sentiment ranking results obtained for different lexicons. In general resource-based lexicons SentiWordNet and SenticNet are outperformed by all the corpus-based lexicons. This is expected, because the vocabulary coverage of these lexicons relevant to social media is limited compared to other lexicons. Furthermore, the results also suggest that the sentiment intensity knowledge captured by the corpus-based lexicons is superior to that of resource-based lexicons.

NRCHashtag lexicon performed significantly better than the remaining baselines and the proposed *EmoSentilex*. The significant performance differences between NRCHashtag lexicon and S140 lexicon and NRCHashtag lexicon and S140-UMM

Table 1 Sentiment ranking result

Method	Spearman's rank correlation coefficient
Baselines (standard sentiment lexicons)	
SentiWordNet	0.479
SenticNet	0.425
S140 lexicon	0.506
NRCHashtag lexicon	0.624
S140-UMM-lexicon	0.517
Proposed methods (emotion-corpus based sentiment lexicons)	
EmoSentiLex	0.572
Sentilex	**0.682**

lexicon clearly suggests the superiority of the NRCHashtag corpus over the S140 corpus in learning transferable lexicons for sentiment intensity prediction. It would be interesting to compare the performance of these lexicons in the sentiment classification tasks.

It is extremely promising to see that the proposed lexicons outperform most of the baselines significantly. Amongst the proposed lexicons, *Sentilex* performed significantly better than *EmoSentilex*. This is not surprising, since *Sentilex* has the ability to incorporate the sentiment-class knowledge of the documents in the learning stage. This exactly follows the findings of earlier research in supervised and unsupervised sentiment analysis.

5.4.2 Sentiment Classification

Sentiment classification results for the S140 data set are shown in Table 2. Here unlike in the sentiment intensity prediction task, SentiWordNet demonstrated comparable performance with that of corpus-based lexicons. However, SenticNet does perform the worst amongst all the lexicons. This suggests that SentiWordNet is better transferable onto social media compared to SenticNet.

The S140 corpus based lexicons significantly outperform NRCHashtag lexicon, given their advantage to train on a corpus, that is similar to the test set. However, the proposed lexicon *Sentilex* recorded the best performance on this data set. once again the superiority of *Sentilex* over *EmoSentilex* is evidenced, given its ability to incorporate sentiment-class knowledge of the documents in the learning stage. The performance improvements of emotion corpus based sentiment lexicons over a majority of baseline lexicons, clearly suggests that emotion knowledge when exploited effectively is very useful for sentiment analysis.

Table 3 summarizes the results for different lexicon on the SemEval-2013 data set. Unlike the previous, this data set has a very skewed class distribution. The impact of this is clearly reflected in the results. Majority of the lexicons recorded strong

Table 2 Sentiment classification results on S140 test data set

Method	Positive F-score	Negative F-score	Overall F-score
Baselines (standard sentiment lexicons)			
SentiWordNet	69.42	67.60	68.51
SenticNet	59.88	59.84	59.86
S140-lexicon	71.55	69.42	70.48
NRCHashtag-lexicon	66.66	64.75	65.70
S140-UMM-lexicon	**75.14**	69.36	72.25
Proposed methods (emotion-corpus based sentiment lexicons)			
EmoSentiLex	67.51	71.14	69.32
Sentilex	72.93	**74.11**	**73.52**

Table 3 Sentiment classification results on SemEval-2013 data set

Method	Positive F-score	Negative F-score	Overall F-score
Baselines (standard sentiment lexicons)			
SentiWordNet	80.14	50.38	65.26
SenticNet	54.95	55.94	55.45
S140-lexicon	80.13	57.87	69.00
NRCHashtag-lexicon	80.25	53.98	67.11
S140-UMM-lexicon	78.87	55.85	67.36
Proposed methods (emotion-corpus based sentiment lexicons)			
EmoSentiLex	64.51	48.37	56.44
Sentilex	**83.06**	**60.98**	**72.02**

performances in classifying positive class documents. Once again SentiWordNet demonstrated that it is better transferable onto social media compared to SenticNet.

Similar to the previous data set, S140 corpus based lexicons performed better than NRCHashtag corpus based lexicon. Overall comparison across the evaluation tasks suggests that S140 corpus based lexicons record better performance in sentiment classification, whereas NRCHashtag lexicon records better performance in sentiment quantification. This offers interesting directions for future work on composing different corpora for learning sentiment lexicons.

The proposed lexicon *EmoSentilex* performed significantly below most of the lexicons on this data set. We believe the inability to learn the document-sentiment relationships, coupled with the skewed class distribution characteristics of the data set resulted in such performance degradation. However, our proposed lexicon *Sentilex* significantly outperformed all the remaining lexicons. The consistent performance of *Sentilex* in all the evaluation tasks, strongly evidences the correlation between emotions and sentiments. We believe that the emotion-sentiment mapping in psychology effectively clusters the emotion corpus into sentiment classes, thereafter the ability of the UMM model to effectively capture the word-sentiment relationships resulted in the performance improvements for *Sentilex*.

6 Conclusions

In this paper we study the mapping proposed in psychology between emotions and sentiments, from a computational modelling perspective in order to establish the role of an emotion corpus for sentiment analysis. By combining a generative unigram mixture model (UMM) with the emotion-sentiment mapping, we propose two different methods to extract lexicons for Twitter sentiment analysis from an emotion labelled Twitter corpus. We comparatively evaluate the quality of the proposed lexicons and standard sentiment lexicons through a variety of sentiment analysis tasks on benchmark Twitter data sets. Our experiments confirm that the proposed sentiment lexicons, yield significant improvements over standard lexicons in sentiment classification and sentiment intensity prediction tasks. It is extremely promising to see the potential of an emotion corpus as a useful knowledge resource for sentiment analysis, especially on social media where emotions and sentiments are widely expressed. Further the cost-effectiveness of the emotion-sentiment mapping to cluster the emotion corpus into positive, negative classes (0.28 million tweets in a second) makes it practically possible to adopt large emotion corpora, in order to extract sentiment lexicons with improved coverage.

References

1. Cambria, E., Olsher, D., Rajagopal, D.: Senticnet 3: a common and common-sense knowledge base for cognition-driven sentiment analysis. In: 28th AAAI Conference on Artificial Intelligence, pp. 1515–1521 (2014)
2. Pang, B., Lee, L.: Opinion mining and sentiment analysis. Found. Trends Inf. Retrieval **2**(1), 1–135 (2008)
3. Hu, M., Liu., B.: Mining and summarizing customer reviews. In: Proceedings of the ACM SIGKDD International Conference on Knowledge Discovery and Data Mining (2004)
4. Stone, P.J., Dexter, D.C., Marshall, S.S., Daniel, O.M.: The general inquirer: a computer approach to content analysis. The MIT Press (1966)
5. Wilson, T., Wiebe, J., Hoffmann, P.: Recognizing contextual polarity in phrase-level sentiment analysis. In: Proceedings of HLT-EMNLP-2005 (2005)
6. Esuli, A., Baccianella, S., Sebastiani, F.: Sentiwordnet 3.0: an enhanced lexical resource for sentiment analysis and opinion mining. In: Proceedings of LREC (2010)
7. Fellbaum, C.: Wordnet and wordnets. In: Encyclopedia of Language and Linguistics, pp. 665–670 (2005)
8. Liu, H., Singh, P.: Conceptnet- a practical commonsense reasoning tool-kit. BT Technol. J. **22**(4), 211–226 (2004)
9. Feng, S., Song, K., Wang, D., Yu, G.: A word-emotion mutual reinformcement ranking model for building sentiment lexicon from massive collection of microblogs. World Wide Web **18**(4), 949–967 (2015)
10. Mohammad, S.M., Kiritchenko, S., Zhu, X.: Nrc-canada: building the state-of-the-art in sentiment analysis of tweets. In: 7th International Workshop on Semantic Evaluation (SemEval 2013), pp. 321–327 (2013)
11. Go, A., Bhayani, R., Huang, L.: Twitter sentiment classification using distant supervision. Processing, pp. 1–6 (2009)
12. Hogenboom, A., Bal, D., Frasincar, F., Bal, M.: Exploiting emoticons in polarity classification of text. J. Web Eng. (2013)
13. Boyd, D., Golder, S., Lotan, G.: Tweet, tweet, retweet: conversational aspects of retweeting on twitter. In: Proceedings of the 43rd Hawaii International Conference on System Sciences (2010)
14. Mohammad, S.M., Turney, P.: Crowdsourcing a word-emotion association lexicon. Comput. Intell. **29**(3), 436–465 (2013)
15. Poria, S., Gelbukh, A., Cambria, E., Hussain, A., Huang, G.B.: Emosenticspace: a novel framework for affective common-sense reasoning. Knowl.-Based Syst. **69**, 108–123 (2014)
16. Rao, Y., Lei, J., Wenyin, L., Li, Q., Chen, M.: Building emotional dictionary for sentiment analysis of online news. World Wide Web **17**, 723–742 (2014)

17. Song, K., Feng, S., Gao, W., Wang, D., Chen, L., Zhang, C.: Build emotion lexicon from microblogs by combining effects of seed words and emoticons in a hetereogeneous graph. In: Proceedings of the 26th ACM Conference on Hypertext and Social Media, pp. 283–292 (2015)
18. Munezero, M., Montero, C.S., Sutinen, E., Pajunen, J.: Are they different? affect, feeling, emotion, sentiment, and opinion detection in text. IEEE Trans. Affect. Comput. **5**(2) (2014)
19. Binali, H., Potdar, V., Wu, C.: Computational approaches for emotion detection in text. In: 4th IEEE International Conference on Digital Ecosystems and Technologies DEST (2010)
20. Ghazi, D., Inkpen, D., Szpakowicz, S.: Hierarchical approach to emotion recognition and classification in texts. In: Proceedings of the 23rd Canadian Conference on Advances in Artificial Intelligence (2010)
21. Wang, W.: Harnessing twitter "big data" for automatic emotion identification. In: Proceedings of the ASE/IEEE International Conference on Social Computing and International Conference on Privacy, Security, Risk and Trust (2012)
22. Hu, X., Tang, J., Gao, H., Liu, H.: Unsupervised sentiment analysis with emotional signals. In: Proceedings of the International World Wide Web Conference (WWW) (2013)
23. Jiang, F., Liu, Y.Q., Luan, H.B., Sun, J.S., Zhu, X., Zhang, M., Ma, S.P.: Microblog sentiment analysis with emoticon space model. J. Comput. Sci. Technol. **30**(5), 1120–1129 (2015)
24. Mohammad, S.M.: #emotional tweets. In: Proceedings of the First Joint Conference on Lexical and Computational Semantics, pp. 246–255 (2012)
25. Bandhakavi, A., Wiratunga, N., Deepak, P., Massie, S.: Generating a word-emotion lexicon from #emotional tweets. In: Proceedings of the 3rd Joint Conference on Lexical and Computational Semantics (*SEM 2014) (2014)
26. Bandhakavi, A., Wiratunga, N., Massie, S., Deepak, P.: Lexicon generation for emotion detection from text. IEEE Intell. Syst. (2017)
27. Ekman, P.: An argument for basic emotions. Cogn. Emot. **6**(3), 169–200 (1992)
28. Plutchik, R.: A general psychoevolutionary theory of emotion. In: Plutchik, R., Kellerman, H. (eds.) Emotion: Theory, Research, and Experience, vol. 1, pp. 3–33 (1980)
29. Parrott, W.: Emotions in Social Psychology. Psychology Press, Philadelphia (2001)
30. Qadir, A., Riloff, E.: Bootstrapped learning of emotion hashtags #hashtags4you. In: the 4th Workshop on Computational Approaches to Subjectivity, Sentiment and Social Media Analysis (WASSA 2013) (2013)
31. Jin, X., Wang, Z.: An emotion space model for recognition of emotions in spoken chinese. In: Proceedings of the First International Conference on Affective Computing and Intelligent Interaction (2005)
32. Binali, H., Potdar, V.: Emotion detection state-of -the-art. In: Proceedings of the CUBE International Information Technology Conference, pp. 501–507 (2012)
33. Nakov, P., Rosenthal, S., Kozareva, Z., Stoyanov, V., Ritter, A., Wilson, T.: Semeval-2013 task2: sentiment analysis in twitter. In: Proceedings of the 7th International Workshop on Semantic Evaluation (SemEval-2013) (2013)
34. Rosenthal, S., Nakov, P., Kiritchenko, S., Mohammad, S.M., Ritter, A., Stoyanov, V.: Semeval-2015: sentiment analysis in twitter. In: Proceedings of the 9th International Workshop on Semantic Evaluation (SemEval-2015) (2015)

Context-Aware Sentiment Detection from Ratings

Yichao Lu, Ruihai Dong and Barry Smyth

Abstract The explosion of user-generated content, especially tweets, customer reviews, makes it possible to build sentiment lexicons automatically by harnessing the consistency between the content and its accompanying emotional signal, either explicitly or implicitly. In this work we describe novel techniques for automatically producing domain specific sentiment lexicons that are optimised for the language patterns and idioms of a given domain. We describe how we use review ratings as sentiment signals. We also describe an approach to recognising contextual variations in sentiment and show how these variations can be exploited in practice. We evaluate these ideas in a number of different product domains.

Keywords Sentiment Analysis · Review Ratings · Context Aware

1 Introduction

Sentiment analysis and opinion mining techniques aim to identify, analyse, and understand the subjective information in textual material provided by users. They have become increasingly popular when it comes to harnessing the explosion of user-generated content from short tweets and status updates to more detailed customer reviews. For example, [7] mine positive and negative sentiment from news articles about stocks and uses this to visualise market sentiment and pricing information for the user. Likewise Tumasjan et al. [19] demonstrated the potential to predict election outcomes by extracting the political sentiment contained within tweets. Elsewhere [15] describes how to help users compare different products by summarising

Y. Lu
Fudan University, Shanghai, China
e-mail: luyc13@fudan.edu.cn

R. Dong (✉) · B. Smyth
Insight Centre for Data Analytics, University College Dublin, Dublin, Ireland
e-mail: Ruihai.Dong@insight-centre.org; Ruihai.dong@ucd.ie

B. Smyth
e-mail: Barry.Smyth@insight-centre.org

M. Bramer and M. Petridis (eds.), *Research and Development in Intelligent Systems XXXIII*, DOI 10.1007/978-3-319-47175-4_6

sentiment information across different product features from customer reviews. And Dong et al. propose an opinionated product recommendation technique that combines similarity and sentiment to estimate review ratings and generate more effective product recommendations [4, 5].

The requirement to handle huge amounts of informal, opinionated, multi-domain, user-generated content has proven to be a challenge for traditional sentiment analysis techniques, which are often based on static dictionaries [10], fixed rules [9] or labeled data [6, 14]. In particular, such static approaches seldom deal with the common idioms that emerge across different domains. And even within a single domain different terms can signal very different sentiment extremes depending on context. For instance, in tablet reviews, the term *high* has positive polarity when modifying the aspect *quality*, but has negative polarity when modifying the aspect *price*.

These challenges speak to the need for a more flexible and dynamic approach to sentiment analysis, perhaps one that can be based on the common language patterns associated with a target domain. One way forward is to harness the sentiment signals (implicit or explicit) that are often associated with user-generated content. For example, users tend to use emoticon symbols such as *:)*, *:(*, or hashtags (#cool, #fail etc.), to express emotional intent. These signals can be used to enhance sentiment learning algorithms; see [2, 8, 16, 17]. For instance, Davidov et al. [2] proposed a supervised classification techniques by using 50 Twitter tags and 15 ‘smileys’ as sentiment labels to avoid intensive manual annotation. Liu et al. [16] present a model (ESLAM) by utilising both manually labeled data and emoticons to train and smooth for twitter sentiment analysis.

For shorter forms of user-generated content a guiding assumption is that positive sentiment words commonly have a closer relationship with positive emotional signals than with negative emotional signals, and vice versa; see for example [11]. Kiritchenko et al. introduced an weakly supervised approach for assessing the sentiment of short informal textual messages on Twitter [13]. Hashtags are used as an indicator of the sentiment polarity of the message, and they use a pointwise mutual information (PMI) technique to measure the relationship between the candidate sentiment words and the polarity of the tweets.

Longer forms of user-generated content, such as blogs or customer reviews, are less amenable to this use of local emotional signals: they are longer and more diverse than others, and typically containing multiple topics and expressing complex and varied sentiment. For example, “*The Thai red curry was delicious, ... , but price is $12, quite expensive*” expresses positive sentiment about the *Thai red curry*, but negative sentiment about *price*. Nevertheless, implicit indications of intent may be available. For example, Bross and Ehrig [1] exploit the information contained in pros and cons summaries as sentiment signals; the assumption is that authors choose positive expressions when describing a product aspect in the pros, whereas negative expressions are common in the cons. Wu and Wen [22] predict the sentiment of nouns by calculating their statistical associations with positive and negative search engine hits, which are obtained by using the search engine Baidu with positive and negative query-patterns.

Explicit ratings are also an important signal for reviews. A review's rating reflects the overall sentiment of a review without necessarily reflecting the sentiment of each individual review topic. For example, Kiritchenko et al. [12] create an domain-specific sentiment lexicon for restaurants by utilising one-, two-star rating as negative signals and four-, five-star ratings as positive signals. But a more interesting challenge is if the ratings can be useful signals to build context-aware sentiment lexicons in these longer forms of user-generated content.

In this paper, we describe a set of novel techniques for automatically building context-sensitive, domain-specific sentiment lexicons. We also use review ratings as emotional signals but describe how context can also be leveraged. We evaluate these ideas in a number of different product domains. We compare our lexicons to conventional one-size-fits-all lexicons using data from TripAdvisor and Amazon and we evaluate the quality of our lexicons by comparing their performance to a number of baselines and state-of-the-art alternatives on a sentiment polarity prediction task.

2 Creating a Domain-Specific Sentiment Lexicon

We describe an approach to automatically constructing a domain specific sentiment lexicon given a set of user-generated reviews which can be reliably separated into positive and negative groupings. For the purpose of this paper we will focus on the domain on hotel reviews from TripAdvisor—and thus creating a travel-related lexicon—but many other domains are amenable to the approach described. To separate our reviews into positive and negative groups we use the corresponding review rating, focusing on 5-star (positive) and 1-star (negative) reviews in this instance. As mentioned in the introduction the basic principle is that words that are frequent in positive reviews but infrequent in negative reviews are more likely to be positive, and vice versa; by comparing a word's positive and negative review frequency we can compute a fine-grained sentiment score.

2.1 Word-Level Sentiment Scoring

We begin with a collection of n positive (5-star) and negative (1-star) reviews and for each word w we calculate the number of times it occurs in positive ($pos(w)$) and negative ($neg(w)$) reviews. On the face of it the relative difference between $pos(w)$ and $neg(w)$ is an indication of the polarity of w. However, we observe that in many online review settings there can be significant differences between the length of positive and negative reviews thus leading to the need for some sort of normalisation approach.

For example, the average length of negative reviews on TripAdvisor is 191 words, significantly higher than the 120 words for the average positive review. Given that we ensure an equal number of positive and negative reviews in our training set we can apply a simple normalisation procedure by dividing word counts by the total review lengths (*POS* and *NEG* for positive and negative reviews, respectively) as in Eqs. 1 and 2.

$$freq^{+}(w) = \frac{pos(w)}{POS} \tag{1}$$

$$freq^{-}(w) = \frac{neg(w)}{NEG} \tag{2}$$

Next we calculate the *polarity difference* (or *PD*) of a word as the relative difference between its positive and negative frequency; see Eq. 3. Thus, a word that is more likely to appear in positive reviews than in negative reviews will have a positive *PD* value, whereas a word that is more likely to be in negative reviews will have a negative *PD* value. *PD* values close to 0 indicate neutral polarity or sentiment whereas PD values at the extremes of –1 or +1 indicate strong negative or positive polarity, respectively.

$$PD(w) = \frac{freq^{+}(w) - freq^{-}(w)}{freq^{+}(w) + freq^{-}(w)} \tag{3}$$

Finally, we compute the sentiment score of a word ($sent(w)$) as the signed square of its PD value; see Eq. 4 where $sign(x)$ returns –1 if $x < 0$ or +1 otherwise. The effect of this transformation is to relatively amplify differences between PD values that are non-neutral.

$$sent(w) = PD(w)^2 * sign(PD(w)) \tag{4}$$

2.2 Dealing with Negation

The approach discussed so far does not consider the issue of negation when it comes to evaluating sentiment. For example the polarity of the sentiment of a word can completely change when it is used in a negative context; for instance the sentiment of a highly positive words like "terrific" is almost completely reversed when used in a negative context. For example, "*the checkin experience was far from terrific*". However dealing with negation is far from straightforward and considerable attention has been paid to this aspect of sentiment analysis; see [20].

In this work, we deal with negation by using a dictionary of common negative terms (e.g. "not") and phrases (e.g. "far from") and look for these terms within a *k_word* distance of a given sentiment word w, within the same sentence. When a negative term is associated in this way with a sentiment word w then we assume w is in a negated context. In order to deal with this negative context we maintain two forms of w. When w is found in a regular (non negated) context it is dealt with in the normal way as described above. However when w is found in a negated context we maintain a separate count for its negated form, which we label and record as not_w.

For example, we may find the word "perfect" occurring in many positive reviews but fewer negative reviews and as such it will be assigned a positive $sent(w)$ value according to Eq. 4. However, we may also find reviews which mention features that are "not perfect" or "far from perfect". In these reviews the sentiment word ("perfect") occurs in a negated context but the counts of these occurrences are stored with reference to *not_perfect* allowing for an independent sentiment calculation for the negated form, $sent(not\_perfect)$.

So as to filter out noise, we set a threshold t of the minimum occurrence. Words that have been found less than t times in the training set will be dropped from the lexicon. The setting of a threshold value guarantees that opinion words included by the lexicon are indeed high-frequency words in the domain and also makes up for the deficiency our sentiment score formula has when dealing with words that only appear a few times in the training set.

3 Creating a Context-Sensitive Sentiment Lexicon

So far we have described an approach to constructing a domain-specific sentiment lexicon; one that captures idiomatic differences between different domains and so provides a more accurate account of sentiment. But this still does not deal with sentiment shifts that can occur at much finer levels of granularity, within a domain. For example describing a level of service as "low" is clearly negative but a "low" price is typically positive. To deal with this we need a context-sensitive lexicon that documents how words assume different sentiment in different contexts and, for example, when applied to different features.

To construct a context-sensitive sentiment lexicon we adopt a similar overall approach to the construction of a domain-specific sentiment lexicon. We begin with two sets of positive and negative reviews, count word occurrences within each of these sets, and deal with negation effects. But this time we document word occurrences with respect to individual *feature words* and keep account of sentiment words that are linked to specific features in positive and negative reviews.

Feature words are assumed to be nouns that occur with above average frequency; as such they are the common aspects that users tend to discuss in a given review domain (e.g. "restaurant" in a hotel review or "display" in a laptop review). To identify feature modifying opinion words (sentiment words) we begin by recognising within-sentence part-of-speech patterns; that is, POS tag sequences with a feature word at one end and a sentiment word on the other. We assume there to be a certain number of fixed collocations when sentiment words are used to modify features. Such collocations can be captured by the corresponding POS tag sequences so that frequent part-of-speech patterns can be employed to identify the relationship between features and opinion words. When looking for patterns, we focus on those that indicate sentiment-features relations.

But at the beginning, for a given feature in a sentence, we neither have its corresponding sentiment words nor POS patterns. Thus the domain-specific lexicon is employed to provide a set of candidate sentiment words whose absolute value of polarity difference is no less than 0.3. We assume that these nearby words with strong polarity difference is used to modify the given feature. Then we scan for and count the sentiment-feature patterns in the training set and mark those with the number of occurrence above the average as valid patterns. Afterwards, we can search for the feature modifying sentiment words by matching the valid patterns. For each sentiment-feature word pairing we count the number of times (w,f) co-occur in positive and negative reviews as $pos(w,f)$ and $neg(w,f)$, respectively. Then we calculate a relative frequency for each (w,f) pairing as in Eqs. 5 and 6, where $pos(f)$ and $neg(f)$ indicates the number of times the feature f occurs in positive and negative reviews.

$$freq^{+}(w,f) = \frac{pos(w,f)}{pos(f)} \tag{5}$$

$$freq^{-}(w,f) = \frac{neg(w,f)}{neg(f)} \tag{6}$$

Next we can calculate a feature-based polarity difference and corresponding sentiment score as in Eqs. 7 and 8, respectively.

$$PD(w,f) = \frac{freq^{+}(w,f) - freq^{-}(w,f)}{freq^{+}(w,f) + freq^{-}(w,f)} \tag{7}$$

$$sent(w,f) = PD(w,f)^{2} * sign(PD(w,f)) \tag{8}$$

In this way, within a given domain, we create a sentiment lexicon in which each sentiment word w is linked to a co-occurring feature word f and associated with a sentiment score that reflects the relative likelihood of the (w,f) pairing occurring in positive or negative reviews. Once again, positive values for $(sent(w,f)$ indicate that w is more likely to co-occur with f in positive reviews, whereas negative values for $sent(w,f)$ mean that w tends to co-occur with f in negative reviews.

We can deal with negation in context-specific lexicon generation in a manner that is exactly analogous to the approach taken when generating domain-specific lexicons. Briefly, sentiment words that appear in a negated context are recorded with a "not_" prefix and their (feature-based) sentiment score are calculated independently from non-negated versions. The threshold of minimum occurrence is set to be the same as we did in the case of domain-independent lexicons.

4 Evaluation

At the beginning of this paper we highlighted two problems with conventional sentiment lexicons—the domain problem and the context problem—which spoke against a one-size-fits-all approach to sentiment labeling. We then went on to describe ways of coping with these problems by automatically generating domain-specific and context-sensitive lexicons from user-generated reviews. In this section we describe the results of a recent study to evaluate the effectiveness of these approaches on real-world datasets from different content domains.

4.1 Datasets

As the basis for our evaluation we collected a large number of reviews from TripAdvisor and Amazon. A summary description of these data is presented in Table 1. For example the TripAdvisor dataset contains a total of 867,644 reviews for hotels. The Amazon dataset includes 90,138 reviews for electronic products in 6 categories (Digital Cameras, GPSes, Laptops, Phones, Printers, and Tablets). Reviews in both datasets are rated on a scale of 1 to 5.

We need a source of user-generated reviews as training data to produce our lexicons, and as test data to evaluate the results. For each dataset we construct our *training set* from a random sampling of the rating-1 (negative) and rating-5 (positive) reviews; in the case of TripAdvisor we select 30,000 positive and negative reviews and for Amazon we select 12,000 each. The remaining reviews (regardless of rating) are used as a source of test data and we randomly select 2,000 from each of the five rating levels for the *test set*.

As a baseline against which to evaluate our algorithmically generated lexicons we choose the popular lexicon Bing Liu opinion lexicon [10]; we will refer to this as *BL* in what follows. This is one of the most widely used in the sentiment analysis literature and is a general purpose sentiment lexicon in which each opinion word is labelled as either positive of negative.

Table 1 Statistics of the TripAdvisor and Amazon review datasets

	Tripadvisor		Amazon	
Rating	#reviews	Avg. Len	#reviews	Avg. Len
1	33,646	191	14,708	143
2	43,372	186	6,703	190
3	112,235	161	18.962	204
4	272,077	133	18,962	208
5	406,314	120	42,100	163

As a second baseline we also use the SO-CAL lexicon [18]. While Bing Liu's opinion lexicon only consists of lists of positive and negative words without sentiment scores, the SO-CAL lexicon differs from it in three respects: (1) The lexicon has been split by the POS tags; (2) The lexicon assigns a sentiment score (rather than a binary label) to each opinion word, and (3) The lexicon deals with negation in a similar manner of ours; i.e. with the "not_" prefix. In other words, the SO-CAL lexicon is structurally more similar to our domain-specific lexicons. We refer to the SO-CAL lexicon as *SOCAL* in what follows.

4.2 Domain-Specific Lexicon Coverage

To begin with, we generate domain-specific lexicons for the TripAdvisor and Amazon domains using the technique described Sect. 2—we will refer to lexicons generated in this way as *DS*—and compare these lexicons to *BL*. We do this by comparing the *DS* and *BL* lexicons in terms of coverage (their term overlap and differences). Note that for the purpose of this evaluation we only consider terms in *BL* and *DS* that are also present in the test dataset.

The results are presented in Fig. 1 as bar charts of lexicon size and venn diagrams for the TripAdvisor and Amazon data. In each case we present 4 regions as follows:

1. *DS+*: the number of positive terms in the domain-specific lexicon;
2. *DS–*: the number of negative terms in the domain-specific lexicon;
3. *BL+*: the number of positive terms in Bing Liu's lexicon;
4. *BL–*: the number of negative terms in Bing Liu's lexicon.

There are a number of observations that can be made at this stage. First of all it is clear that the *DS* lexicons tend to be larger (overall and with respect to positive and negative terms) than *BL* lexicon. Figure 1 graphs the number of positive and negative terms in the *DS* and *BL* lexicons for the TripAdvisor and Amazon datasets. For example, the TripAdvisor *DS* lexicon includes 4,963 terms overall (2,049 positive terms and 2,914 negatives) compared to the 3,429 terms in the *BL* lexicon (1,211

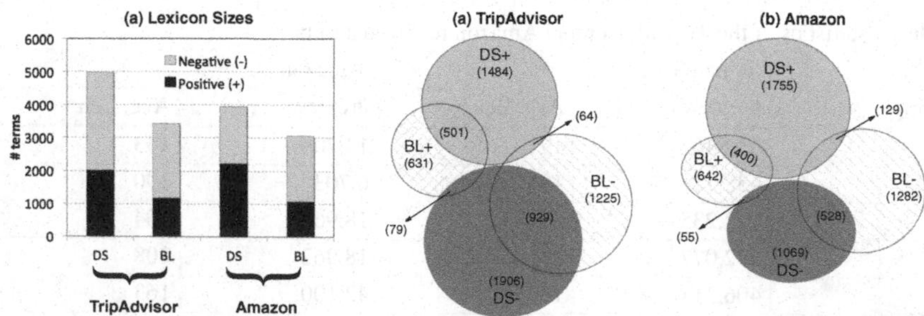

Fig. 1 A coverage analysis of the *DS* and *BL* lexicons for the TripAdvisor and Amazon datasets

positive and 2,218 negatives); similar proportions are evident for the Amazon data. Clearly there are many words in TripAdvisor and Amazon reviews, which can be linked to either positive or negative opinions, but that are absent from the more conventional *BL* lexicon.

A second observation is that there is a relatively small degree of overlap between the *DS* and *BL* lexicons. For TripAdvisor the total overlap is only 1,567 terms and for Amazon it is only 1,112 terms. Perhaps more significantly, there are also large number of terms in each lexicon that are absent from the other. For example, the TripAdvisor *DS* lexicon includes 1,906 positive and 1,484 negative terms that are unique to it, that is absent from the *BL* lexicon. These are terms that are commonly associated with positive or negative hotel reviews but are not typically included in conventional lexicons. A good example of such terms includes words like *manager* or *carpets* in the hotel domain. In each case the term is commonly associated with a negative review; travellers rarely talk about a hotel manager in a review unless it is negative and when they mention a hotel's carpets it is almost always to complain about their lack of cleanliness. On the other hand the 631 positive and 1,225 negative terms that are unique to *BL* (and missing from *DS*) simply do not occur as common sentiment words in hotel reviews. We note a very similar pattern for the Amazon data, although obviously many of the terms will be different.

Thirdly, while both lexicons agree on the polarity of the majority of their overlapping terms there are also overlapping terms where there is clear disagreement. For instance, in TripAdvisor we see that 79 terms that are classified as negative by *DS* but that are listed as positive in the *BL* lexicon; words like "portable" usually have a negative connotation in a hotel review (for example, guests often complain about a portable TV) but *BL* classifies it as a positive word. Conversely there are 64 terms that are classified as positive by *DS* but that are considered to be negative by *BL*. For example, words such as "extravagant" or "sunken" are considered to be negative by *BL* but are invariably positive in hotel reviews. Once again we see a similar pattern in the Amazon results.

So far we have demonstrated that our approach to generating domain-specific lexicons has the potential to identify a greater variety of sentiment words within a corpus of positive and negative user-generated reviews; it is worth remembering that these lexicons include the same terms as our context-sensitive lexicons. While some of these words match those in conventional sentiment lexicons, a majority do not. And for those that are found in conventional sentiment lexicons, some are found to have conflicting sentiment polarity according to our domain-specific approach. Of course none of this speaks to the quality of the *DS* lexicons nor have we considered the impact of adding context-sensitivity. Thus the next section we consider a more objective test of lexicon quality and consider these matters more fully.

4.3 Sentiment Polarity Prediction

As an independent test of lexicon quality we propose to use review content and sentiment data to automatically predict the overall polarity of the review. To do this we represent each review as a sentiment vector and train classifiers to predict the overall review polarity. We will evaluate a number of different conditions by varying the lexicons and the classifiers. We will also separately evaluate the influence of dealing with negation and then benefits of our context-sensitive approach.

4.3.1 Setup

To begin with, the testing data from the previous analysis will be used as our overall dataset for this evaluation. This is necessary because our previous training data is used to build the lexicons and we need to ensure that it does not participate in this experiment's classifier training or testing. For the purpose of this classification task we label the 1-star and 2-star rated reviews (4,000 in all in each domain) as negative and the 4-star and 5-star reviews as positive (again 4,000 review in all).

In terms of lexicons we will compare our *DS* approach to *BL* and *SOCAL* as examples of conventional one-size-fits-all lexicons. In addition we will include also include another alternative which we will designate as *PMI*, based on the work of [13]. The significance of this *PMI* lexicon is that it is generated in a similar manner to our *DS* approach, but uses a mutual information score to estimate the sentiment polarity of words. Thus it provides baseline domain-specific lexicon.

In terms of classification approach we will consider Bayes Nets (*BN*) [21], a rules-based approach in the form of *JRip* [21], and random forests (*RF*) [21]. We will also compare the classifier performance with and without any negation testing, as per Sect. 2. The implementation of negation for *BL* is carried out by simply switching the original polarity of sentiment in terms of negation. And finally, we will consider the impact of our context-sensitive approach (Sect. 3) by applying it to the *DS* and *PMI* lexicons. Applying our approach to *PMI* is analogous to the way in which we describe context-sensitivity in Sect. 3 as the the mutual information scores can be separately calculated with respect to individual, in-scope feature words. However, the conventional binary sentiment lexicon *BL* is not amenable to this context-sensitivity approach.

4.3.2 Methodology

Our basic classification task begins by representing each review as a simple feature vector. To do this we classify each word in the review by its part-of-speech tag (adjective, adverb, noun, verb) and sum the sentiment scores (–1 or +1 for *BL* versus real-valued scores for the other lexicons). Thus each review is represented by a 4-element sentiment vector and associated with a binary class label (positive or negative). We perform a 10-fold cross-validation, splitting the data into training and testing sets, build the appropriate classifiers with the training data, and evaluate them against the test data. In this experiment, we use WEKA[1] to run the classifications. In what follows we focus on presenting f-measure scores for each condition.

4.3.3 Results

The results are presented in Table 2 for both the TripAdvisor data and the Amazon data. In the first column we list the basic classification techniques used based on the underlying sentiment lexicon and the classification algorithm; for example, $DS - BN$ refers to our domain-specific lexicon with a Bayes Net classifier. The next three columns present the precision (recall, f-measure) scores classifiers built using *no negation*, *with negation*, and *negation + context*.

4.3.4 Discussion

We can see that the domain-specific lexicons (*DS* and *PMI*) perform very similarly, with f-measure scores around 0.93, significantly higher than the performance of the one-size-fits-all *BL* and *SOCAL* lexicons which has an average f-measure in the range of 0.81 to 0.85. We can also see that by explicitly dealing with negation during lexicon generation we can improve our classification performance overall. Moreover, by incorporating context into the *DS* and *PMI* lexicons—we cannot easily apply this approach to the *BL* and *SOCAL* lexicon—we can further improve classification performance, albeit more marginally in the case of TripAdvisor data.

The results for the Amazon domain are broadly similar although this time we see that *DS* performs marginally better than *PMI* across the various conditions. The introduction of negation and context also drives a greater f-measure improvement (compared with TripAdvisor), suggesting that negation and context play a more significant role in Amazon's reviews. That being said, negation has little or no effect when applied to the *BL* lexicon.

In this analysis we have focused on the f-measure results, but similar effects have been noted for precision and recall measures also. Overall it is clear that there is a significant classification performance benefit when we compare the domain specific, context sensitive lexicons to the more traditional one-size-fits-all lexicons. For example, according to Table 2 the best performing TripAdvisor condition is for *DS* using a Bayes Net classifier with negation and context to produce the f-measure score of approximately 0.954. By comparison the corresponding score for *BL* arise only 0.881. Likewise, in the Amazon domain, we see best performance for *DS* using a Bayes Net classifier with f-measure score of about 0.861, as compared with only 0.762 for *BL*. These differences—between domain specific, context sensitive results

[1]. http://www.cs.waikato.ac.nz/ml/weka/.

Table 2 Precision, Recall, F-measure for TripAdvisor and Amazon dataset; The differences between domain specific, context sensitive results are significant at the 0.05 level

		No negation			With negation			Negation + context		
	Method	Pre.	Recall	F-measure	Pre.	Recall	F-measure	Pre.	Recall	F-measure
TripAdvisor	DS-BN	0.935	0.935	0.934	0.946	0.946	0.946	**0.954**	**0.954**	**0.954**
	PMI-BN	0.933	0.933	0.933	0.941	0.941	0.941	0.949	0.949	0.949
	BL-BN	0.857	0.856	0.856	0.882	0.882	0.881	–	–	–
	SOCAL-BN	0.837	0.837	0.837	0.858	0.858	0.858	–	–	–
	DS-JRip	0.928	0.928	0.928	0.942	0.942	0.942	0.948	0.948	0.948
	PMI-JRip	0.93	0.93	0.93	0.938	0.938	0.938	0.945	0.945	0.945
	BL-JRip	0.853	0.853	0.853	0.882	0.882	0.881	–	–	–
	SOCAL-JRip	0.842	0.842	0.842	0.862	0.862	0.862	–	–	–
	DS-RF	0.932	0.932	0.932	0.943	0.943	0.943	0.95	0.95	0.95
	PMI-RF	0.927	0.927	0.927	0.941	0.941	0.941	0.949	0.949	0.949
	BL-RF	0.839	0.839	0.838	0.868	0.867	0.867	–	–	–
	SOCAL-RF	0.814	0.814	0.814	0.845	0.845	0.845	–	–	–
Amazon	DS-BN	0.82	0.82	0.82	0.853	0.853	0.853	**0.861**	**0.861**	**0.861**
	PMI-BN	0.819	0.818	0.817	0.846	0.846	0.845	0.857	0.857	0.856
	BL-BN	0.726	0.726	0.726	0.762	0.762	0.762	–	–	–
	SOCAL-BN	0.703	0.703	0.703	0.73	0.73	0.73	–	–	–
	DS-JRip	0.801	0.801	0.801	0.847	0.847	0.847	0.859	0.859	0.859
	PMI-JRip	0.802	0.802	0.801	0.839	0.839	0.839	0.854	0.854	0.854
	BL-JRip	0.738	0.738	0.738	0.76	0.76	0.76	–	–	–
	SOCAL-JRip	0.715	0.715	0.715	0.731	0.731	0.731	–	–	–
	DS-RF	0.802	0.799	0.799	0.848	0.848	0.848	0.859	0.859	0.859
	PMI-RF	0.797	0.793	0.793	0.84	0.84	0.84	0.851	0.85	0.85
	BL-RF	0.709	0.709	0.708	0.743	0.743	0.743	–	–	–
	SOCAL-RF	0.678	0.678	0.678	0.707	0.707	0.707	–	–	–

(a) Positive. (b) Negative.

Fig. 2 Domain-specific adjectives and adverbs from the TripAdvisor dataset

and *BL* and *SOCAL* results—were found to be significant at the 0.05 level using a paired t-test [3].

To make more concrete, we also generate word-clouds for the *DS* lexicons for TripAdvisor and Amazon datasets in order to evaluate the quality of the lexicons intuitively. For reasons of space, we only display domain-specific adjectives and adverbs from the TripAdvisor dataset in Fig. 2 and context-aware sentiment words in *screen* context from the Amazon dataset in Fig. 3. In these figures, size represents the sentiment strength, and brightness represents the frequency. From Fig. 2, we can see that *home-made*, *paved*, *dreamy*, *well-lit*, *good-sized*, etc. have strong positive polarity, but *poorest*, *shoddy*, and *unanswered* etc. have strong negative polarity. From Fig. 3, we can see that *sharp* is often used to describe *screen* and has the strongest positive polarity under *screen* context, however, it has strong negative polarity when describing *corner of the table* in hotel reviews. We also see that *unresponsive*, *blank*, *bad* etc. have strong negative polarity for *screen*.

(a) Positive. (b) Negative.

Fig. 3 Context-aware sentiment words in *screen* context from the Amazon dataset

5 Conclusion

One-size-fits-all sentiment lexicons have proven to be problematic when dealing with idiomatic and contextual word-usage differences that are commonplace in user-generated content. Words that are reliably positive in one setting can have negative sentiment elsewhere, and vice versa. In this paper we have described a novel technique for automatically building context-sensitive, domain-specific sentiment lexicons by employing review ratings as signals. Evaluation results based on TripAdvisor and Amazon review data have demonstrated the effectiveness of these lexicons compared to more conventional baselines. Our domain-specific lexicons capture many new sentiment terms compared to conventional lexicons and by attending to context they are better able to deal with the occasional sentiment shifts that can occur within domains.

Acknowledgments This work is supported by Science Foundation Ireland under Grant Number SFI/12/RC/2289.

References

1. Bross, J., Ehrig, H.: Automatic construction of domain and aspect specific sentiment lexicons for customer review mining. In: Proceedings of the 22nd ACM International Conference on Information and Knowledge Management, pp. 1077–1086. ACM (2013)
2. Davidov, D., Tsur, O., Rappoport, A.: Enhanced sentiment learning using twitter hashtags and smileys. In: Proceedings of the 23rd International Conference on Computational Linguistics: Posters, COLING '10, pp. 241–249. Association for Computational Linguistics, Stroudsburg, PA, USA (2010). http://dl.acm.org/citation.cfm?id=1944566.1944594
3. Dietterich, T.G.: Approximate statistical tests for comparing supervised classification learning algorithms. Neural Comput. **10**(7), 1895–1923 (1998)
4. Dong, R., Schaal, M., O'Mahony, M.P., Smyth, B.: Topic extraction from online reviews for classification and recommendation. In: Proceedings of the 23rd International Joint Conference on Artificial Intelligence, IJCAI '13. AAAI Press, Menlo Park, California (2013)
5. Dong, R., OMahony, M.P., Smyth, B.: Further experiments in opinionated product recommendation. In: Proceedings of the 22nd International Conference on Case-Based Reasoning, ICCBR '14, pp. 110–124. Springer (2014)
6. Esuli, A., Sebastiani, F.: Sentiwordnet: a publicly available lexical resource for opinion mining. In: Proceedings of LREC, vol. 6, pp. 417–422. Citeseer (2006)
7. Feldman, R., Rosenfeld, B., Bar-Haim, R., Fresko, M.: The stock sonarsentiment analysis of stocks based on a hybrid approach. In: Twenty-Third IAAI Conference (2011)
8. Go, A., Bhayani, R., Huang, L.: Twitter sentiment classification using distant supervision. CS224N Project Report, Stanford **1**, 12 (2009)
9. Hatzivassiloglou, V., McKeown, K.R.: Predicting the semantic orientation of adjectives. In: Proceedings of the 35th Annual Meeting of the Association for Computational Linguistics and Eighth Conference of the European Chapter of the Association for Computational Linguistics, pp. 174–181. Association for Computational Linguistics (1997)
10. Hu, M., Liu, B.: Mining opinion features in customer reviews. In: Proceedings of the 19th National Conference on Artifical Intelligence, AAAI'04, pp. 755–760. AAAI Press (2004). http://dl.acm.org/citation.cfm?id=1597148.1597269

11. Hu, X., Tang, J., Gao, H., Liu, H.: Unsupervised sentiment analysis with emotional signals. In: Proceedings of the 22nd International Conference on World Wide Web, pp. 607–618. International World Wide Web Conferences Steering Committee (2013)
12. Kiritchenko, S., Zhu, X., Cherry, C., Mohammad, S.M.: Nrc-canada-2014: Detecting aspects and sentiment in customer reviews. In: Proceedings of the 8th International Workshop on Semantic Evaluation (SemEval 2014), pp. 437–442 (2014)
13. Kiritchenko, S., Zhu, X., Mohammad, S.M.: Sentiment analysis of short informal texts. J. Artif. Intell. Res. 723–762 (2014)
14. Liu, B.: Sentiment analysis and opinion mining. Synth. Lect. Hum. Lang. Technol. **5**(1), 1–167 (2012)
15. Liu, B., Hu, M., Cheng, J.: Opinion observer: analyzing and comparing opinions on the web. In: Proceedings of the 14th International Conference on World Wide Web, WWW '05, pp. 342–351. ACM, New York, NY, USA (2005). doi:10.1145/1060745.1060797. http://doi.acm.org/10.1145/1060745.1060797
16. Liu, K.L., Li, W.J., Guo, M.: Emoticon smoothed language models for twitter sentiment analysis. In: AAAI (2012)
17. Lu, Y., Castellanos, M., Dayal, U., Zhai, C.: Automatic construction of a context-aware sentiment lexicon: An optimization approach. In: Proceedings of the 20th International Conference on World Wide Web, WWW '11, pp. 347–356. ACM, New York, NY, USA (2011). doi:10.1145/1963405.1963456. http://doi.acm.org/10.1145/1963405.1963456
18. Taboada, M., Brooke, J., Tofiloski, M., Voll, K., Stede, M.: Lexicon-based methods for sentiment analysis. Comput. linguist. **37**(2), 267–307 (2011)
19. Tumasjan, A., Sprenger, T.O., Sandner, P.G., Welpe, I.M.: Predicting elections with twitter: what 140 characters reveal about political sentiment. ICWSM **10**, 178–185 (2010)
20. Wiegand, M., Balahur, A., Roth, B., Klakow, D., Montoyo, A.: A survey on the role of negation in sentiment analysis. In: Proceedings of the workshop on negation and speculation in natural language processing, pp. 60–68. Association for Computational Linguistics (2010)
21. Witten, I., Frank, E.: Data Mining: Practical machine learning tools and techniques. Morgan Kaufmann (2005)
22. Wu, Y., Wen, M.: Disambiguating dynamic sentiment ambiguous adjectives. In: Proceedings of the 23rd International Conference on Computational Linguistics, COLING '10, pp. 1191–1199. Association for Computational Linguistics, Stroudsburg, PA, USA (2010). http://dl.acm.org/citation.cfm?id=1873781.1873915

Recommending with Higher-Order Factorization Machines

Julian Knoll

Abstract The accumulated information about customers collected by the big players of internet business is incredibly large. The main purpose of collecting these data is to provide customers with proper offers in order to gain sales and profit. Recommender systems cope with those large amounts of data and have thus become an important factor of success for many companies. One promising approach to generate capable recommendations are Factorization Machines. This paper presents an approach to extend the basic 2-way Factorization Machine model with respect to higher-order interactions. We show how to implement the necessary additional term for 3-way interactions in the model equation in order to retain the advantage of linear complexity. Furthermore, we carry out a simulation study which demonstrates that modeling 3-way interactions improves the prediction quality of a Factorization Machine.

Keywords Higher Order · Factorization Machine · Simulation Study · Collaborative Filtering · Factorization Model · Machine Learning

1 Introduction

Recommender systems were invented to counteract the problem of information overflow. They provide a user with relatively few pieces of information that might be relevant to him or her, selected from the mass of data. One way to produce recommendations is through the use of collaborative filtering (CF). This very intuitive approach is often used as a way to illustrate the idea of recommender systems. CF algorithms make recommendations based on data relating to the attitude (e.g., ratings) or to the

J. Knoll (✉)
Technische Hochschule Nürnberg Georg Simon Ohm,
Keßlerplatz 12, 90489 Nuremberg, Germany
e-mail: Julian.Knoll@th-nuernberg.de

J. Knoll
Friedrich-Alexander-Universität Erlangen-Nürnberg,
Lange Gasse 20, 90403 Nuremberg, Germany

M. Bramer and M. Petridis (eds.), *Research and Development in Intelligent Systems XXXIII*, DOI 10.1007/978-3-319-47175-4_7

behavior (e.g., mouse clicks) of a user concerning items he or she might be interested in. The data CF uses are usually referred to as the user-item matrix (UI matrix) since they are represented as a table with users in rows and items in columns. Based on the UI matrix, the algorithm selects a list of potentially interesting items for a specific user (item prediction) or predicts the rating a specific user might give to a specific item (rating prediction).

It is hardly surprising that approaches have been developed that make use of data beyond the UI matrix. These data can relate to characteristics of the respective user (rich side information of users), to properties of the respective item (rich side information of items), or to the relation of a specific user to a specific item (interaction-associated information). The idea of these approaches is easy to understand: increasing the amount of data a recommender algorithm uses should lead to a gain in quality of the recommendations the algorithm suggests [15].

A Factorization Machine (FM) produces recommendations in that way. FMs are general predictors—like many other machine learning approaches (e.g., Decision Trees, Support Vector Machines)—that can be applied to most prediction tasks, such as regression, binary classification, and ranking. Basic FMs can be defined by a model equation with an intercept, linear weights, and 2-way interactions. Due to the exponential growth of 2-way interactions with the increase of variables, each 2-way interaction is represented by the cross-product of two vectors of a matrix. This allows for a decrease in memory usage, linear instead of exponential time complexity, and an intended effect of the generalization of data [10].

In this paper, we analyze how to extend the basic FM model used by Rendle [10] with respect to interactions of a higher-order. We make the following main contributions. First, we extend the basic FM model by adding 3-way interactions. Second, we present an approach for implementing 3-way interactions to keep linear time complexity. And third, we carry out a simulation study demonstrating that modeling 3-way interactions in FMs makes sense with respect to the quality of recommendations.

The remainder of this work is structured as follows. In Sect. 2, we give an overview of the related work. The theoretical background concerning FMs and the extension of the FM model is presented in Sect. 3. After describing the simulation study conducted in Sect. 4, we discuss its results in Sect. 5. Concluding remarks and an outlook on our future work are contained in Sect. 6.

2 Related Work

2.1 Overview of Recommender Systems

According to Desrosiers et al. [4], recommender systems are software tools that suggest objects (items) that are of interest for a specific user (active user). Following Su and Khoshgoftaar [16] in their survey of CF techniques, there are three ways to generate recommendations—the first two defined as being in contraposition to each other, and the third combining both addressed approaches to benefit from their advantages:

1. *Content-based filtering*: Content-based recommender systems make use of data related to the characteristics of users (rich side information of users) or the properties of items (rich side information of items). A content-based algorithm generates recommendations by locating items with similar properties (e.g., color of the item) or searching for users with the same characteristics as the active user (e.g., age of the user) [16].
2. *Collaborative filtering*: CF assumes that users with the same preference for specific items show a similar behavior with respect to these items. Thus, a CF algorithm makes recommendations based on the similarity between items or users calculated from data about explicit behavior (e.g., ratings) or implicit actions related to items (e.g., mouse clicks) [2]. Therefore, Goldberg coined the phrase "wisdom of the crowd" to describe the working principle of CF [5].
3. *Hybrid approaches*: Both addressed approaches have disadvantages—while CF only makes use of information of the UI matrix, content-based filtering only exploits information relating to item or user properties. Thus, hybrid recommender techniques have been developed to utilize information from both data sources. For example, content-boosted CF [4] combines these approaches by processing both algorithms individually and composing the results. More sophisticated approaches attempt to include every available information into the recommendation process and also take advantage of more complex methods. Shi et al. [15] identify tensor factorization, graph-based approaches, and FMs as state-of-the-art algorithms that benefit from different interaction-associated information during the process of generating recommendations.

Since FMs are one of these state-of-the-art approaches, it is worth analyzing the existing algorithm concerning improvements. In the following subsections we will give an overview of the current literature regarding FMs.

2.2 Articles Introducing Factorization Machines

An early pre-stage of FMs can be seen in the tensor factorization approach of Rendle et al. [14]. In 2010 Rendle introduced the term "Factorization Machine" [10] in an article which explained an approach combining Support Vector Machines and factorization models. In this first article regarding FMs, he proposed stochastic gradient descent (SGD) as the learning method for the training of FMs. In the following year Rendle et al. [13] illustrated that FMs outperform Tucker tensor factorization.

Later, Rendle published the paper "Factorization Machines with libFM" [11] which describes FMs in more detail and comments on more sophisticated optimization methods, such as alternating least-squares (ALS) and Markov Chain Monte

Carlo (MCMC). Furthermore, he released the open source software library libFM—an implementation of the FM approach.

In both papers Rendle mentions the possibility of defining higher-order FMs without going into detail. Consequently, it might be beneficial to analyze FMs of a higher order regarding improvements related to the basic approach. In Sect. 3, after describing Rendle's approach (2-way FMs), we will detail how to set up higher-order FMs (3-way FMs).

2.3 Articles Extending Factorization Machines

Since FMs hold advantages in performance and precision compared to other existing machine learning algorithms, it is not surprising that there are several improvements related to the basic approach. First of all, Rendle [12] extended the FM approach so that it scales to relational data by providing learning algorithms that are not based on design matrices.

Sun et al. [17] improved the FM approach with regards to its application in parallel computing. They developed a FM algorithm based on the map reduce model that splits the whole problem set into smaller ones which can be solved parallel, thereby requiring less time for the computation.

Another extension was invented by Blondel et al. [1]. They formulated a convex FM model and imposed fewer restrictions on the learned model. In line with that, they presented a globally-convergent learning algorithm to solve the corresponding optimization problem.

Though literature provides some extensions to the basic FM approach, currently articles related to higher-order FMs—as we propose the in this paper—do not exist.

2.4 Articles Applying Factorization Machines

Since Rendle [10] initially described the FM approach as a recommender algorithm, there are plenty of applications in this context, e.g., simultaneous CF on implicit feedback (mouse clicks) and explicit feedback (ratings) [9], predicting user interests and individual decisions in Twitter [6], and cross-domain CF (between songs and videos) [7]. Further, Yan et al. [18] used the FM approach for the RecSys 2015 Contest and achieved the 3rd best result involving an e-commerce item recommendation problem.

As a machine learning algorithm, FMs can be used to solve non recommender problems as well. Chen et al. [3], for example, employ FMs to exploit social media data for stock market prediction and Oentaryo et al. [8] use FMs for the prediction of response behavior in mobile advertising.

UI matrix

u_1	NA	4	2	NA
u_2	3	NA	1	NA
u_3	NA	5	NA	5

design matrix X / target

x_1	x_2	x_3	x_4	x_5	x_6	x_7	y
u_1	u_2	u_3	i_1	i_2	i_3	i_4	
1	0	0	0	1	0	0	4
1	0	0	0	0	1	0	2
0	1	0	1	0	0	0	3
0	1	0	0	0	1	0	1
0	0	1	0	1	0	0	5
0	0	1	0	0	0	1	5

Fig. 1 Exemplary transformation of a UI matrix to its design matrix X and target vector $\mathbf{y}$

The multifaceted areas of application demonstrate that FMs are a state-of-the-art algorithm which can be employed on different problem sets. All those areas can benefit from improving the FM approach. Extending the approach to a higher order to receive better results can be one such improvement.

3 Theoretical Background

3.1 Data Preparation

Unlike basic CF algorithms, such as user-based CF or item-based CF (which engage a UI matrix to suggest recommendations), FMs require the data to appear in the form of features. In the first step, we ignore side information (like user age) or interaction-associated information (like tags). Let us assume we have n ratings of a users regarding b movies. Basic CF algorithms would use a UI matrix with a rows and b columns in which n elements contain ratings a specific user gave to the respective item.

In contrast, FMs work with a design matrix X with n rows and $a + b = p$ columns and a target vector $\mathbf{y}$ that contains the n ratings corresponding to each row of X. Thus, each row h of X represents one case of the prediction problem with $\mathbf{x}^{(h)}$ as input values and y_h as prediction target. In each $\mathbf{x}^{(h)}$, two elements (the specific user and the respective item) are equal to 1 and all other elements are 0. To simplify the notation, in the following the vector $\mathbf{x}^{(h)}$ will be referred to as $\mathbf{x}$ with its elements being x_1 to x_p and the target value being y.

Consequently, we transform the usually sparse UI matrix into an even more sparse design matrix X and a target vector $\mathbf{y}$, as illustrated in Fig. 1. Remember, p describes the number of features, so if we have more data than pure ratings, we can also include this side information or interaction-associated information into the matrix X by adding features. Let us assume we have data regarding c tags (that a respective user gave to a corresponding item), then the matrix X has $a + b + c = p$ columns.

In this way we include all available data in the matrix X, thus the prediction problem can be described by the design matrix $X \in \mathbb{R}^{n \times p}$—which contains all features—and a vector $\mathbf{y} \in \mathbb{R}^n$—which contains the corresponding target values [11].

Based on X and $\mathbf{y}$, FMs estimate a statistical model with a global intercept, a weight for every feature, and interactions between the features. The main characteristic that makes FMs so successful in solving recommendation problems is the possibility to include interactions. These interactions describe the situation in which two or more features influence the target by more than the sum of the involved linear weights and lower-order interactions. For instance, a 2-way interaction appears when a specific user rates a specific item higher or lower than the global intercept, the specific weight of the user, and the specific weight of the item would suggest. Accordingly, when a specific user rates a specific item which belongs to a specific genre, a 3-way interaction captures all information that is not included by the global intercept; the linear weights of the user, the item, and the genre; and all 2-way interactions between the user, the item, and the genre.

3.2 Basic 2-Way Factorization Machines

3.2.1 The 2-Way Factorization Machine Model

The basic 2-way FM model factorizes the weights of all 2-way interactions by using a matrix V. While $w_0 \in \mathbb{R}$, $\mathbf{w} \in \mathbb{R}^p$, and $V \in \mathbb{R}^{p \times k}$ are the model parameters of the 2-way FM, the corresponding FM model equation is defined as [11]:

$$\hat{y}(x) := w_0 + \sum_{j=1}^{p} w_j x_j + \sum_{j=1}^{p} \sum_{j'=j+1}^{p} x_j x_{j'} \sum_{f=1}^{k} v_{j,f} v_{j',f} \tag{1}$$

The hyperparameter $k \in \mathbb{N}^{+0}$ determines how many values factorize the weight of a 2-way interaction and therefore expresses how strong the FM generalizes. Thus, k specifies the so-called number of factors, or 2-way factors, to be more precise. Consequently, a 2-way FM is able to capture all single and pairwise interactions between the features [10]: w_0 as global intercept, w_j as weight of the feature j, and $\hat{v}_{j,j'} = \sum_{f=1}^{k} v_{j,f} v_{j',f}$ as the estimate for the interaction between feature j and j'.

3.2.2 Complexity of 2-Way Factorization Machines

One important advantage of FMs is that the model can be computed in linear time. Rendle showed this by reformulating the 2-way interactions term as follows [10]:

$$\sum_{j=1}^{p}\sum_{j'=j+1}^{p} x_j x_{j'} \sum_{f=1}^{k} v_{j,f} v_{j',f}$$
$$= \frac{1}{2}(\sum_{j=1}^{p}\sum_{j'=1}^{p} x_j x_{j'} \sum_{f=1}^{k} v_{j,f} v_{j',f}) - \frac{1}{2}(\sum_{j=1}^{p} x_j x_j \sum_{f=1}^{k} v_{j,f} v_{j,f})$$
$$= \frac{1}{2}\sum_{f=1}^{k}(\sum_{j=1}^{p} x_j v_{j,f} \sum_{j'=1}^{p} x_{j'} v_{j',f}) - \frac{1}{2}\sum_{f=1}^{k}(\sum_{j=1}^{p} x_j^2 v_{j,f}^2)$$
$$= \frac{1}{2}\sum_{f=1}^{k}\Big((\sum_{j=1}^{p} x_j v_{j,f})^2 - (\sum_{j=1}^{p} x_j^2 v_{j,f}^2)\Big)$$

This equation proves that the 2-way term—and therefore the whole 2-way FM model—has linear complexity in k and p and, thus, can be calculated in $O(kp)$.

According to Rendle [10], the model parameters w_0, $\mathbf{w}$, and V can be learned by stochastic gradient descent (SGD) for different loss functions (e.g., square, logit, or hinge loss). As the name suggests, SGD is based on the gradient of the model equation. The gradient of the 2-way FM model is as follows:

$$\frac{\partial}{\partial\theta}\hat{y}(x) = \begin{cases} 1, & \text{if } \theta \text{ is } w_0 \\ x_j, & \text{if } \theta \text{ is } w_j \\ x_j(\sum_{j'=1}^{p} v_{j',f} x_{j'}) - v_{j,f} x_j^2, & \text{if } \theta \text{ is } v_{j,f} \end{cases} \tag{2}$$

As we can see, the sum $\sum_{j'=1}^{p} v_{j',f} x_{j'}$ is independent of j and thus can be precomputed when computing $\hat{y}(x)$. Therefore, using the SGD method all parameter updates for one case can be calculated in $O(kp)$.

More sophisticated learning methods have been developed since FMs were initially defined, such as ALS and MCMC, which have some advantages in comparison to SGD—especially faster convergence [11]. However, in this paper we focus on SGD as the learning method because it is a basic method that is easy to understand and implement. Nevertheless, the approach shown in this paper can easily be applied to the named methods.

3.3 Higher-Order Factorization Machines

The 2-way FM as well as the 3-way FM are special cases of the generalized d-way FM model. Rendle [10] defined the following model equation without going into detail (e.g. about implementation or expected results):

$$\hat{y}(x) := w_0 + \sum_{j=1}^{p} w_j x_j + \sum_{l=2}^{d}\sum_{j_1=1}^{p} \dots \sum_{j_d=j_{d-1}+1}^{p} \Big(\prod_{z=1}^{l} x_{j_z}\Big)\Big(\sum_{f=1}^{k_l}\prod_{z=1}^{l} v_{j_z,f}^{(l)}\Big) \tag{3}$$

The model parameters existing in the d-way FM model are a global intercept, a weight for every feature, and an estimation for every interaction from 2-way interactions to d-way interactions with several matrices $V^{(l)} \in \mathbb{R}^{p \times k_l}$, $k_l \in \mathbb{N}^{+0}$.

3.3.1 The 3-Way Factorization Machine Model

To extend the basic FM model to 3-way interactions we need another matrix defined as $U \in \mathbb{R}^{p \times m}$ (beside the model parameters $w_0 \in \mathbb{R}$, $\mathbf{w} \in \mathbb{R}^p$, and $V \in \mathbb{R}^{p \times k}$). When we include the 3-way interactions the FM model equation changes as follows:

$$\hat{y}(x) := w_0 + \sum_{j=1}^{p} w_j x_j + \sum_{j=1}^{p} \sum_{j'=j+1}^{p} x_j x_{j'} \sum_{f=1}^{k} v_{j,f} v_{j',f} + \sum_{j=1}^{p} \sum_{j'=j+1}^{p} \sum_{j''=j'+1}^{p} x_j x_{j'} x_{j''} \sum_{f=1}^{m} u_{j,f} u_{j',f} u_{j'',f} \tag{4}$$

Just like the 2-way FM model, this model includes a global intercept (w_0), a weight for every feature (w_j), and an estimation for all 2-way interactions ($\hat{v}_{j,j'} = \sum_{f=1}^{k} v_{j,f} v_{j',f}$). In addition, the model equation contains an estimate for all 3-way interactions $\hat{u}_{j,j',j''} = \sum_{f=1}^{m} u_{j,f} u_{j',f} u_{j'',f}$.

3.3.2 Complexity of 3-Way Factorization Machines

By including the 3-way interactions we do not lose the advantage of linear complexity. In analogy to Sect. 3.2.2 we can show that the term estimating the 3-way interactions has linear complexity as well:

$$\begin{aligned}
&\sum_{j=1}^{p} \sum_{j'=j+1}^{p} \sum_{j''=j'+1}^{p} x_j x_{j'} x_{j''} \sum_{f=1}^{m} u_{j,f} u_{j',f} u_{j'',f} = \\
&= \frac{1}{6} \Big(\sum_{j=1}^{p} \sum_{j'=1}^{p} \sum_{j''=1}^{p} x_j x_{j'} x_{j''} \sum_{f=1}^{m} u_{j,f} u_{j',f} u_{j'',f} \Big) \\
&\quad - \frac{1}{2} \Big(\sum_{j=1}^{p} \sum_{j'=1}^{p} x_j x_j x_{j'} \sum_{f=1}^{m} u_{j,f} u_{j,f} u_{j',f} \Big) + \frac{1}{3} \Big(\sum_{j=1}^{p} x_j x_j x_j \sum_{f=1}^{m} u_{j,f} u_{j,f} u_{j,f} \Big) \\
&= \frac{1}{6} \sum_{f=1}^{m} \Big(\sum_{j=1}^{p} x_j u_{j,f} \sum_{j'=1}^{p} x_{j'} u_{j',f} \sum_{j''=1}^{p} x_{j''} u_{j'',f} \Big) \\
&\quad - \frac{1}{2} \sum_{f=1}^{m} \Big(\sum_{j=1}^{p} x_j^2 u_{j,f}^2 \sum_{j'=1}^{p} x_{j'} u_{j',f} \Big) + \frac{1}{3} \sum_{f=1}^{m} \Big(\sum_{j=1}^{p} x_j^3 u_{j,f}^3 \Big)
\end{aligned}$$

$$= \sum_{f=1}^{m} \Big(\frac{1}{6} (\sum_{j=1}^{p} x_j u_{j,f})^3 - \frac{1}{2} \Big(\sum_{j=1}^{p} x_j^2 u_{j,f}^2 \sum_{j=1}^{p} x_j u_{j,f} \Big) + \frac{1}{3} \Big(\sum_{j=1}^{p} x_j^3 u_{j,f}^3 \Big) \Big)$$

This equation proves that the 3-way term has linear complexity in m and p and thus can be calculated in $O(mp)$. Therefore, the whole model equation, including a global intercept, a weight for every feature, 2-way interactions, and 3-way interactions can be computed in linear time.

For the same reasons that SGD works with 2-way FMs, it can also serve as a learning method for 3-way FMs. The gradient of the 3-way FM model is:

$$\frac{\partial}{\partial \theta} \hat{y}(x) = \begin{cases} 1, & \text{if } \theta \text{ is } w_0 \\ x_j, & \text{if } \theta \text{ is } w_j \\ x_j (\sum_{j'=1}^{p} v_{j',f} x_{j'}) - v_{j,f} x_j^2, & \text{if } \theta \text{ is } v_{j,f} \\ \frac{1}{2} x_j (\sum_{j'=1}^{p} u_{j',f} x_{j'})^2 - u_{j,f} x_j^2 (\sum_{j'=1}^{p} u_{j',f} x_{j'}) & \\ \quad - \frac{1}{2} x_j (\sum_{j'=1}^{p} u_{j',f}^2 x_{j'}^2) + u_{j,f}^2 x_j^3, & \text{if } \theta \text{ is } u_{j,f} \end{cases} \tag{5}$$

We obtain four terms in the 3-way interaction gradient by applying the product rule while building the derivation. The derivation of the other terms of the gradient is straightforward.

3.4 Regularization of Factorization Machines

Typically, the number of parameters of a FM is relatively large to allow for the consideration of plenty of interactions. This means that FMs tend to overfit during the learning process and require a kind of regularization. Usually L2 regularization is applied, which means that high parameter values get penalized by adding a corresponding term to the loss function. The regularization value λ determines the influence of regularization.

Though every parameter could have its own λ, it is commonly set for every parameter depending on the order. So in a 2-way FM there are two hyperparameters—λ_1 for the weights per variable (w_1 to w_p) and λ_2 for the factors modeling the 2-way interactions ($v_{1,1}$ to $v_{p,k}$). In case of a 3-way FM, there is one more hyperparameter λ_3 regularizing the factors modeling the 3-way interactions ($u_{1,1}$ to $u_{p,m}$).

While using the learning method SGD entails a manual determination of the regularization values, enhanced learning methods (e.g. MCMC) automatically cope with regularization.

4 Simulation Study

To demonstrate the advantage of using 3-way FMs we conducted a simulation study based on the MovieLens 100 k data set. We chose this data set due to the included side information and the better performance resulting from the manageable amount of ratings. The 100k data set contains about 100,000 ratings from 943 users, relating to 1,682 movies. All users selected by MovieLens had rated at least 20 movies. Besides the ratings, the data set includes side information related to users (age, gender, occupation, and zip code) and to items (movie title, release date, video release date, and genre information).

Generating recommendations based only on the rating data from the data set only allows for having 2-way interactions—in this case using a 3-way FM does not make sense. To have a database that allows for 3-way interactions, we enriched the ratings with the data of the 18 genres included in the item side information. As a result, in total we got a matrix X containing $943 + 1,682 + 18 = 2,643$ columns.

The simulation study's aim was to analyze the influence of including 3-way interactions in the FM model. Since we discovered in a preexamination that the regularization values have a huge influence on the results, we chose the setup pictured in Fig. 2. In simulation 1 we employed five-fold cross validation. This means that we sampled five different training (80 %) and test (20 %) data sets (which are disjunct), trained the FM model with the training data set, and determined the performance by applying the model to the corresponding test data set. For simulation 2 we sampled 500 cross validation data sets in order to draw valid conclusions. In both simulations the estimation quality was measured with the metric Root Mean Squared Error (RMSE), which Rendele [10] used it in his examination.

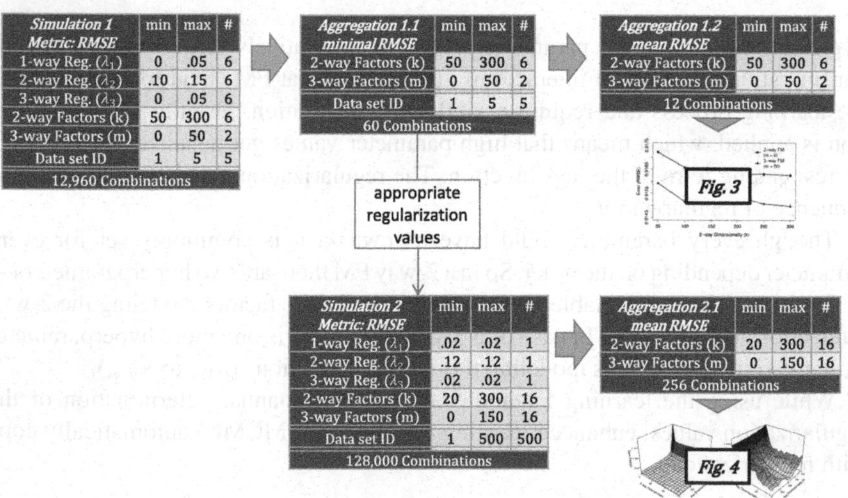

Fig. 2 Basic setup of simulation study

During simulation 1 we identified acceptable regularization values for simulation 2. Further, we compared 2-way FMs (with different k) and 3-way FMs (with different k and $m = 50$). Thus, we first conducted a grid search over the three hyperparameters λ_1, λ_2, and λ_3 in a feature space we determined in a preexamination. This grid search was applied to a different number of 2-way factors (k) on the 5 different cross validation data sets we sampled from the MovieLens 100k data set.

Afterwards, we identified the minimum prediction error of every constellation regarding k, m, and the cross validation data set. These identified minima determined the acceptable regularization values we used during simulation 2. Furthermore, we calculated the means of every combination over the cross validation data sets. As a consequence, we were able to compare the mean results of a specific 2-way FM, either with the 3-way FM with additional parameters, or with the 3-way FM with the same number of parameters in total (see Fig. 3).

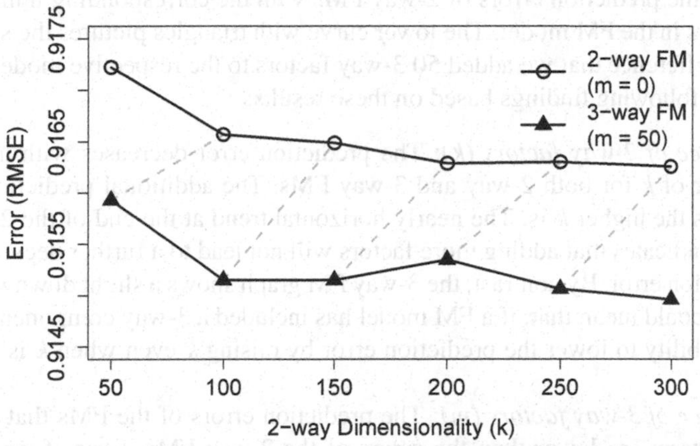

Fig. 3 Performance of 3-way FMs versus 2-way FMs with optimized regularization, *dashed grey* edges connect models with an equal number of parameters

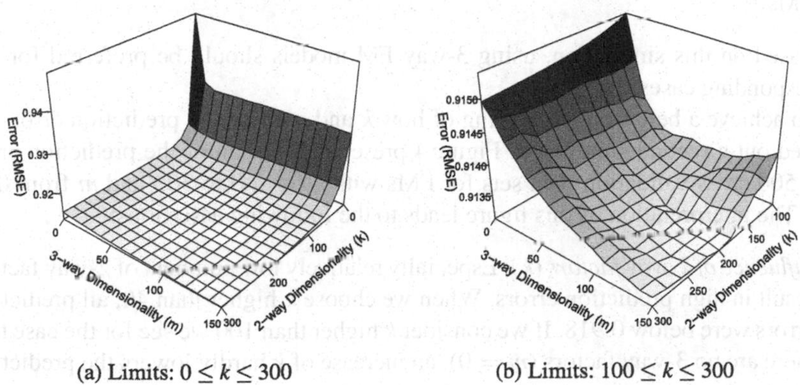

(a) Limits: $0 \leq k \leq 300$

(b) Limits: $100 \leq k \leq 300$

Fig. 4 Performance of mixed FM models regarding 2-way and 3-way dimensionality (with $\lambda_1 = 0.02$, $\lambda_2 = 0.12$, and $\lambda_3 = 0.02$, determined in simulation 1)

In simulation 2 we wanted to find out how different combinations of 2-way factors and 3-way factors perform in this setting. Therefore, we simulated with constant regularization values (captured in simulation 1) and varied the 2-way dimensionality k and the 3-way dimensionality m. In this simulation we used 500 cross validation data sets and learned overall 128,000 FMs in order to obtain valid results. Afterwards, we computed the mean over the data sets and obtained results for each of the 256 combinations of k and m we simulated. Based on these results, we drew two 3D-plots that show the connection between 2-way and 3-way factors (see Fig. 4).

5 Results

Figure 3 shows the prediction quality for different FMs. The upper curve with circles embodies the prediction errors of 2-way FMs with the corresponding number of 2-way factors in the FM model. The lower curve with triangles pictures the same FMs with the difference that we added 50 3-way factors to the respective model. We can reveal the following findings based on these results:

1. *Influence of 2-way factors (k)*: The prediction error decreases with a growing number of k for both 2-way and 3-way FMs. The additional prediction benefit reduces the higher k is. The nearly horizontal trend at the end of the 2-way FM graph indicates that adding more factors will not lead to a further decrease of the prediction error. By contrast, the 3-way FM graph shows a slight downward trend which could mean that, if a FM model has included a 3-way component, there is a possibility to lower the prediction error by raising k even when k is relatively large.
2. *Influence of 3-way factors (m)*: The prediction errors of the FMs that include a 3-way term are lower than the errors of the 2-way FMs. Even if we compare models with the same number of parameters $(k + m)$—connected with dashed grey edges—the FMs with 3-way factors provide better results than the 2-way FMs.

Based on this simulation, using 3-way FM models should be preferred for the corresponding cases.

To achieve a better understanding of how k and m drive the prediction error, we carried out a second simulation. Figure 4 presents the mean of the prediction error over 500 cross validation data sets for FMs with k from 0 to 300 and m from 0 to 150. The interpretation of this figure leads to the following assertions:

1. *Influence of 2-way factors (k)*: Especially relatively low numbers of 2-way factors result in high prediction errors. When we choose k higher than 40, all prediction errors were below 0.918. If we consider k higher than 100 we see for the case that there are no 3-way factors $(m = 0)$, an increase of k hardly lowers the prediction error.

2. *Influence of 3-way factors (m)*: For $k = 0$, even the use of 10 3-way factors lowers the prediction error from 0.949 to 0.931. In such cases where we simulated a relatively low number of 2-way factors ($k < 100$, see Fig. 4a), the effect of the additional 3-way factors is covered by the huge effect of the 2-way interactions. Nevertheless, adding a 3-way term improves the results in these cases. For $k \geq 100$ (see Fig. 4b) we can see that raising m takes a much higher effect on the results than increasing k. This indicates that there is a saturation limit regarding 2-way interactions. When we reach that point it is much more effective to capture 3-way interactions in the model than to raise the number of 2-way factors.

Hence, both simulations suggest that using 3-way FMs can bring more precise prediction results and thus should be taken into consideration. To confirm these results, a broader study should be conducted regarding the influence of k and m, which will be our subject of future research. Nevertheless, this simulation study demonstrates that there are cases in which 3-way FMs provide better results than 2-way FMs.

6 Conclusions and Future Work

In this paper, we showed how a FM model equation can include 3-way interactions and how it can be implemented with linear complexity. Furthermore, we illustrated that in specific cases 3-way FMs outperform 2-way FMs in prediction problems. This suggests that greater attention should be paid to the development of higher-order FMs.

The next steps of our research will be devoted to investigating how ALS and MCMC can be implemented for 3-way FMs. Moreover, we plan to explore how 2-way and 3-way dimensionality influence each other and how to find the optimal values for k and m.

References

1. Blondel, M., Fujino, A., Ueda, N.: Convex factorization machines. In: Appice, A., Rodrigues, P.P., Santos Costa, V., Gama, J., Jorge, A., Soares, C. (eds.) Machine Learning and Knowledge Discovery in Databases, pp. 19–35. Springer International Publishing, Cham (2015)
2. Bobadilla, J., Ortega, F., Hernando, A., Gutiérrez, A.: Recommender systems survey. Knowl.-Based Syst. **46**, 109–132 (2013)
3. Chen, C., Dongxing, W., Chunyan, H., Xiaojie, Y.: Exploiting social media for stock market prediction with factorization machine. In: Proceedings of the International Joint Conferences on Web Intelligence, pp. 142–149. IEEE Computer Society (2014)
4. Desrosiers, C., Karypis, G.: A comprehensive survey of neighborhood-based recommendation methods. In: Ricci, F., Rokach, L., Shapira, B., Kantor, P.B. (eds.) Recommender Systems Handbook, pp. 107–144. Springer, Boston (2011)
5. Goldberg, K., Roeder, T., Gupta, D., Perkins, C.: Eigentaste: a constant time collaborative filtering algorithm. Inf. Retrieval **4**, 133–151 (2001)

6. Hong, L., Doumith, A., Davison, B.: Co-factorization machines: modeling user interests and predicting individual decisions in Twitter. In: Proceedings of the 6th ACM International Conference on Web Search and Data Mining, pp. 557–566. ACM, Rome, Italy (2013)
7. Loni, B., Shi, Y., Larson, M., Hanjalic, A.: Cross-domain collaborative filtering with factorization machines. In: Rijke, M., Kenter, T., Vries, A.P., Zhai, C., Jong, F., Radinsky, K., Hofmann, K. (eds.) Proceedings of the 36th European Conference on Information Retrieval Research, pp. 656–661. Springer International Publishing, Cham (2014)
8. Oentaryo, R., Lim, E., Low, J., Lo, D., Finegold, M.: Predicting response in mobile advertising with hierarchical importance-aware factorization machine. In: Proceedings of the 7th ACM International Conference on Web Search and Data Mining, pp. 123–132. ACM, New York, New York, USA (2014)
9. Pan, W., Liu, Z., Ming, Z., Zhong, H., Wang, X., Xu, C.: Compressed knowledge transfer via factorization machine for heterogeneous collaborative recommendation. Knowl. Based Syst. **85**, 234–244 (2015)
10. Rendle, S.: Factorization machines. In: Proceddings of the 10th International Conference on Data Mining (2010)
11. Rendle, S.: Factorization machines with libFM. ACM Trans. Intell. Syst. Technol. **3**, 1–22 (2012)
12. Rendle, S.: Scaling factorization machines to relational data. Proc. VLDB Endow. **6**, 337–348 (2013)
13. Rendle, S., Gantner, Z., Freudenthaler, C., Schmidt-Thieme, L.: Fast context-aware recommendations with factorization machines. In: Proceedings of the 34th international ACM SIGIR Conference on Research and Development in Information Retrieval, pp. 635–644. ACM, Beijing, China (2011)
14. Rendle, S., Marinho, L., Nanopoulos, A., Schmidt-Thieme, L.: Learning optimal ranking with tensor factorization for tag recommendation. In: Proceedings of the 15th ACM SIGKDD international conference on Knowledge discovery and Data Mining, pp. 727–736. ACM, Paris, France (2009)
15. Shi, Y., Larson, M., Hanjalic, A.: Collaborative filtering beyond the user-item matrix: a survey of the state of the art and future challenges. ACM Comput. Surv. **47**, 1–45 (2014)
16. Su, X., Khoshgoftaar, T.: A survey of collaborative filtering. Adv. Artif. Intell. **10**, 1–19 (2009)
17. Sun, H., Wang, W., Shi, Z.: Parallel factorization machine recommended algorithm based on MapReduce. In: Proceedings of the 10th International Conference on Semantics, Knowledge and Grids (2014)
18. Yan, P., Zhou, X., Duan, Y.: E-Commerce item recommendation based on field-aware factorization machine. In: Proceedings of the 2015 International ACM Recommender Systems Challenge, pp. 1–4. ACM, Vienna, Austria (2015)

Machine Learning

Multitask Learning for Text Classification with Deep Neural Networks

Hossein Ghodrati Noushahr and Samad Ahmadi

Abstract Multitask learning, the concept of solving multiple related tasks in parallel promises to improve generalization performance over the traditional divide-and-conquer approach in machine learning. The training signals of related tasks induce a bias that helps to find better hypotheses. This paper reviews the concept of multitask learning and prior work on it. An experimental evaluation is done on a large scale text classification problem. A deep neural network is trained to classify English newswire stories by their overlapping topics in parallel. The results are compared to the traditional approach of training a separate deep neural network for each topic separately. The results confirm the initial hypothesis that multitask learning improves generalization.

Keywords Machine Learning · Neural Networks · Text Classification

1 Introduction

A divide-and-conquer approach has established itself as the de-facto standard to solve complex problems in machine learning. Difficult tasks are broken down into easier sub-tasks. This approach, however, omits a rich source of information in order to fit to the specific sub-task. *Multitask learning (MTL)*, a concept of learning to solve multiple tasks in parallel promises to result in improved generalization performance by taking into account an inductive bias. This bias is present if multiple related tasks are solved in parallel.

The aim of this paper is to review the concept of MTL and to evaluate it experimentally against the traditional divide-and-conquer approach. This is done on a large scale *text classification* problem. A *deep neural network (DNN)* is trained jointly to

H. Ghodrati Noushahr (✉) · S. Ahmadi
Centre for Computational Intelligence, School of Computer Science and Informatics,
De Montfort University, The Gateway, Leicester LE1 9BH, UK
e-mail: hossein@ghodrati.net

S. Ahmadi
e-mail: sahmadi@dmu.ac.uk

M. Bramer and M. Petridis (eds.), *Research and Development in Intelligent Systems XXXIII*, DOI 10.1007/978-3-319-47175-4_8

solve four related classification tasks. This model is compared against four separate DNN that solve only one of the classification tasks separately.

The remainder of this paper is structured as follows: in Sect. 2 we review the concept of MTL and show up related work. DNN and relevant elements related to them are reviewed in Sect. 3. A brief overview about text classification is given in Sect. 4. Neural language models as an alternative to traditional language models are reviewed in Sect. 5. The experimental setup and the evaluation criteria are presented in Sect. 6; we also describe briefly the Reuters RCV1-v2 corpus which has been used for the text classification task, in this section. The model architecture of the multitask and singletask learning models is presented in Sect. 7. The results of the experiments are listed in Sect. 8 and a conclusion drawn in Sect. 9.

2 Multitask Learning

The common approach in supervised machine learning is to split a complex problem into easier tasks. These tasks are often binary classification problems, or the classification of multiple, mutually exclusive classes. But this approach ignores substantial information that is available in a mutually non-exclusive classification problem. A simplified example should help to illustrate these two approaches. An insurance company could be interested in cross-selling life insurances to its customers and also observe if good customers are at risk of churn. The traditional approach would be to create two separate models, one for the cross-selling of the life insurance, and one for the churn prediction. This approach is referred to as *singletask learning (STL)* as each classifier is solving exactly one specific task. Both tasks could however be solved in parallel, but a *softmax* classifier would not be suited here as the two classes are not mutually exclusive. A customer could have a real demand for a life insurance while also being at risk of churn. *Multitask learning (MTL)* is about solving these mutually non-exclusive classification problems jointly.

Caruana [1] compared STL and MTL on three different problems with MTL achieving consistently better results than STL. He argued that MTL performs an inductive transfer between different tasks that are related to each other. The different training signals represent an inductive bias which helps the classifier to prefer hypotheses that explain more than just one task.

Caruana [1] conducted an experiment to prove this. The two tasks T and T' should be solved with a neural network. For task T, the two local minima *A* and *B* exist. *A* and *C* represent the two local minima for task T'. Caruana found that if trained jointly on both tasks, the hidden layer representation of the neural network fall into the local minimum *A*. This is illustrated in Fig. 1.

Remains the question what can be defined as relatedness of tasks. Caruana [1] identified the lack of a common definition and defined the concept as follows: if two tasks are the same function, but with independent noise added, the tasks are related [1] (see Definition 1). If two tasks are to predict different aspects of an individual, event,

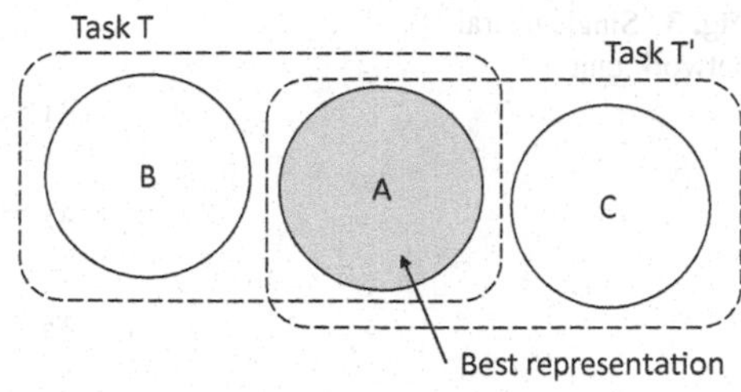

Fig. 1 Hidden layer representations findable by backpropagation [1]

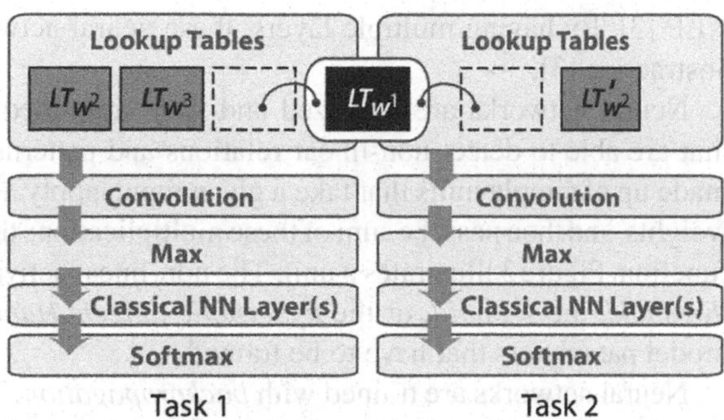

Fig. 2 Collobert and Westons multitask CNN [2]

or object, the tasks are related. But, if they were to predict aspects from different individuals, events, or objects, the tasks are not related.

Definition 1 Tasks that are the same function of inputs, but with independent noise processes added to the task signals, are related [1].

Collobert and Weston [2] applied MTL to solve multiple *natural language processing (NLP)* tasks in parallel. These task include *parsing*, *chunking*, *part-of-speech tagging* or *named entity recognition*. They trained a *convolutional neural network (CNN)* jointly on all NLP tasks in a unified architecture. Figure 2 illustrates this architecture. The joint training is achieved by sharing the lookup table LT_{w^1}. The jointly trained CNN was able to solve several NLP tasks in parallel while improving the generalization performance. Also, state-of-the-art performance was achieved on a *semantic role labelling* task with the unified architecture.

3 Deep Neural Networks

Deep neural networks (DNN) are one fundamental building block of deep learning alongside with large amounts of labelled training data, very flexible models, and a lot of computing power [3]. These neural networks have multiple hidden layers wherefore they are characterized as deep. They had a significant impact on challenging problems, such as image and speech recognition, and more recently also in

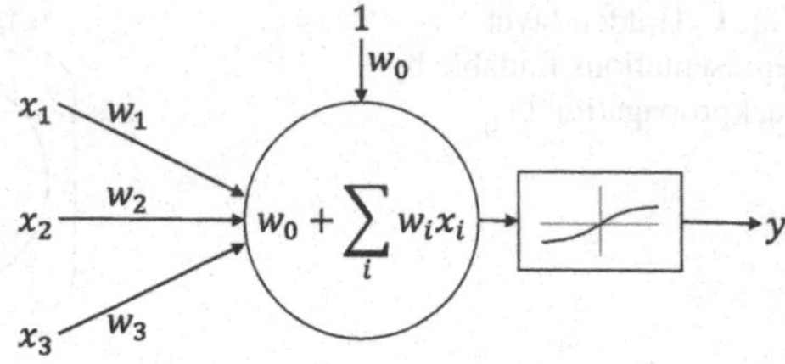

Fig. 3 Single neural network unit

NLP [3]. By having multiple layers, these neural networks learn multiple levels of abstraction [3].

Neural networks are idealized and very simplified models of the human brain that are able to detect non-linear relations and patterns. These neural networks are made up of simple units that take a given input, apply a multiplication with trainable weights, and then pass the sum of these multiplications through a non-linear activation function. Figure 3 illustrates a unit. The non-linear activation function can be a *binary threshold*, the *sigmoid*, or the *hyperbolic tangent (tanh)* function. The weights are model parameters that have to be trained.

Neural networks are trained with *backpropagation*. Training examples are passed through the neural network and an error is calculated. The error depends on a loss function such as the *mean squared error*, *mean absolute error*, *binary crossentropy*, or *categorical crossentropy*. The error is then converted into an *error derivative* that is passed backwards through the network. At each hidden layer, the error derivative is computed from the error derivatives in the layer above. Instead of calculating the error after a whole pass through the whole training data set, it is more effective and efficient to train the neural network with a small random sample; this is called *mini-batch training* [4].

The weights should be initialized by drawing randomly from a distribution such as the uniform distribution. Inputs should be standardized by calculating z-values or by normalizing to a 0–1 range and also decorrelated with techniques such as *principal component analysis (PCA)* [4]. To avoid overfitting, a technique called *dropout* is recommended [5]. Dropout removes randomly units and their connections within a neural network during training. Units are prevented from co-adapting too much with this technique [5].

One of the early challenges of training DNN is the *vanishing gradient problem*. Vanishing gradients occur when higher layer units in a DNN are nearly saturated at -1 or 1 using the tanh, and at 0 or 1 using the sigmoid activation function; the gradients in the lower layer units are approx. 0 [6]. This causes slow optimization convergence, and often the convergence happens to a poor local minima [6]. This effect was investigated by Hochreiter [7] and Bengio et al. [8] based on *recurrent neural networks* which are deep once unfolded over time, but the same holds also true for deep feedforward neural networks [8].

A new activation function helped to overcome this problem. So called *rectified linear units (ReLU)* are defined as:

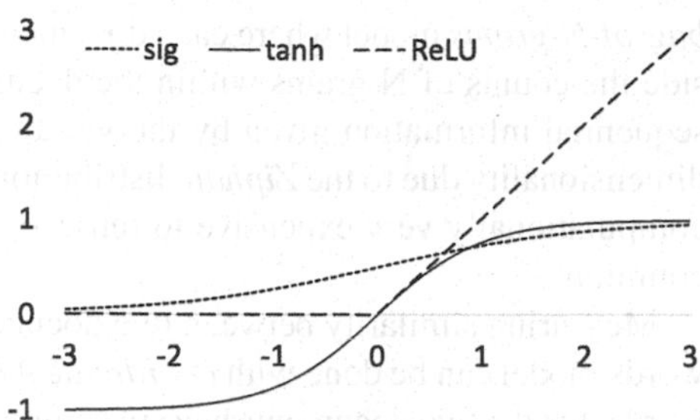

Fig. 4 Rectified linear function in comparison to sigmoid and tanh functions

$$f(x) = \max(0, x) = \begin{cases} x & \text{if } x > 0 \\ 0 & \text{else} \end{cases} \tag{1}$$

If a ReLU is in the active area above 0, the partial derivative is 1 and vanishing gradients do not exist in DNN along active units [6]. If a ReLU is in the inactive area, the gradient is 0. While this might look at first glance like a disadvantage, it allows a neural network to obtain a sparse representation [9]. A neural network that has been initialized from a uniform distribution, has around half of its hidden unit output values as zeros after being trained [9]. This is not only biologically more plausible, but also mathematically more favourable [9].

ReLU made unsupervised pre-training [10] with *stacked autoencoder* [11, 12] obsolete and a whole new family of activation functions based on ReLU [13] led to state-of-the-art results on the *ImageNet* image classification task surpassing human-level performance [14]. Figure 4 illustrates ReLU together with sigmoid and tanh activation functions.

4 Text Classification

Text classification assigns documents that contain natural language to pre-defined categories based on their content [15]. More broadly, text classification falls into the area of text mining which is a shallower technique more used for *information retrieval* and driven by the need to process large amounts of 'real world' text documents [15]. Other text mining techniques are *text summarization*, *document clustering*, or *language identification* [15].

Due to the unstructured nature of natural language text, intensive data preprocessing techniques are required for text classification. This includes *tokenization*, *stopword removal*, *lemmatizing*, and part-of-speech tagging. Also, a decision on a representation must be made. Two simple, but effective models are the *bag-of-words* model and *N-grams*. The bag-of-words model represents each document by a set of words they contain along with the frequency within each document. N-grams are probabilistic language models that find sequences of *N* consecutive words that have high occurrence probabilities [15]. These two models can also be combined to a *bag-of-N-grams* model where each document is a set of N-grams they contain alongside the counts of N-grams within the document. The bag-of-words model ignores sequential information given by the word order and also suffers from the curse of dimensionality due to the *Zipfian* distribution of words. The N-gram model in turn is computationally very expensive to retrieve; N-grams with N greater than 5 are not common.

Measuring similarity between two documents that are represented with a bag-of-words model can be done with *coordinate matching*. That is, counting the number of words that the two documents have in common. A more effective way is to weight the words with *term frequency—inverse document frequency (TF-IDF)* and subsequently measure the cosine similarity between two documents. First, words are weighted by the inverse number of documents in which the word appears (IDF) assuming that less frequent words bear more information than frequent words. Then, words are weighted by the number of occurrences within a document (TF). More formally, TF-IDF is calculated as follows:

$$\mathrm{IDF} = \log \frac{N}{n_t} \tag{2}$$

where N is the total number of documents in scope, and n_t the number of documents that contain the term t.

$$\mathrm{TF} = \log\left(1 + f_{t,d}\right) \tag{3}$$

where $f_{t,d}$ is the raw frequency of term t in document d.

$$\text{TF-IDF} = \mathrm{TF} \times \mathrm{IDF} \tag{4}$$

5 Neural Language Models

A new approach to model language are *neural language models* that are based on neural networks. They overcome the shortcomings of the bag-of-words model mentioned in the previous section by finding dense representations that preserve semantic and syntactic information. In the following, we introduce continuous vector embeddings for words, and two models introduced by Mikolov et al. [16] to compute these word embeddings.

The atomic representation of words bears two major drawbacks: absence of semantic information and high dimensionality. A dense representation of words, often referred to as word embeddings, overcomes these issues. Each word is mapped into a vector in a N-dimensional space. By doing so, it is possible to measure multiple levels of similarity between words. Also, many linguistic regularities and patterns are encoded in the dense vector embeddings [17]. That way, vector calculations such as $V(Madrid) - V(Spain) + V(France) = V(Paris)$ make it possible to answer analogy questions. The similarity between two vectors is measured by the cosine

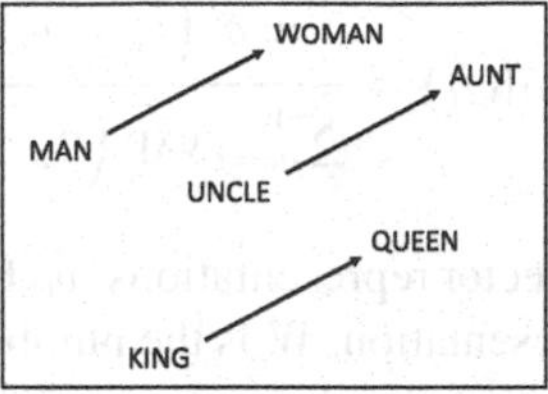

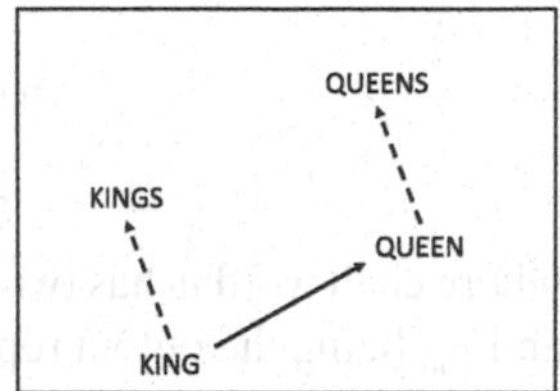

Fig. 5 Illustration of gender relation on the *left panel* and singular/plural relation on the *right panel* [18]

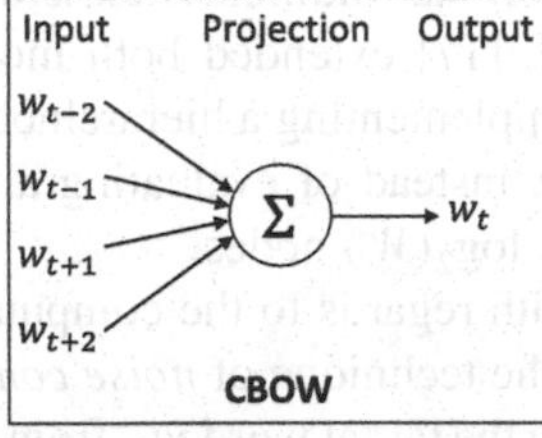

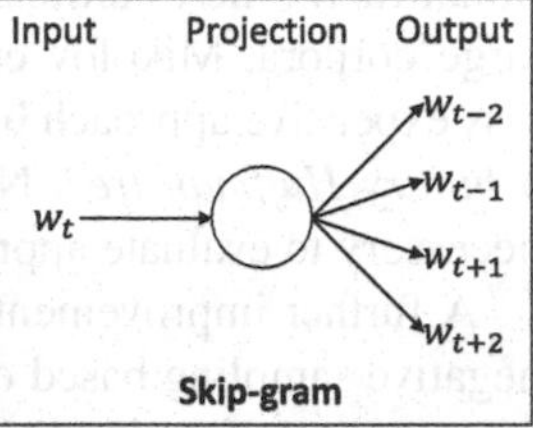

Fig. 6 CBOW and Skip-gram model architectures [16]

similarity measure. To summarize, word embeddings not only capture attributional, but also relational similarity between words as illustrated on Fig. 5. Woman, aunt, and queen are similar with regards to their attribute gender. The word pairs (man,woman), (uncle,aunt), and (king,queen) are similar with regards to their male-female relation.

The *continuous bag-of-words (CBOW)* model learns dense representations of words in a self-supervised manner by training a shallow neural network with a large unlabelled text corpus. Words in the text corpus are mapped into a N-dimensional vector space. The initial vector values are set by drawing from a random distribution. A sliding window passes over the whole text corpus predicting each time for a given context the center word. The context consists of a number of c preceding and following words. The left panel on Fig. 6 illustrates the architecture of CBOW. The vector representations of the context words are either summed or averaged.

The *Skip-gram* model is very similar to the CBOW model except for the difference that it predicts the surrounding context for a given word. The right panel on Fig. 6 illustrates the model architecture for Skip-gram. Formally, Skip-gram maximizes the average log probability

$$\frac{1}{T}\sum_{t=1}^{T}\sum_{-c\leq j\leq c, j\neq 0} \log p\left(w_{t+j}|w_t\right) \tag{5}$$

for a given word sequence $w_1, w_2, w_3, \ldots, w_T$, with c being the context window size [16]. Larger c will lead to more training examples which can lead to higher accuracy at the expense of computational cost. The softmax function defines $p(w_{t+j}|w_t)$ as

$$p(w_O|w_I) = \frac{\exp\left({v'_{w_O}}^{\top} v_{w_I}\right)}{\sum_{w=1}^{W} \exp\left({v'_w}^{\top} v_{w_I}\right)} \tag{6}$$

where each word w has two vector representations: v_w being the input representation, and v'_w being the output representation. W is the number of words in the vocabulary.

As the computational cost of this formulation $\nabla \log p(w_O|w_I)$ grows proportional to W, the traditional softmax multiclass classifier becomes impractical with large corpora. Mikolov et al. [17] extended both models with a computationally less expensive approach by implementing a hierarchical softmax classifier based on a *binary Huffman tree*. Now, instead of evaluating all W output nodes, it is only necessary to evaluate approx. $\log_2(W)$ nodes.

A further improvement with regards to the computational cost was achieved by negative sampling based on the technique of *noise contrastive estimation*. The task now consists of distinguishing the target word w_O from k negative samples randomly drawn from a noise distribution $P_n(w)$. The objective of Skip-gram with negative sampling changes to

$$\log \sigma\left({v'_{w_O}}^{\top} v_{w_I}\right) + \sum_{i=1}^{k} \mathbb{E}_{w_i \sim P_n(w)} \left[\log \sigma\left({v'_{w_i}}^{\top} v_{w_I}\right)\right] \tag{7}$$

The simplest way to find a continuous vector representation for a sequence of words, such as a sentence, paragraph, or a whole document is to average the vector embeddings across all words in the sequence. This is referred to as a *neural bag-of-words* model. We use this representation in addition to the TF-IDF weighted bag-of-words model for our experiments.

6 Experimental Setup and Evaluation

We investigate MTL with a DNN for text classification and compare it to four separate neural networks that are trained with STL. Both MTL and STL are also compared against a logistic regression model representing the baseline model. We use the Reuters RCV1-v1 text corpus [19] for the experiments. It consists of over 800,000 manually categorized newswire stories from Reuters that were written between 1996 and 1997 in English language [19]. Each story is categorized by topic, industry and region. The categories overlap and vary in granularity. The topic categories are organized in four hierarchical groups: *CCAT (Corporate/Industrial)*, *ECAT (Economics)*, *GCAT (Government/Social)* and *MCAT (Markets)* [19]. The industry categories are organized in ten hierarchical groups, such as *I2 (Metals and Minerals)* or *I5 (Construction)* [19]. The region codes are not organized hierarchically; they contain information about geographic locations and economic/political groupings [19]. We use the four root topic categories CCAT, ECAT, GCAT and MCAT for the text classification in the course of this research. The first 23,149 stories represent the training data set and the last 781,265 stories the test set. In the following, one of the newswire stories is shown. It falls into the CCAT topic, is assigned to the industry *I34420 (Electrical Instruments, Control Systems)*, and marked with the region code *USA*.

> The stock of Tylan General Inc. jumped Tuesday after the maker of process-management equipment said it is exploring the sale of the company and added that it has already received some inquiries from potential buyers. Tylan was up $2.50 to $12.75 in early trading on the Nasdaq market. The company said it has set up a committee of directors to oversee the sale and that Goldman, Sachs & Co. has been retained as its financial adviser.

We evaluate the performance based on the F_1 metric [20] being defined as:

$$F_1 = 2 \times \frac{\text{precision} \times \text{recall}}{\text{precision} + \text{recall}} \tag{8}$$

with *precision* being defined as:

$$\text{precision} = \frac{\text{true positives}}{\text{true positives} + \text{false positives}} \tag{9}$$

and with *recall* being defined as:

$$\text{recall} = \frac{\text{true positives}}{\text{true positives} + \text{false negatives}} \tag{10}$$

We report *macro* F_1 values which are unweighted averages of all four F_1 values for the root topic categories CCAT, ECAT, GCAT and MCAT. Macro F_1 accounts better for imbalances between categories.

7 Models

We train two DNN with MTL for the classification of the Reuters RCV1-v2 corpus. We use both TF-IDF weighted bag-of-words and 300-dimensional neural bag-of-words representations for these two DNN. The baseline models are two logistic regression models: one with the bag-of-words, and one with the neural bag-of-words representation.

The architecture of the bag-of-words DNN is illustrated on Fig. 7. It has four hidden layers with units varying between 10,000 in the first hidden layer, and 50 in the last hidden layer. The input consists of 47,236 features based on a bag-of-words representation that have been weighted with TF-IDF.

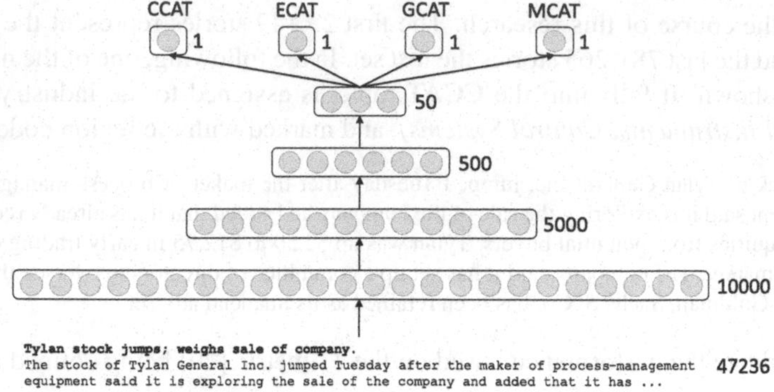

Fig. 7 Multitask learning model with bag-of-words representation

The neural bag-of-words DNN has ten hidden layers, starting with 1,200 units in the first hidden layer, and decreasing down to 10 units in the last hidden layer. This network is less complex compared to the previous DNN. Although it has ten hidden layers, the input consists of only 300 units making it a 'narrower', but deeper network. This network has only 1.9 million trainable weights compared to the 524.9 million weights of the previous DNN. The architecture of the neural bag-of-words DNN is shown on Fig. 9.

The architecture of both models is determined by a heuristic: gradually increase the number of hidden layers while at the same time decreasing the hidden layer nodes until the model starts to overfit. As the scope of this paper is to investigate MTL rather than identifying the optimal network architecture, we think that this is an acceptable approach for this paper.

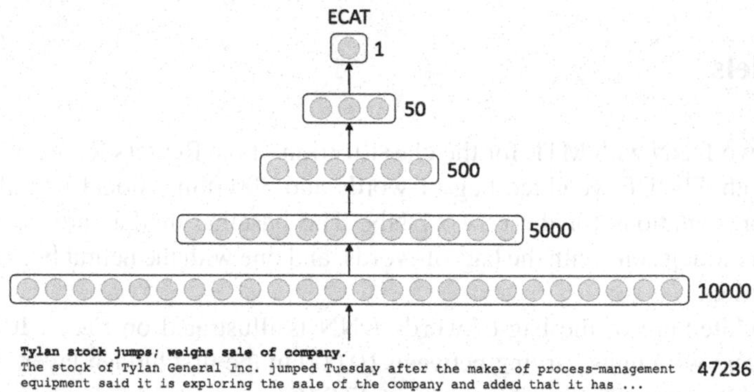

Fig. 8 Singletask learning model for ECAT topic category with bag-of-words representation

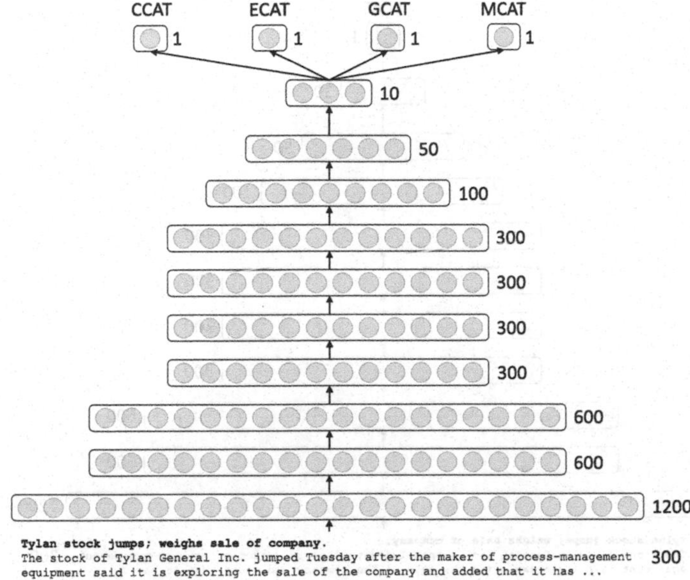

Fig. 9 Multitask learning model with neural bag-of-words representation.eps

Dropout is applied to 20 % of the units in each layer to avoid overfitting. The activation functions in all hidden layers are ReLU. The output layer consists of four units for the four root categories CCAT, ECAT, GCAT, and MCAT. The activation function in the output layer is the sigmoid with binary crossentropy being the loss function. A softmax layer cannot be used here as the four tasks are mutually not exclusive. Instead, the error of all four output units is summed and propagated back through the network. The bag-of-words network is trained with mini-batches of size 256 for 10 *epochs*, and the neural bag-of-words network is trained similarly with mini-batches of size 256 for 30 epochs.

The same architectures and hyper-parameters have been used for the four separate DNN that are trained with STL for each of the four root topic categories, both with the bag-of-words and the neural bag-of-words representation. Figures 8 and 10 show one of the networks for each representation that classifies the ECAT topic. The other three models are identical. The main difference compared to the MTL model is that each DNN has only one final output unit for the topic category that has to be predicted.

The logistic regression baseline models are trained with scikit-learn [21]. Theano [22] and Keras [23] are used for modelling and training of the neural networks.

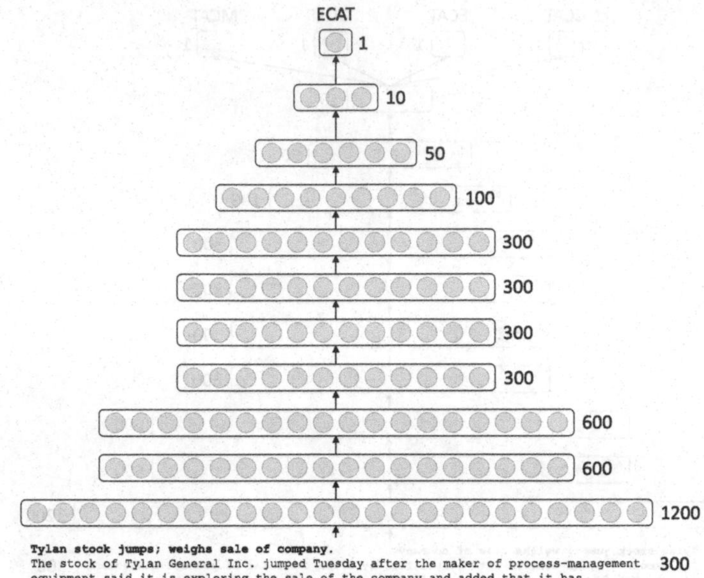

Fig. 10 Singletask learning model for ECAT topic category with neural bag-of-words representation

8 Results

The results of the experiments are shown in the Tables 2 and 3. We compare the F_1 measures of the DNNs that are trained with MTL against the the four separate DNN that are trained with STL for both representations: bag-of-words and neural bag-of-words. The DNNs with the neural bag-of-words representation are trained and evaluated ten times with different random seeds. Mean and standard deviation of all ten iterations are provided. Due to long training times, the DNNs with the bag-of-words representation are only trained and evaluated once. In addition to that, the results are compared against a logistic regression baseline (see Table 1).

MTL achieves the best results evaluated with the F_1 metric across all four separate classification tasks for the four root topic categories CCAT, ECAT, GCAT and

Table 1 Results of baseline models (reported as F_1 metric)

Model	CCAT	ECAT	GCAT	MCAT	Macro F_1 [a]
Log. regression[b]	0.9244	0.7291	0.9194	0.9029	0.8690
Log. regression[c]	0.9037	0.7080	0.9060	0.8877	0.8514

[a]Macro F_1 is the unweighted average of all separate F_1 metrics. [b]Logistic regression with bag-of-words representation. [c]Logistic regression with neural bag-of-words representation.

Table 2 Results with bag-of-words representation (reported as F_1 metric)

Learning	CCAT	ECAT	GCAT	MCAT	Macro F_1 [a]
Singletask	0.9128	0.7637	0.9062	0.9004	0.8708
Multitask	0.9144	0.7808	0.9158	0.9125	0.8809
Improvement (%)	0.18	2.24	1.07	1.35	1.17

[a]Macro F_1 is the unweighted average of all separate F_1 metrics.

Table 3 Results[a] with neural bag-of-words representation (reported as F_1 metric)

Learning	CCAT	ECAT	GCAT	MCAT	Macro F_1 [b]
Singletask	0.8785 ± 0.11	0.7155 ± 0.09	0.8628 ± 0.09	0.8632 ± 0.14	0.8300 ± 0.04
Multitask	0.9356 ± 0.00	0.8150 ± 0.01	0.9182 ± 0.01	0.9224 ± 0.00	0.8978 ± 0.00

[a]Mean and standard deviation of ten iterations with different random seeds. [b]Macro F_1 is the unweighted average of all separate F_1 metrics.

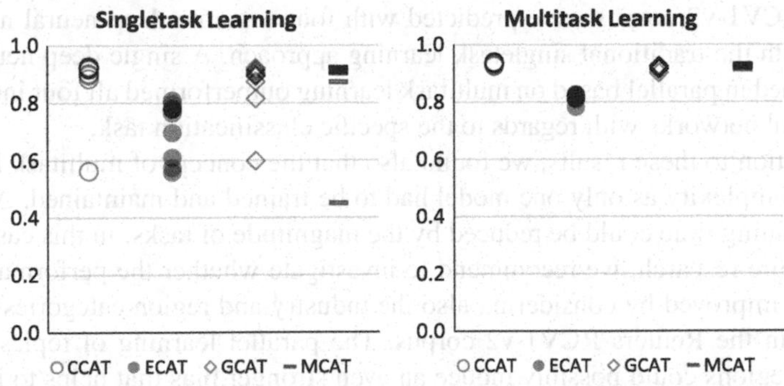

Fig. 11 F_1 results of ten iterations with STL and MTL based on the DNN with a neural bag-of-words representation

MCAT with both representations. This holds also true for the macro F_1 metric. Here, the performance could be improved by 1.17 % for the bag-of-words representation compared to STL. For the neural bag-of-words representation, the tenfold iteration reveals the true generalization performance of MTL. The mean macro F_1 metric could be improved by 8.16 % from 0.8300 to 0.8978. Some STL iterations result in very poor F_1 results which can be explained by overfitting to the training data. This effect can be seen on Fig. 11: very poor F_1 results as low as 0.45 for MCAT, and 0.56 for CCAT were obtained. With MTL however, the standard deviation across all ten iteration results is very low which is also visible on Fig. 11. The joint training of multiple tasks avoids overfitting in any of the individual tasks. Both DNNs trained with MTL outperform also the logistic regression baseline models (see Table 1).

In addition to the improved results, MTL also yields the benefit of complexity reduction. In this case, only one model needs to be trained instead of four separate ones. Also, the training time is dramatically reduced as only one model needs to be trained instead of four separate ones. The DNN based on the neural bag-of-words representation achieved better results than the DNN with the TF-IDF weighted bag-of-words representation.

9 Conclusion

We reviewed the concept of multitask learning for deep neural networks in this paper. The joint training of a neural network with related tasks induces a bias that helps to improve generalization. This hypothesis could be confirmed with experiments based on a large scale text classification problem. The four root topic categories of the Reuters RCV1-v2 corpus were predicted with four separate deep neural networks trained with the traditional singletask learning approach. A single deep neural network trained in parallel based on multitask learning outperformed all four individual deep neural networks with regards to the specific classification task.

In addition to these results, we found also that the concept of multitask learning reduced complexity as only one model had to be trained and maintained. Also, the overall training time could be reduced by the magnitude of tasks: in this case four.

For future research, we recommend to investigate whether the performance can be further improved by considering also the industry and region categories that are available in the Reuters RCV1-v2 corpus. The parallel learning of topics, industries and regions could possibly induce an even stronger bias that helps to improve generalization.

Acknowledgments We thank Microsoft Research for supporting this work by providing a grant for the Microsoft Azure cloud platform.

References

1. Caruana, R.: Multitask learning. Mach.Learn. **28**(1), 41–75 (1997)
2. Collobert, R., Weston, J.: A unified architecture for natural language processing. In: Proceedings of the 25th International Conference on Machine learning—ICML '08, pp. 160–167. ACM Press, New York, USA (2008)
3. Hinton, G.E., Bengio, Y., Lecun, Y.: Deep Learning: NIPS 2015 Tutorial (2015)
4. LeCun, Y., Bottou, L., Orr, G.B., Müller, K.R.: Efficient BackProp. In: Neural Networks: Tricks of the Trade, vol. 1524, chap. Efficient BackProp, pp. 9–50. Springer, Berlin, Heidelberg (1998)
5. Srivastava, N., Hinton, G.E., Krizhevsky, A., Sutskever, I., Salakhutdinov, R.: Dropout : a simple way to prevent neural networks from overfitting. J. Mach. Learn. Res. (JMLR) **15**, 1929–1958 (2014)

6. Maas, A.L., Hannun, A.Y., Ng, A.Y.: Rectifier nonlinearities improve neural network acoustic models. In: ICML Workshop on Deep Learning for Audio, Speech and Language Processing, vol. 28, p. 1 (2013)
7. Hochreiter, S.: Recurrent neural net learning and vanishing gradient. Int. J. Uncertainity Fuzziness Knowl. Based Syst. **6**(2), 8 (1998)
8. Bengio, Y., Simard, P., Frasconi, P.: Learning long term dependencies with gradient descent is difficult. IEEE Trans. Neural Netw. **5**(2), 157–166 (1994)
9. Glorot, X., Bordes, A., Bengio, Y.: Deep Sparse Rectifier Neural Networks. In: Proceedings of the Fourteenth International Conference on Artificial Intelligence and Statistics (AISTATS-11), vol. 15, pp. 315–323 (2011). (Journal of Machine Learning Research—Workshop and Conference Proceedings)
10. Bengio, Y., Lamblin, P., Popovici, D., Larochelle, H.: Greedy layer-wise training of deep networks. Adv. Neural Inf. Process. Syst. **19**(1), 153–160 (2007)
11. Hinton, G.E.: Learning multiple layers of representation. Trends Cogn. Sci. **11**(10), 428–434 (2007)
12. Hinton, G.E., Salakhutdinov, R.R.: Reducing the dimensionality of data with neural networks. Science **313**(5786), 504–507 (2006)
13. Jin, X., Xu, C., Feng, J., Wei, Y., Xiong, J., Yan, S.: Deep Learning with S-shaped Rectified Linear Activation Units. arXiv preprint arXiv:1512.07030 (2015)
14. He, K., Zhang, X., Ren, S., Sun, J.: Delving deep into rectifiers: surpassing human-level performance on imagenet classification. In: Proceedings of the IEEE International Conference on Computer Vision, pp. 1026–1034. IEEE (2015)
15. Witten, I.H.: Text mining. Practical Handbook of Internet Computing, pp. 14–1 (2005)
16. Mikolov, T., Chen, K., Corrado, G., Dean, J.: Efficient estimation of word representations in vector space. In: Proceedings of the International Conference on Learning Representations (ICLR 2013), pp. 1–12 (2013)
17. Mikolov, T., Sutskever, I., Chen, K., Corrado, G., Dean, J.: Distributed representations of words and phrases and their compositionality. In: Advances in Neural Information Processing Systems, pp. 3111–3119 (2013)
18. Mikolov, T., Yih, W.t., Zweig, G.: Linguistic regularities in continuous space word representations. In: Proceedings of NAACL-HLT, pp. 746–751 (2013)
19. Lewis, D.D., Yang, Y., Rose, T.G., Li, F.: RCV1: a new benchmark collection for text categorization research. J. Mach. Learn. Res. **5**, 361–397 (2004)
20. Powers, D.M.W.: Evaluation: from precision, recall and f-measure to roc, informedness, markedness and correlation. J. Mach. Learn. Technol. **2**(1), 37–63 (2011)
21. Pedregosa, F., Varoquaux, G., Gramfort, A., Michel, V., Thirion, B., Grisel, O., Blondel, M., Prettenhofer, P., Weiss, R., Dubourg, V., Vanderplas, J., Passos, A., Cournapeau, D., Brucher, M., Perrot, M., Duchesnay, É.: Scikit-learn: machine learning in python. J. Mach. Learn. Res. **12**, 2825–2830 (2012)
22. Theano Development team: theano: a python framework for fast computation of mathematical expressions. arXiv e-prints (2016)
23. Chollet, F.: Keras. https://github.com/fchollet/keras (2015)

An Investigation on Online Versus Batch Learning in Predicting User Behaviour

Nikolay Burlutskiy, Miltos Petridis, Andrew Fish, Alexey Chernov and Nour Ali

Abstract An investigation on how to produce a fast and accurate prediction of user behaviour on the Web is conducted. First, the problem of predicting user behaviour as a classification task is formulated and then the main problems of such real-time predictions are specified: the accuracy and time complexity of the prediction. Second, a method for comparison of online and batch (offline) algorithms used for user behaviour prediction is proposed. Last, the performance of these algorithms using the data from a popular question and answer platform, *Stack Overflow*, is empirically explored. It is demonstrated that a simple online learning algorithm outperforms state-of-the-art batch algorithms and performs as well as a deep learning algorithm, Deep Belief Networks. The proposed method for comparison of online and offline algorithms as well as the provided experimental evidence can be used for choosing a machine learning set-up for predicting user behaviour on the Web in scenarios where the accuracy and the time performance are of main concern.

Keywords Online Learning · Deep Learning · Classification

N. Burlutskiy (✉) · M. Petridis · A. Fish · A. Chernov · N. Ali
The University of Brighton, Brighton, UK
e-mail: N.Burlutskiy@brighton.ac.uk; nburlutsky@hotmail.com

M. Petridis
e-mail: M.Petridis@brighton.ac.uk

A. Fish
e-mail: Andrew.Fish@brighton.ac.uk

A. Chernov
e-mail: A.Chernov@brighton.ac.uk

N. Ali
e-mail: N.Ali2@brighton.ac.uk

M. Bramer and M. Petridis (eds.), *Research and Development in Intelligent Systems XXXIII*, DOI 10.1007/978-3-319-47175-4_9

1 Introduction

The era of the Internet and Big Data has provided us with access to a tremendous amount of digital data generated as a result of user activities on the Web. Recently, many researchers have achieved promising results in using Machine Learning (ML) for predicting user activity at home [1, 2], in online forums [3], social media [4], and customer preference [5].

Ideally, a prediction provides accurate results delivered in real time. However, in the real world predictive models cannot provide 100 % accuracy as well as instantaneous predictions. Also, a complex model which provides accurate predictions can be unacceptably slow [6]. Thus, there is a need to find a ML set-up for a trade-off in the accuracy and the speed of a prediction, especially in the cases when the speed of the prediction is crucial.

There are two modes of learning commonly used in machine learning, *online* and *offline/batch* learning. In most scenarios, online algorithms are computationally much faster and more space efficient [7]. However, there is no guarantee that online learning provides results as accurate as batch learning results [1, 6]. Thus, choosing the learning mode is a challenge.

Choosing a ML algorithm for both batch and online learning modes is a challenge as well. Recently deep learning showed promising results in terms of the accuracy but training such models may take significant time [1]. In contrast, simple linear classifier can provide fast but probably less accurate results [3]. The following question is addressed in this paper:

1. How do online and batch (offline) learning modes compare in terms of accuracy and time performance?

In order to answer this question, we compare five *batch* algorithms, including a deep learning algorithm, to three *online* learning algorithms in terms of their *accuracy* and *time* performance. Due to the conceptual difference in online and offline learning, this comparison is not straightforward. Thus, we propose a method for such comparison. We investigate the performance of these two learning modes and the chosen algorithms for Stack Overflow, the largest Q&A forum. We predict user behaviour, more specifically users' response time to questions at this forum.

To sum up, our intended contribution is as follows:

- To provide a method to compare online and offline learning algorithms in terms of accuracy and time performance;
- To compare the efficiency of online algorithms to offline state-of-the-art algorithms including a deep learning algorithm for user behaviour on a Q&A forum;
- To find and propose a ML set-up for a fast and accurate prediction of user behaviour on a Q&A forum.

The remainder of this paper is organised as follows. Section 2 summarises related works. In Sect. 3 we introduce ML options for predicting user behaviour. Then in Sect. 4, we describe our method for comparing the efficiency of such predictions. In

Sect. 5 we empirically evaluate offline and online algorithms for an efficient prediction. Finally, we conclude the paper in Sect. 6.

2 Related Work

Predicting user behaviour has been broadly researched in the literature. This behaviour includes but is not limited to predicting one's location at a particular time [8], customer preferences [5], user activities at home [1, 2], and behaviour on social media [3, 9]. As a result, a variety of statistical models have been used for predicting user behaviour. For example, a predictive model of time-varying user behaviour based on smoothing and trends of user activities was proposed in [10]. Generally speaking, advanced machine learning algorithms along with intelligent feature engineering demonstrates higher accuracy but the complexity of such models and features negatively affects the time performance of the models [11, 12]. Nevertheless, analysis and evaluation of the time needed for training and predicting user behaviour is often omitted [1, 10].

Predicting user behaviour on social media has included, for example, predicting the posting time of messages in Q&A forums [11, 12], churn of users [9], response time to a tweet [4]. For example, the authors in [9] focused on churn prediction in social networks where the authors proposed using a modified Logistic Regression (LR) model. A similar prediction was accomplished in [4] where a novel approach of activity prediction was proposed. The problem formulated by the authors of [4] was "given a set of tweets and a future timeframe, to extract a set of activities that will be popular during that timeframe". However, in these papers the authors did not address the problem of how to build a fast and accurate model for the prediction of user behaviour.

The authors in [2] proposed an approach to predict the start time of the next user's daily activity. The approach was based on using continuous normal distribution and outlier detection. In another paper [5], the authors showed that neural networks can provide accurate predictions of customer restaurant preference. Both works were focused on providing an accurate prediction but the complexity of the approach as well as the time required to build the model were not analysed.

At the moment of writing this paper, Deep Learning (DL) was an evolving topic attracting many ML researchers and, as a result, we found some promising results of applying DL for prediction of user behaviour. For instance, in [1] the researchers used Deep Belief Networks (DBN) for predicting user behaviour in a smart home environment. They demonstrated that DBN-based algorithms outperformed existing methods, such as a nonlinear Support Vector Machines (SVM) and k-means. A similar advantage of DBN in terms of accuracy was shown in [3] where the authors showed that a DBN-based algorithm outperformed other existing methods such as LR, Decision Trees (DT), k Nearest Neighbours (k-NN), and SVM.

Online prediction is a hot research topic due to the need of the Big Data world to provide real-time results. For example, predicting daily user activity for the next few seconds or minutes prohibits the user of batch training unless a model is very simple. The authors of the paper [1] compared the accuracy of prediction using online, mini-batch, and batch training. However, they did not evaluate the time performance of the prediction.

A comparison of training and testing times for different ML algorithms was performed in [13]. However, the authors did not compare online and offline settings.

Even though many works have shown high accuracy for different prediction tasks, the time and complexity of the trained models are usually omitted. In our paper, we evaluate both the accuracy and time efficiency of online and batch learning modes across several state-of-the-art algorithms including a deep learning algorithm. First, we introduce a method for comparison of these two modes of learning. Second, we compare five batch algorithms, LR, DT, *k*-NN, SVM, and DBN with three online learning algorithms, namely Stochastic Gradient Decent (SGD), Perceptron, and Passive-Aggressive (PA) algorithms. Last, we provide a guideline how to choose ML algorithms for real-time prediction of user behaviour on the Web.

3 Machine Learning for Predicting User Behaviour

We formulate the problem of predicting user behaviour as a classification task. This task is defined by an input feature set X and a class label set Y where the goal is to match labels from the set Y to instances in the set X. The input feature set X is formed from factors potentially affecting the accuracy of predicting user behaviour which are identified in the literature [3, 11, 12, 14–17]. Then the class label Y is formulated. Supervised ML algorithms can be divided into two large groups, namely *batch* or *offline* learning and *online* learning algorithms. The batch training can also be divided in full-batch learning where a model is trained over the entire training data and mini-batch learning when the model is trained and then updated after some number m of training instances. On the contrary, online learning means that a model is updated after every new instance.

3.1 Batch Prediction

A batch learning algorithm uses a training set D_{tr} to generate an output hypothesis, which is a function F that maps instances of an input set X to a label set Y. Thus, batch learners build a statistical assumption on a probability distribution over the product space $X \times Y$. The batch learning algorithm is expected to generalise, in the sense that its output hypothesis predicts the labels Y of previously unseen examples X sampled from the distribution [18].

3.2 Online Prediction

Online prediction is based on a training algorithm in which a learner operates on a sequence of data entries. At each step t, the learner receives an example $x_t \in X$ in a d-dimensional feature space, that is, $X = \mathbb{R}^d$. The learner predicts the class label $\hat{y}_t$ for each example x_t as soon as it receives it:

$$\hat{y}_t = sgn(f(x_t, w_t)) \in Y \tag{1}$$

where $\hat{y}_t$ is the predicted class label, x_t is an example, w_t is the weight assigned to the example, $sgn()$ is the sign function returning $\{0, 1\}$, f is a function mapping x_t, w_t into a real number $r \in \mathbb{R}$, and Y is a class label. Then the true label $y_t \in Y$ is revealed which allows the calculation of the loss $l(x_t, y_t, w_t)$ which reflects the difference between the learner's prediction and the revealed true label y_t. The loss is used for updating the classification model at the end of each learning step.

A generalised online learning algorithm is shown in Algorithm 1 which was described in [19]. This algorithm can be instantiated by substituting the prediction function f, loss function l, and update function Δ.

Algorithm 1 Online Learning

Initialise: $w_1 = 0$
for t=1,2,..,T **do**
    The learner receives an incoming instance: $x_t \in X$;
    The learner predicts the class label: $\hat{y}_t = sgn(f(x_t, w_t))$;
    The true class label is revealed from the environment: $y_t \in Y$;
    The learner calculates the suffered loss: $l(w_t, (x_t, y_t))$;
    if $l(w_t, (x_t, y_t)) > 0$ **then**
        The learner updates the classification model: $w_{t+1} \leftarrow w_t + \Delta(w_t, (x_t, y_t))$;

Online algorithms process data sample by sample which is natural for predicting user behaviour since the records of user activities are often in a chronological order. In this paper. we chose Stochastic Gradient Descent (SGD), Perceptron, and Passive-Aggressive (PA) as threc of the most popular instances of the algorithm [19].

- *Stochastic Gradient Descent*: the function f in Algorithm 1 for this learner is the dot product: $f(x_t, w_t) = w_t \cdot x_t$ and the function Δ is calculated as follows: $\Delta = \eta(y_t - \hat{y}_t)x_t$ where η is the learning rate. For simplicity, we chose hinge loss as the loss function: $l(y_t) = \max(0, 1 - y_t(w_t \cdot x_t))$.
- *Perceptron*: the difference to SGD is in the way the learner's loss is calculated: $l(y_t) = \max(0, -y_t(w_t \cdot x_t))$.
- *Passive-Aggressive*: the family of these algorithms includes a regularisation parameter C. The parameter C is a positive parameter which controls the influence of the slack term on the objective function. The update function is: $\Delta = \tau y_t x_t$ where $\tau = min(C, \frac{l_t}{||x_t||^2})$ and the loss function l is hinge loss. It was demonstrated that larger values of C imply a more aggressive update step [20].

3.3 Time Complexity of ML Algorithms

The complexity of ML algorithms for both training and testing times varies significantly across different families of algorithms and depends on the number of samples, number of features, and algorithm parameters. For some algorithms, a formal time complexity analysis has been performed [21–24]. This analysis can be used in choosing an algorithm for a prediction. For example, if time performance is crucial then it might be advantageous to choose an algorithm with lesser time complexity.

In this paper, we decided to choose algorithms with different time complexity. As a result, we chose five algorithms, namely a Decision Tree Algorithm (DT), Logistic Regression (LR), Support Vector Machines (SVM), *k* Nearest Neighbors (*k*-NN), and Deep Belief Networks (DBN), and three online algorithms, namely Stochastic Gradient Descent (SGD), Perceptron, and Passive-Aggressive (PA). The time complexity of training these models is in Table 1.

For the DT learning algorithm C4.5, the time complexity is $O(mn^2)$ where m is the number of features and n is the number of samples [24]. The time complexity of Logistic Regression (LR) is $O(mn)$ but might be worse depends on the implementation of the optimisation method [23]. For SVM, the time complexity, again, depends on the optimisation method but for one of the most popular implementations, in LibSVM, the time complexity is $O(n^3)$ [22]. For a k-NN learner implemented as a kd-tree, the training cost is $O(n \log(n))$ in time and the predicting complexity is $O(k \log(n))$, where k is the number of nearest neighbors and n is the number of instances. For more complicated models, such as Deep Belief Networks (DBNs), a formal complexity analysis is hard since there is no measure to evaluate the complexity of functions implemented by deep networks [21].

The training cost for all linear online learners is $O(kn\bar{p})$, where n is the number of training samples, k is the number of iterations (epochs), $\bar{p}$ is the average number of non-zero attributes per sample [25]. Although kernel-based online learners potentially can achieve better accuracy compared to linear online learners, kernel-based online learners require more memory to store the data, as well as more computational effort, which makes them unsuitable for large-scale applications. Thus, we considered only linear online learners in this paper.

Table 1 Time complexity of the algorithms

Algorithm	Implementation	Complexity	Reference
DT	C4.5	$O(mn^2)$	[24]
SVM	LibSVM	$O(n^3)$	[22]
k-NN	kd-tree	$O(k \log(n))$	–
LR	LM BFGS	$O(mn)$	[23]
DBN	m hidden layers	high	[21]
SGD, P, PA	Algorithm 1	$O(kn\bar{p})$	[25]

4 The Method to Compare Online and Offline Learning Modes

Due to the conceptual difference in online and offline learning, it is hard to make a comparison between them. Thus, we propose a method for comparing the accuracy and the time performance of training and testing the learners. The standard performance measurements for offline learning are accuracy, precision, recall, and F1-measure [26].

The performance of online learners is usually measured by the cumulative loss a learner suffers while observing a sequence of training samples. In order to compare online and offline learning algorithms, we introduced a method which is a modification of mini-batch training (see Fig. 1).

Algorithm 2 Calculating the accuracy of learners

Initialise training and testing times: $T_{tr} = 0$, $T_{tt} = 0$
Sort the data D chronologically from the oldest to the latest;
Divide the data D into m equal batches D_i
for i=1,2,..,m **do**
  Divide each batch D_i into a training set D_{itr} and test set D_{itt};
  Train a learner l_i on D_{itr};
  Record the time taken for the training T_{itr};
  Update the training time $T_{tr} = T_{tr} + T_{itr}$;
  Test l_i on D_{itr};
  Record the time taken for the testing T_{itt};
  Update the testing time $T_{tt} = T_{tt} + T_{itt}$;
  Calculate the average accuracy A_i over D_{itr};
Calculate the average accuracy A_{av} over all A_i

First, we used a classical 5-fold validation for training and testing the models. We divided the whole set into five parts (Fig. 1), trained a model on any four parts and then tested on the fifth part. Then we changed the parts for training and testing and repeated the process until we eventually used all the five possible combinations

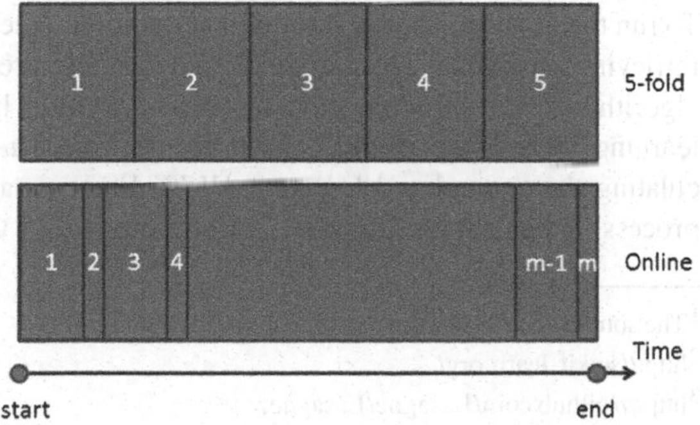

Fig. 1 Proposed training and testing scheme for comparison of online and offline learning algorithms

of the parts for training and testing. Then we calculated the average accuracy A_{av} and the time performance for training T_{tr} and testing T_{tt} for these five tests. We repeated the same process for both online and offline algorithms.

In the second set-up, we applied Algorithm 2 to online learners l_{on} and then we repeated the same algorithm for offline learners l_{off}. The purpose of applying the same aforementioned algorithm to both online and offline learners was to compare the accuracy A_{av} and the performance of online and offline algorithms in terms of the training time T_{tr} and testing time T_{tt} on the same data D in the same set-up.

5 Empirical Study

In order to compare offline to online learning, and to evaluate the performance of a deep learning algorithm, DBN, we conducted an empirical study to predict user behaviour in a Q&A forum. Q&A forums are the platforms where users ask and answer questions. We chose *Stack Overflow* website for the experiments since it is one of the most popular and fastest growing Q&A platforms (see Table 2).

We formulated the prediction task as a binary classification task – we predicted users' response time to asked questions [3]. *Stack Overflow* allows an asker to accept a satisfactory answer explicitly by clicking a button 'accept an answer'. In our experiments, we used this information to predict users' response time - the time when the first accepted answer for a question will be received. We created the label for prediction by assigning '1' to response times $t < T_m$ and '0' to response times $t \geq T_m$ where $T_m = 26$ min was the median time. Since we chose the median time for binarisation, the class for prediction was balanced.

The raw data dump (September 2014 data dump) of *Stack Overflow* consists of eight xml files: Users, Posts, Badges, Posts History, Post Links, Comments, Tags, and Votes (~96.6 GB). We imported Users, Posts, Badges, Comments, Tags, and Votes into an sqlite database (~41.3 GB). Then we extracted features influencing user behaviour on the Web. These features potentially affecting the accuracy of predicting user behaviour were identified from the literature [3, 11, 12, 14–17]. As a result, we extracted 32 features (see Table 3) for 1,537,036 samples (~1.5 GB).

System set-up

To run the experiment, we designed and implemented a software system for storing, retrieving, visualisation, statistical analysis, and prediction using machine learning algorithms.[1] We built the system using a Python library *scikit-learn* for machine learning.[2] For Deep Belief Networks, we used *Theano Lasagne* library.[3] For calculating the text-related features, *NLTK* library was deployed.[4] As a machine for processing this data (~96.6 GB), a computer with 8 GB RAM and 4 CPUs was used.

[1] The source code is at https://github.com/Nik0l/UTemPr.

[2] http://scikit-learn.org/.

[3] https://github.com/Lasagne/Lasagne.

[4] http://www.nltk.org/.

Table 2 The statistics on the *Stack Overflow* dataset used in the experiments

Forum	Users	Questions		Question response time						
				Temporal statistics				Answered within		
		Total	Answered	Mean, days	Med, min	Min, min	Max, days	1 h	1 day	1 month
Stack overflow	3,472,204	7,990,488	4,596,829	5.7	26	0.20	2,087	61	84	96

5.1 Experiments

In our experiment, we trained and tested three online (SGD, Perceptron, PA) and five offline (LR, k-NN, DT, SVM, DBN) algorithms on the same dataset with $1,537,036$ samples of 32 features (see Table 3). We scaled the data to [0, 1] since the SGD algorithm is sensitive to feature scaling. We did not shuffle the training data since the records of user activities were recorded in a chronological order and we wanted to preserve that order.

Online algorithms: we chose the number of iterations equal to one since it was enough for the convergence of SGD. We did not use any regularisation methods.[5] The learning rate η was set to be equal '1'.

Batch algorithms: DBN was trained with 10 epochs, the hidden layer was represented as a Restricted Boltzmann Machine (RBM) with 300 nodes. We chose $k = 3$ nearest neighbors for the k-NN learner.

In the first experiment we use the proposed method for comparison of online and offline learning modes. This set-up has mini-batchs with 5000 samples each (see Table 4 columns '1' and '2' correspondingly). In the second experiment we use a classic 5-fold validation. Since there is variance in prediction models trained on the same data, each algorithm was executed ten times and then we calculated the average of these ten independent runs. The results of these runs for the two experiments are shown in Table 4.

5.2 Results and Discussions

The results for the prediction of users' response time are in Table 4. DBN slightly outperformed other machine learning algorithms in all set-ups by 3–9 % in terms of the accuracy. There was no significant difference in the accuracy of the batch learning algorithms compared to online learning algorithms. However, the time spent on training and testing the batch models was significant especially compared to DBN and SVM.

The fastest batch learning algorithm, LR, showed almost 3 % poorer accuracy and was 13 times slower in terms of the training time. SGD showed the highest accuracy of 63.8 % out of the online learners (see Table 4). On contrary, a Perceptron learner showed relatively poor performance with the accuracy of 57.7 % probably due to the fact that it is sensitive to badly labelled examples even after training on many examples whereas the SGD and Passive-Aggressive algorithms are more robust to badly labelled examples. Also, the aspect that the used online algorithms put different importance on each training instance over time might have influenced the results. Also, the online learners had less variation in the accuracy over the process of training the models compared to the batch learners.

[5] In our experiments, we tried L_1 and L_2 regularisation but we did not find any significant improvements in the results compared to the results without regularisation reported in this paper.

Table 3 Features for the prediction of users' response time

F_i	Description	Comments	Reference
F_1	The total number of question asked by a user	This number is calculated from the time a user registered at a website	[12, 15, 17]
F_2	The total number of answers given by a user	This number is calculated from the time a user registered at a website	[12, 15, 17]
F_3	The total number of user's profile views by other users	Some websites provide with the information on how many times user's profile has been viewed	[14]
F_4	The number of user's answers accepted by other users	*Stack Exchange* websites provide a functionality for an asker to accept an answer	[3]
F_5	The reputation of the user	*Stack Exchange* provides with user's reputation	[14, 17]
F_6	The number of user's upvotes	Some websites provide this feature	[15, 17]
F_7	The number of user's downvotes	Some websites provide this feature	[15, 17]
F_8	The location of a user	For example, *USA, FL, Miami*	[3]
F_9	User's latitude	For example, 101.67	[3]
F_{10}	User's longitude	For example, 20.27	[3]
F_{11}	User's time zone	For example, UTC+2	[3]
F_{12}	The number of days a user has been registered	$T_{now} - T_{registered}(u_i)$ where T_{now} is the current time and $T_{registered}$ is the time when a user u_i registered	[17]
F_{13}	The average answering time	The time is calculated for all questions	[3]
F_{14}	12 entries vector: the number of posts for each month in last year	$(p_1(u_i), ..., p_{12}(u_i))$ where p_i is the number of posts by a user u_i in a particular month j	[3]
F_{15}	24 entries vector: the average number of posts for each hour	$(p_1(u_i), ..., p_{24}(u_i))$, where p_j is the average number of posts by a user u_i at a particular hour j	[11, 12]
F_{16}	The number of questions asked by a user during a time interval	In this paper, we used the number of questions asked by a user during the last week	[27]

(continued)

Table 3 (continued)

F_i	Description	Comments	Reference
F_{17}	The number of answers given by a user during a time interval	In this paper, we used the number of answers given by a user in the last week	[27]
F_{18}	The length of the question title	We ignored spaces in the titles	[15, 16]
F_{19}	The length of the question body	We ignored spaces in the text	[15, 16]
F_{20}	# of 'wh' words in the title	'Wh' words are question words, i.e. 'who', 'what'	[11, 16]
F_{21}	# of 'wh' words in the body	Wh' words are question words, i.e. 'who', 'what'	[11, 16]
F_{22}	# of active verbs in the title	Active verbs include such verbs as 'do', 'make'	[16]
F_{23}	# of active verbs in the body	Active verbs include such verbs as 'do', 'make'	[16]
F_{24}	# of times a user mentioned himself, (for ex., 'we', 'I', 'me')	If a user mentioned themselves in a question means they tried to solve the problem	[16]
F_{25}	The number of url links	For example, the number *href* words	[11]
F_{26}	The number of images	For example, the number of *img* words in the text	[16]
F_{27}	The total number of times a question was viewed	A question can be viewed by both registered and not registered users	[3]
F_{28}	The number of tags in a question	Usually questions are tagged with some words, for example, 'computers' or 'linux'	[12, 17]
F_{29}	The popularity of the tags	Frequency of the tags in the questions	[16]
F_{30}	The number of popular tags	The tags are divided into popular and non-popular	[16]
F_{31}	The 'togetherness' of tags a question is tagged with	$\frac{p(x,y)}{p(x)p(y)}$, $p(x, y)$ is the probability of tags x and y occur together, whereas $p(x)$ and $p(y)$—independently	[15, 16]
F_{32}	The number of comments for a question during a time interval	In this paper, we used the number of comments and replies for a question during the last day	[3]

Table 4 The results for online and offline learning in comparison (1,537,036 samples) for 5-fold validation (5 fold) any mini-batches of 5,000 samples (batch)

Algorithm	Accuracy, %		Training time, s		Testing time, s	
	5 fold	Batch	5 fold	Batch	5 fold	Batch
k-NN	57.4 ± 0.72	57.7 ± 0.72	819.7	98.2	2813	944.3
DT	60.0 ± 0.20	59.2 ± 0.43	193.8	60.8	0.568	0.993
LR	63.7 ± 0.35	61.1 ± 0.69	**85.3**	**18.1**	**0.068**	**0.356**
SVM	62.4 ± 0.46	61.9 ± 0.46	2,933.3	2,333.3	380.6	360.6
DBN	**64.5 ± 0.82**	**66.5 ± 0.52**	2,812.2	3,009.2	705.3	725.7
Perceptron	55.6 ± 0.61	57.7 ± 0.61	4.8	1.4	0.066	**0.002**
PA	59.6 ± 0.52	63.0 ± 0.52	**4.2**	**0.7**	**0.052**	0.003
SGD	**61.2 ± 0.42**	**63.8 ± 0.82**	4.7	1.3	0.056	**0.002**

The time performance of batch algorithms corresponds to their time complexity evaluated in Sect. 3.3 earlier. We can notice that the time complexity of DBN is similar to the time complexity of SVM since both DBN and SVM performed the worst and relatively similar in terms of the training time.

The time performance of online algorithms in the conducted experiments corresponds to their time complexity evaluated in Sect. 3.3 as well. However, training a Perceptron algorithm took slightly longer than the other algorithms.

We expected online learning to perform faster than batch learning in both training and the prediction. The reason is in the nature of these algorithms - in online learning the weights of the model are updated at each step compared to the computationally expensive training in the batch mode. However, we did not expect such significant differences. Even though the experimental results are very hardware dependent, in our experiments the online training and predicting took 13-3000 times less time than batch training and predicting.

It is important to mention that most learners, both online and batch except DT, SVM, and LR, showed poorer accuracy in the results for 5-fold validation (see Table 4) whereas DT, SVM, and LR showed almost the same accuracy. One possible cause of such drop in the accuracy can be the sequential nature of user behaviour and, as a result, the models must be trained in a sequential order rather than shuffling parts of data as it is done in 5-fold validation.

6 Conclusions

We investigated the problem of comparing online and batch ML set-ups in the context of user behaviour on the Web. This comparison allows to choose a ML set-up which provides a fast and accurate prediction. Predicting user behaviour on the Web often requires fast predictions performed in units of seconds or minutes. Thus, as a case study, we predicted the time when a user will answer a question on the largest technical Q&A forum, *Stack Overflow*. Our experiment showed that even though a deep learning algorithm demonstrated slightly more accurate results, the time for both training and predicting was several magnitudes higher compared to the simplest online learning algorithms. Thus, in the world of Big Data, online learning can serve as a simple solution for providing a fast, near real-time prediction. Even though we conducted the experiment only for *Stack Overflow*, we are planning to explore more datasets and prediction tasks for improving the generalisation of our findings beyond the context of user behaviour on the Web. In the future, these findings will be useful for choosing a ML setup for predictions when there are tough requirements on computational complexity as well as the accuracy of a real time prediction system.

Acknowledgments The authors are grateful for illuminating discussions to Dr Yuri Kalnishkan's team in the project "On-line Self-Tuning Learning Algorithms for Handling Historical Information" (funded by the Leverhulme Trust).

References

1. Choi, S., Kim, E., Oh, S.: Human behavior prediction for smart homes using deep learning. In: 2013 IEEE RO-MAN, pp. 173–179 (2013)
2. Nazerfard E., Cook, D.: Using Bayesian Networks for Daily Activity Prediction (2013)
3. Burlutskiy, N., Petridis, M., Fish, A., Ali, N.: Prediction of users' response time in Q&A communities. In: ICMLA'15, International Conference on Machine Learning and Applications (2015)
4. Weerkamp, W., De Rijke, M.: Activity prediction: a twitter-based exploration. In: Proceedings of TAIA'12 (2012)
5. Zheng, B., Thompson, K., Lam, S.S., Yoon, S.W., Gnanasambandam, N.: Customers behavior prediction using artificial neural network. In: Industrial and Systems Engineering Research Conference (ISERC), pp. 700–709. Institute of Industrial Engineerings (2013)
6. Loumiotis, I., Adamopoulou, E., Demestichas, K., Theologou, M.: On trade-off between computational efficiency and prediction accuracy in bandwidth traffic estimation. Electron. Lett. **50**(10), 754–756 (2014)
7. Liang, N.Y., Huang, G.B., Saratchandran, P., Sundararajan, N.: A fast and accurate online sequential learning algorithm for feedforward networks. IEEE Trans. Neural Netw. **17**(6), 1411–1423 (2006)
8. Sadilek, A., Krumm, J.: Predicting long-term human mobility. In: AAAI, Far out (2012)
9. Zhu, Y., Zhong, E., Pan, S.J., Wang, X., Zhou, M., Yang, Q.: Predicting user activity level in social networks. In: Proceedings of the 22Nd ACM International Conference on Information and Knowledge Management, CIKM '13, pp. 159–168, New York, NY, USA. ACM (2013)
10. Radinsky, K., Svore, K., Dumais, S., Teevan, J., Bocharov, A., Horvitz, E.: Modeling and predicting behavioral dynamics on the web. In: Proceedings of the 21st International Conference on World Wide Web, WWW '12, pp. 599–608, New York, NY, USA. ACM (2012)
11. Dror, G., Maarek, Y., Szpektor, I.: Will my question be answered? predicting "question answerability" in community question-answering sites. In: Blockeel, H. (ed.) Machine Learning and Knowledge Discovery in Databases. Lecture Notes in Computer Science, vol. 8190, pp. 499–514. Springer, Berlin Heidelberg (2013)
12. Yang, L., Bao, S., Lin, Q., Wu, X., Han, D., Su, Z., Yu, Y.: Analyzing and predicting not-answered questions in community-based question answering services. In: Burgard, W. (ed.) AAAI. AAAI Press (2011)

13. Lim, T.S., Loh, W.Y., Shih, Y.S.: A comparison of prediction accuracy, complexity, and training time of thirty-three old and new classification algorithms. Mach. Learn. **40**(3), 203–228 (2000)
14. Anderson, A., Huttenlocher, D., Kleinberg, J., Leskovec, J.: Discovering value from community activity on focused question answering sites: a case study of stack overflow. In: Proceedings of the 18th ACM SIGKDD International Conference on Knowledge Discovery and Data Mining, pp. 850–858, New York, USA. ACM (2012)
15. Asaduzzaman, M., Mashiyat, A.S., Roy, C.K., Schneider, K.A.: Answering questions about unanswered questions of stack overflow. In: Proceedings of the 10th Working Conference on Mining Software Repositories, pp. 97–100. Piscataway, NJ, USA (2013)
16. Bhat, V., Gokhale, A., Jadhav, R., Pudipeddi, J., Akoglu, L.: Min(e)d your tags: analysis of question response time in stack overflow. In: 2014 IEEE/ACM International Conference on Advances in Social Networks Analysis and Mining (ASONAM), pp. 328–335 (2014)
17. Lezina, C.G.E., Kuznetsov, A.M.: Predict Closed Questions on Stack Overflow (2012)
18. Dekel, O.: From online to batch learning with cutoff-averaging. In: Koller, D., Schuurmans, D., Bengio, Y., Bottou, L. (eds.) Advances in Neural Information Processing Systems 21, pp. 377–384. Curran Associates, Inc. (2009)
19. Hoi, S.C., Wang, J., Zhao, P.: Libol: a library for online learning algorithms. J. Mach. Learn. Res. **15**, 495–499 (2014)
20. Crammer, K., Dekel, O., Keshet, J., Shalev-Shwartz, S., Singer, Y.: Online passive-aggressive algorithms. J. Mach. Learn. Res. **7**, 551–585 (2006)
21. Bianchini, M., Scarselli, F.: On the complexity of neural network classifiers: a comparison between shallow and deep achitectures. IEEE Trans. Neural Netw. Learn. Syst. **25**(8), 1553–1565 (2014)
22. Chapelle, O.: Training a support vector machine in the primal. Neural Comput. **19**(5), 1155–1178 (2007)
23. Minka, T.P.: A Comparison of Numerical Optimizers for Logistic Regression. Technical report (2003)
24. Su, J., Zhang, H.: A fast decision tree learning algorithm. In: Proceedings of the 21st National Conference on Artificial Intelligence—Volume 1, AAAI'06, pp. 500–505. AAAI Press (2006)
25. Bottou, L.: Proceedings of COMPSTAT'2010: 19th International Conference on Computational StatisticsParis France, August 22–27, 2010 Keynote, Invited and Contributed Papers, chapter Large-Scale Machine Learning with Stochastic Gradient Descent, pp. 177–186. Physica-Verlag HD, Heidelberg (2010)
26. Mohri, M., Rostamizadeh, A., Talwalkar, A.: Foundations of Machine Learning. The MIT Press (2012)
27. Cai, Y., Chakravarthy, S.: Answer quality prediction in Q&A social networks by leveraging temporal features. IJNGC **4**(1) (2013)

A Neural Network Test of the Expert Attractor Hypothesis: Chaos Theory Accounts for Individual Variance in Learning

P. Chassy

Abstract By positing that complex, abstract memories can be formalised as network attractors, the present paper introduces chaos theory in the field of psychological learning and, in particular, in the field of expertise acquisition. The expert attractor hypothesis is that the cortical re-organisation of biological networks via neural plasticity leads to a stable state that implements the memory template underpinning expert performance. An artificial neural network model of chess players' strategic thinking, termed Templates for Expert Knowledge Simulation, was used to simulate, in 500 individuals, the learning of 128 positions which belong to 8 different chess strategies. The behavioural performance of the system as a novice, as an expert, and its variance in learning, are all in line with psychological findings. Crucially, the distribution of weights, the learning curves, and the evolution of the distribution of weights support the attractor hypothesis. Following a discussion on the psychological implications of the simulations, the next step towards unravelling the chaotic features of the human mind are evoked.

Keywords Learning · Chaos · Attractor · Expertise

1 The Expert Attractor Hypothesis

The classical view of psychologists conceptualises learning as a storage mechanism. That is, new information is added to a database of existing knowledge. In this context, long-term memory (LTM) is defined in psychology as the repository of knowledge and skills. Unfortunately, this classic notion of memory has been challenged by biological and neuroimaging research. At the cellular level, learning relates to the construction of biological terminals that re-structure the network. The process, termed neural plasticity [13], requires the activation of memory-specific genes that will create new axons and synapses to increase (excitatory) or decrease (inhibitory)

P. Chassy (✉)
Mathematical Cognition Research Group, Department of Psychology,
Liverpool Hope University, Liverpool L16 9JD, UK
e-mail: chassyp@hope.ac.uk

M. Bramer and M. Petridis (eds.), *Research and Development in Intelligent Systems XXXIII*, DOI 10.1007/978-3-319-47175-4_10

the influence of one neuron to the next. By this process, neurons get fused into cell assemblies that code for concepts [11]. A typical feature of expertise acquisition is the development of highly rich and complex memory structures, termed templates [9], that encode the essential features of problems and directly connect the problem to potential solutions. The central role of rich memory structures in high level performance has been demonstrated behaviourally and has received considerable support from neuroimaging studies. It is thus understandable that concepts of high complexity such as those used by experts require a huge amount of rewiring. The estimate of the time required to achieve expert status is about 10,000 h of dedicated training or around 10 years of deliberate practice [7]. A question that remains difficult to answer is how to account for the wide variance in the time necessary to acquire expertise. A study with chess players indicates that the time, though still with an average of around 10,000 h, can require as much as 24,000 h or as little as 4,000 h. These figures imply that for some individuals, acquiring expertise takes 6 times more training than others.

A longstanding issue in psychology and education has been to account for individual variances in learning. Many factors which play a role in the quality and speed of learning have been identified. Among the factors, those which have been most explored are the amount of deliberate practice [7], intelligence [10], genetics [3], motivation, and pedagogy. Deliberate practice is the most common factor cited to account for individual differences in performance. It has been shown to hold for many types of expertise such as music [16], medicine [6], and sports [28]. It is worth noting that the neural structure in charge of supporting the creation of complex memories, the hippocampus [27], is directly connected to the neural networks implementing the emotional system [25]. In consequence, it is not surprising that emotional responses and motivation influence learning speed. There is no doubt that genetic mechanisms and emotions play a role, respectively, in supporting and facilitating learning [3]. Analysing learning by considering exclusively the main neural circuits, such as the amygdala-hippocampus complex [25], emphasises learning processes common to all individuals. Yet, since learning takes place at a cell level [13], considering only large scale structures erases individual differences in the micro structure of neural networks [15, 24]. Strikingly, the fact that the brain displays a huge individual variance in connectivity seems to have been completely neglected. It is reasonable to consider that the initial state of the network, regardless of other parameters, should influence its ability to restructure the connections so as to store new, complex information. For instance, the same lecture will impact differently on the neural networks of two students due to the fact that they are initially wired differently (i.e., cognitive styles).

The question is thus how much the initial state influences the speed at which the complex concepts will be learned. It is reasoned that if the learning process in an artificial neural network reflects the restructuration of biological neural networks through neural plasticity, then the initial state of the artificial network will influence how neural plasticity rewires the system after each learning session (or epoch). It is worth noting at this stage that the variance in connectivity is also a recognised biological phenomenon [1]. Typically, learners with slightly different past experiences, as implemented in non-significant variations in the weight matrix, will develop very differently and achieve expertise at different times. Elaborated memories underpinning expert performance have been shown to include core elements which are fixed and flexible elements, instantiated to adapt to the problem at hand. These elaborated memories are the hallmark of expertise acquisition in many disciplines. They encapsulate the abstract concepts that help experts make good decisions quickly [4]. Regardless of the learner's past, the abstract concepts will encapsulate the essence of the domain under study. The networks of different learners will be rewired to code the same abstract concept. It is assumed that there is convergence towards a definite distribution of neurons and connections. Various initial configurations of long-term memory networks will require different amounts of time for encoding the abstract concept. Learning speed is thus sensitive to initial conditions. The abstract concept is formalised as a network attractor. That learning is sensitive to initial conditions and that abstract concepts can be formalised as attractors makes human memory exhibit properties in line with chaos theory [26].

At this stage, it is reasonable to draw the parallel between biological and artificial neural networks. The original research by McCulloch and Pitts [20], in formalising biological processes mathematically, has paved the way for the artificial reproduction of biological life. Yet, the initial conception of the neural system as a simple information processing system has received considerable support [8]. Over the last decades neurobiology has unravelled the processes underpinning learning at the cell and molecular levels [14], and computer sciences have developed equations that reflect such biological processes [21, 22]. Both carbon and silicon-based agents evolve from an initial state, the novice state, to a final state, the expert state, by rewiring the network's connections. Similar to the biological system, the change in the artificial network can be formalised as the change of influence or weight between two neurons. Likewise, similarly to the biological system, learning in artificial systems takes place in the module processing the information. Finally, in both cases the system stores information in its connections but does not create new units. The present paper used a neural network architecture that simulates expert classification of strategies in chess [2], termed Templates for Expert Knowledge Simulation (TEKS), to explore the influence of initial network connectivity on learning speed. The model appropriately simulates both novice and expert chess players' performance in recognising chess strategies and as such is a reliable neural network model to experiment on strategy learning.

2 Templates for Expertise Knowledge Simulation

TEKS is a two-module artificial neural network that simulates novice and expert ability to recognise chess strategies. It has been developed in line with current biological findings in the field of learning and expertise acquisition.[1] The visual template store (VTS) implements long-term memory processes. It performs recognition and, crucial

[1] See Chassy (2013) for details about the neuroscience background.

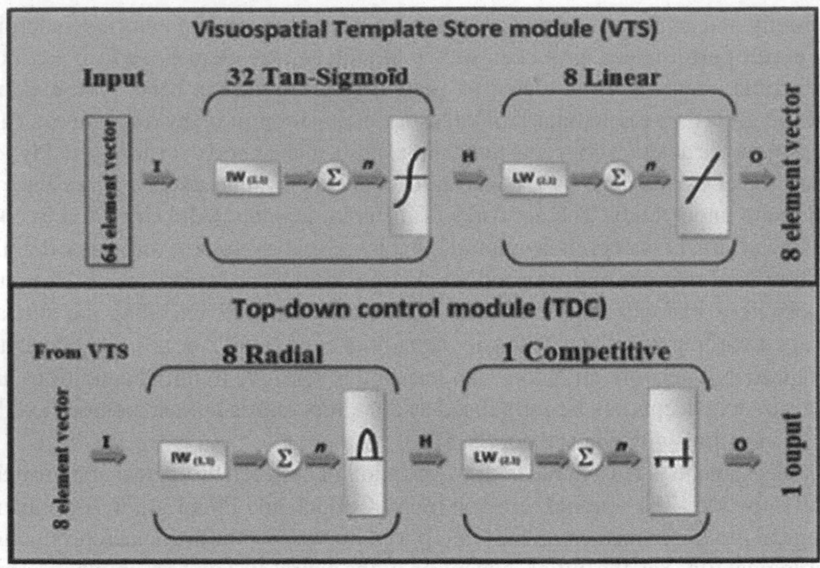

Fig. 1 Detailed architecture of the VTS and TDC modules

to this study, constitutes the target of learning. A top-down control (TDC) module ensures attentional filtering to simulate the competition between cognitions in the perceptual system. When a position is presented to TEKS, the signal is processed sequentially. VTS establishes the degree of match with the memory templates and then TDC forwards the best match to conscious recollection. The neural architecture of the two modules is detailed in Fig. 1.

VTS is a feed-forward network made of a 64 input neuron, a hidden layer of 32 tan sigmoid neurons, and an output of 8 linear neurons. The output of VTS is fed into TDC module. TDC is a probabilistic neural network made of 8 input neurons, a hidden layer of 8 radial basis neurons, and 1 output unit. Equations defining the behaviour of TEKS in each of the four layers are reported below:

$$\text{VTS hidden layer: H} = 2/[(1 + e^{-2n}) - 1)] \quad (1)$$

$$\text{VTS output layer: O} = \text{L } W_{i,j}{}^{*}n \quad (2)$$

$$\text{TDC hidden layer: H} = e^{-n^2} \quad (3)$$

$$\text{TDC output: O} = \max(n_1, \ldots, n_8) \quad (4)$$

At the end of a simulation, TEKs indicates which type of strategy has been used by a player. This element is crucial in terms of theory of expertise since the identification of strategies provides players with typical manoeuvres and tactical options.

3 Methods

The network had to learn to classify 8 types of chess strategies. These strategies are well-known systems played by chess experts [5]. Each system is based on strategic principles that guide, if not dictate, the distribution of the pieces over the board, hence creating patterns that display some variance but whose core implements the strategic idea of the system. To examine novices' variance in learning, the learning of 500 individuals was simulated using TEKS. For each novice, the weights were randomly initialised with a value $[-1; 1]$ before training took place. The performance criterion c to stop performance was $c = 0.01$.

3.1 Coding of Chess Positions

In TEKS, chess positions are coded as 64-element input vectors. There is a one-to-one mapping between a board location and a vector position, hence topological constancy is maintained. Pieces are coded with a numerical equivalence detailed in Table 1. 128 positions were submitted to numerical recoding, simulating the retinal input of the system and its subsequent transcoding to a neural signal. Though 128 positions seems a small number for an artificial simulation, these positions have been used in previous research; making results psychologically meaningful.

3.2 Training Cycle

As indicated in the above, learning takes place in the long-term visual store. Due to its high success in classifying, learning was performed with the Levenberg-Marquardt algorithm ([17, 19], see Eq. 5).

$$x_{k+1} = x_k - [J^t J + \mu I]^{-1} J^T e \tag{5}$$

Table 1 Numerical code used to code pieces as integers

Piece	Colour white	Colour black
King	6	−6
Queen	5	−5
Rook	4	−4
Bishop	3	−3
Knight	2	−2
Pawn	1	−1
Empty	0	0

The cycle of presenting an input vector, calculating the output, and adjusting the weight matrix, is termed an epoch. In the context of this paper, we would consider an epoch to represent several training sessions. In line with human data showing an average of 10,000 h of training to reach high expert status, the correction of the weights is gradual and so learning requires several epochs. Epochs will thus be used to measure the individual variance in learning the classification of chess strategies.

3.3 Network Reorganisation and Connection Distribution

As indicated in the above, learning is a slow process that requires time to create new axons and synapses in the biological neural network. As a proxy, for the artificial neural network, we consider that a weight of zero is the absence of a synapse and a weight different than zero is a synapse. If the weight is positive the synapse is excitatory and if the weight is negative the synapse is inhibitory. As a proxy, the weight matrix thus reflects the distribution of axons and number of synapses in the network. VTS represents long-term memory; its connectivity will inform the overall average change in weight generated by the learning process.

To assess the influence of a network's reorganisation due to learning, we analysed the impact of learning on the density of the two layers of VTS. The layer connecting the input to the 32 Tan sigmoid neurons, referred to as layer 1, has 2048 connections. The layer connecting the hidden neurons to the 8 linear output neurons, referred to as layer 2, has 256 connections. To make comparisons in rate of change due to learning, weight was averaged per connection. The equations determining density of Layer 1 and Layer 2 are presented in Eqs. 6 and 7 respectively.

$$\text{Density}_{\text{L1}} = \Sigma w_{i,j}/2248 \text{ with } i = 1 \text{ to } 64 \text{ and } j = 1 \text{ to } 32. \quad (6)$$

$$\text{Density}_{\text{L2}} = \Sigma w_{i,j}/256 \text{ with } i = 1 \text{ to } 32 \text{ and } j = 1 \text{ to } 8. \quad (7)$$

Various predictions are derived from the chaos theory approach introduced in this paper. Should the chaos theory apply to learning chess strategies, then TEKS should display the following features:

- TEKS should replicate the learning curve of human learning
- TEKS should display variance in the time to reach expertise (as indicated by the number of epochs to reach performance criterion)
- TEKS density should increase with learning
- TEKS should reveal an expert attractor: that is, a specific distribution of weights that underpin expert performance in identifying strategies
- Finally, TEKS should display increased density change when individuals are further away from the expert attractor.

4 Results

4.1 TEKS Simulation of Human Characteristics

TEKS displays four crucial aspects of human learning. First, the 500 novices performed poorly in categorising chess strategies as indicated by their performance close to chance. Second, experts performed well above chance. These findings replicate previous results with TEKS [2]. The third result of interest is that the learning curve displayed by TEKS across the 500 simulations replicates the one found in studies with human subjects. Regressing the number of epochs to the average performance, as measured by mean error, provides an equation that significantly accounts for 95.6 % of variance, *Performance* = $\text{epoch}^{-1.973} + 3.065$, $F(1, 10) = 215.66$, $p < .001$. The last, and most important result of these simulations from a behavioural point of view, is that TEKS replicates human variance in the speed at which it attains expert stage. All 500 simulated learners reached expert stage with a mean number of epochs of 9.664 (SD = 3.805), but the number of epochs varies between 5 and 23 and is not normally distributed (see Fig. 2). This is a critical result for psychology, education, and neuroscience. It shows that differences in learning speed are not solely due to genetic or motivational factors, but simply stem from individual differences in neural wiring. TEKS simulates the behavioural characteristics displayed by human subjects.

4.2 Chaotic-Like Properties of Learning

The variance in time to achieve expertise brings empirical evidence that a slight variation in the network's weight distribution before learning impacts drastically

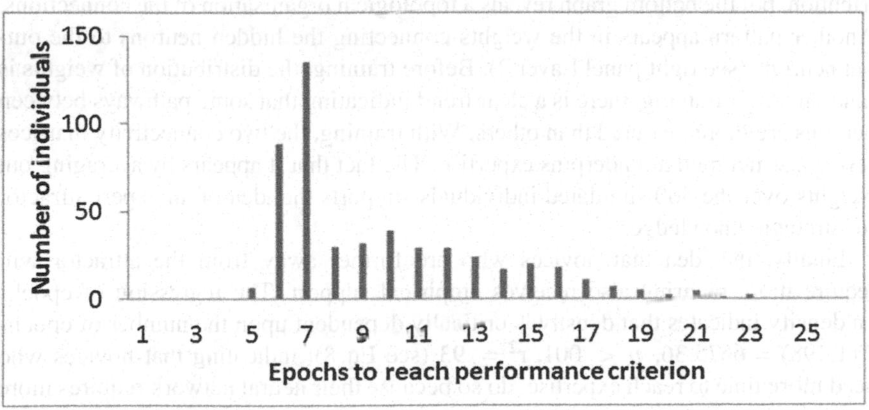

Fig. 2 Variance in speed to reach expert attractor

Table 2 Average densities (SD) per layer and per learning stage

	Novice	Expert	Learning
Density L1	0.500 (0.006)	1.478 (0.702)	0.978 (0.701)
Density L2	0.499 (0.01823)	0.419 (0.120)	−0.080 (0.122)

on learning time. The variance in epochs reflects the difficulty of the network to rewire itself to attain the stable expert attractor. We shall now examine density as an indicator of the degree of reorganisation of the network. Table 2 below reports the main descriptive statistics for the 500 individuals at both the Novice and Expert stage.

The mean weight of the novice network was lower ($M = .50$; $SD = .006$) than the mean weight of the expert network ($M = 1.36$; $SD = .61$), $t(499) = 31.48$, $p < .001$. Hence, the expert network is denser than the non-expert network. However, learning had the opposite effect on the two weights matrices. The weight matrix connecting the input to the hidden layer, referred to as Layer 1, saw a significant increase in positive connections $t(499) = -31.206$, $p < .001$. Inversely, the weight matrix connecting the hidden layer to the output, referred to as Layer 2, saw an increase in negative connections $t(499) = 14.629$, $p < .001$. The change in density distribution due to learning is significant.

The weight matrix of VTS (long-term memory module) was averaged out across all 500 simulated individuals. Since positive weights (excitatory synapses) and negative weights (inhibitory synapses) reflect the existence of axons and synaptic terminals, absolute weights were used to map the existence of an attractor. Figure 3 shows a striking reorganisation in connectivity after expertise has been acquired. The connectivity map before training (novice stage) and after training (expert stage) reveals a radical change. The left column reports the connectivity map between the input and the hidden layer. The weight matrix in the top graph reflects a random distribution, but the bottom graph reveals a topological organisation of the connections. Another pattern appears in the weights connecting the hidden neurons to the output neurons (see right panel Layer 2). Before training, the distribution of weights is random. After training, there is a clear trend indicating that some pathways between neurons are more favoured than others. With training, the two connectivity matrices reveal a structure that underpins expertise. The fact that it appears by averaging out weights over the 500 simulated individuals supports the idea of an expert attractor for strategic knowledge.

Finally, the idea that novices who are further away from the attractor will require more rewiring also receives empirical support. The regression of epochs on density indicates that density is critically dependent upon the number of epochs $F(1{,}498) = 6675.36$, $p < .001$, $r^2 = .93$ (see Eq. 8); indicating that novices who need more time to reach expertise, do so because their neural network requires more rewiring.

$$\text{Density} = 0.155 * \text{Epoch} - 0.142 \tag{8}$$

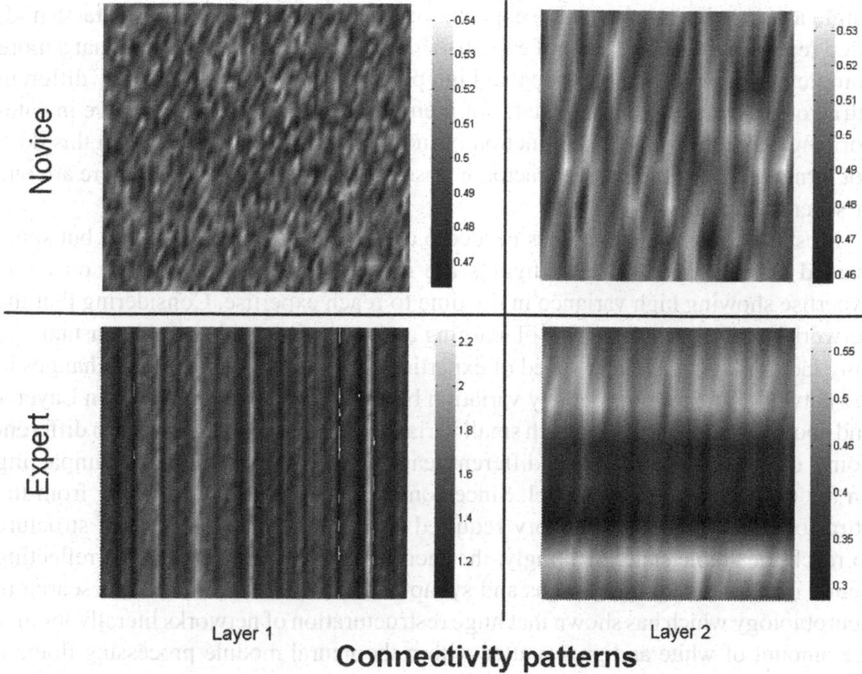

Fig. 3 Topological maps reporting the average weights of the network in the two layers of long-term memory (VTS module)

5 Discussion

The present paper tested (a) whether TEKS account for human learning behavioural features and (b) whether slight variance in connectivity influences speed to achieve expert performance. The findings have revealed that the model replicates several characteristics of human learning and displays features of chaotic dynamical systems. By replicating the central psychology finding that learning curves take the form of a power function [23], TEKS demonstrates its ability to simulate human learning.

Even though novice network connectivity was made of randomly distributed weights, learning has led all networks to achieve a specific distribution of densities that seems constant across all individuals. Hence, learning makes networks converge towards one stable state that underpins expert performance. The central finding of the present paper is that the analysis of the expert weight matrix reveals the existence of a stable state that implements high-level performance. The time to reach expertise is thus a function of the novice's position in the attractor landscape. The number of epochs is determined by the distance to the attractor. The present simulations support the idea that the initial state of the neural network, encoding a specific form of knowledge, has a significant influence on the time necessary to acquire the said knowledge. The convergence of the topological properties of the trained network argues for a single attractor but an alternative explanation is not totally ruled out. The task modelled represents only one facet of expertise in chess players. It is possible that a more complex task would demonstrate that high performance can be reached by different attractors. Several equilibrium states of memory might be equally effective in categorising perceptual input as a function of more abstract concepts. Though this does not tarnish the idea of expert attractor, it poses the question of whether there are one or several attractors.

Most of the simulated novices needed 6 or 7 epochs to reach expertise but some needed as much as 24. These figures are in perfect line with research on chess expertise showing high variance in the time to reach expertise. Considering that the network parameters such as rate of learning and type of algorithm were constant, the only factor playing a role in speed of expertise acquisition was the initial changes in the network. The average density variation before training was about .01 in Layer 1 and about .02 in Layer 2. Yet, such small variations, by putting the novices in different points of the state space, led to different learning trajectories, ultimately impacting on the time to reach expert level. Since some novices were further away from the attractor, their long-term memory required more changes to its internal structure to reach the stable state. Strikingly, the increase in the average weights, reflecting neural construction of new axons and synaptic terminals, is in line with research in neurobiology which has shown that huge restructuration of networks literally modify the amount of white and grey matter within the neural module processing domain specific information. Taxi drivers for example have been shown to have a bigger hippocampus than controls [18]. Similarly, expert musicians have more connections than non-musicians in the motor cortices controlling their instrument [12].

Two factors impose a limitation upon the scope of the findings revealed in this artificial network exploration of the chaotic features of human learning. Learners in any discipline come from different backgrounds which can be formalised as random variations in their network. It should be noted that the assumption of random variation entails that the learner has no knowledge at all about the topic to be learned, or else the network (whether biological or artificial) would display some significant useful connections that would orient the learning process. In this respect it might be that past experience in domains akin to the one being learned pre-wire the networks thus leading to an easier restructuration later on. The apparent facility of some learners of this type might be mistakenly taken as talent. The second point relates to the method used to reveal the attractor presented in Fig. 3. The fact that weights were averaged does not permit to rule out the possibility of rare, alternative states of the system that would perform similarly to the main attractor. The question of multiple attractors to account for human memory should be addressed in future research.

To dissipate a potential misinterpretation, it is worth mentioning at this stage that the present study is not in contradiction with previous research highlighting the role of genetic information in acquiring expert level [3]. It is established by animal and human data that the neurobiological processes underpinning the creation of new neural networks are controlled by a series of genes [13]. The ability to rewire the network to store new information would thus partly be inherited. This might mislead to the view that expertise acquisition is solely inherited. Yet, the present paper highlights the role of environmental factors by pointing that the speed of learning does not necessarily result from a cognitive impairment but might only reflect the lack of experience of the individual in processing a given type of information. Hence, the two views, genetic and chaotic, are complementary rather than contradictory in accounting for variance in expertise acquisition. Future research should be undertaken to simulate the relative importance of both factors.

Our results call for a debate on the nature of expertise and of the representation of human memory in AI. The idea that expert knowledge can be represented by an attractor accounts for of all the findings in the present article and is in line with all of the main findings in psychology and neuroscience [29]. The present paper crucially contributes to our understanding of human learning processes by introducing chaos theory in the field of expertise acquisition. It has demonstrated, through a Monte Carlo simulation of a psychological model of expertise acquisition in chess, that the behavioural and neuroscientific features of expertise are accounted for by chaos theory.

References

1. Alexander-Bloch, A., Giedd, J.N., Bullmore, E.: Imaging structural co-variance between human brain regions. Nat. Rev. Neurosci. **14**(5), 322–336 (2013)
2. Chassy, P.: The role of memory templates in experts' strategic thinking. Psychol. Res. **3**(5), 276–289 (2013)
3. Chassy, P., Gobet, F.: Speed of expertise acquisition depends upon inherited factors. Talent Dev. Excell. **2**, 17–27 (2010)
4. Chassy, P., Gobet, F.: A hypothesis about the biological basis of expert intuition. Rev. General Psychol. **15**(3), 198–212 (2011)
5. Chassy, P., Gobet, F.: Risk taking in adversarial situations: civilization differences in chess experts. Cognition **141**, 36–40 (2015)
6. Ericsson, K.A.: Deliberate practice and the acquisition and maintenance of expert performance in medicine and related domains. Acad. Med. **79**(10), S70–S81 (2004)
7. Ericsson, K.A., Krampe, R.T., Tesch-Römer, C.: The role of deliberate practice in the acquisition of expert performance. Psychol. Rev. **100**(3), 363 (1993)
8. Gobet, F., Chassy, P., Bilalic, M.: Foundations of Cognitive Psychology. McGraw-Hill (2011)
9. Gobet, F., Simon, H.A.: Templates in chess memory: a mechanism for recalling several boards. Cogn. Psychol. **31**, 1–40 (1996)
10. Grabner, R.H., Neubauer, A.C., Stern, E.: Superior performance and neural efficiency: the impact of intelligence and expertise. Brain Res. Bull. **69**(4), 422–439 (2006)
11. Hebb, D.O.: The Organization of Behavior: A Neuropsychological Approach. Wiley (1949)
12. Imfeld, A., et al.: White matter plasticity in the corticospinal tract of musicians: a diffusion tensor imaging study. Neuroimage **46**(3), 600–607 (2009)
13. Kandel, E.R.: The molecular biology of memory storage: a dialogue between genes and synapses. Science **294**(5544), 1030–1038 (2001)
14. Kandel, E.R., Dudai, Y., Mayford, M.R.: The molecular and systems biology of memory. Cell **157**(1), 163–186 (2014)
15. King, A.V., et al.: Microstructure of a three-way anatomical network predicts individual differences in response inhibition: a tractography study. Neuroimage **59**(2), 1949–1959 (2012)
16. Krampe, R.T., Ericsson, K.A.: Maintaining excellence: deliberate practice and elite performance in young and older pianists. J. Exp. Psychol.: General **125**(4), 331 (1996)

17. Levenberg, K.: A method for the solution of certain non-linear problems in least squares. Q. Appl. Math. **2**, 164–168 (1944)
18. Maguire, E.A., et al.: Navigation-related structural change in the hippocampi of taxi drivers. Proc. Natl. Acad. Sci. **97**(8), 4398–4403 (2000)
19. Marquardt, D.W.: An algorithm for least-squares estimation of nonlinear parameters. J. Soc. Ind. Appl. Math. **11**(2), 431–441 (1963)
20. McCulloch, W.S., Pitts, W.: A logical calculus of the ideas immanent in nervous activity. Bull. Math. Biophys. **5**(4), 115–133 (1943)
21. Mishra, D., et al.: Levenberg-Marquardt learning algorithm for integrate-and-fire neuron model. Neural Inf. Process.-Lett. Rev. **9**(2), 41–51 (2005)
22. Moré, J.J.: The Levenberg-Marquardt algorithm: implementation and theory. In: Numerical Analysis, pp. 105–116. Springer (1978)
23. Newell, A., Rosenbloom, P.S.: Mechanisms of skill acquisition and the law of practice. Cogn. Skills Acquis. **1**, 1–55 (1981)
24. Niogi, S.N., McCandliss, B.D.: Left lateralized white matter microstructure accounts for individual differences in reading ability and disability. Neuropsychologia **44**(11), 2178–2188 (2006)
25. Phillips, R., LeDoux, J.: Differential contribution of amygdala and hippocampus to cued and contextual fear conditioning. Behav. Neurosci. **106**(2), 274 (1992)
26. Prigogine, I.: Les lois du chaos. Flammarion (1994)
27. Squire, L.R.: Memory and the hippocampus: a synthesis from findings with rats, monkeys, and humans. Psychol. Rev. **99**(2), 195 (1992)
28. Starkes, J.L., Hodges, N.J.: Team sports and the theory of deliberate practice. J. Sport Exerc. Psychol. **20**, 12–24 (1998)
29. Wills, T.J., et al.: Attractor dynamics in the hippocampal representation of the local environment. Science **308**, 873–876 (2005)

AI Techniques

A Fast Algorithm to Estimate the Square Root of Probability Density Function

Xia Hong and Junbin Gao

Abstract A fast maximum likelihood estimator based on a linear combination of Gaussian kernels is introduced to represent the square root of probability density function. It is shown that, if the kernel centres and kernel width are known, then the underlying problem can be formulated as a Riemannian optimization one. The first order Riemannian geometry of the sphere manifold and vector transport are explored, and then the well-known Riemannian conjugate gradient algorithm is used to estimate the model parameters. For completeness the k-means clustering algorithm and a grid search are applied to determine the centers and kernel width respectively. Illustrative examples are employed to demonstrate that the proposed approach is effective in constructing the estimate of the square root of probability density function.

Keywords Density Estimation · Riemannian optimization · Maximum likelihood

1 Introduction

The Riemannian optimisation algorithms have been recently developed on many types of matrix manifolds such as the Stiefel manifold, Grassmann manifold and the manifold of positive definite matrices, see Sect. 3.4 of [1]. Since Riemannian optimisation is directly based on the curved manifolds, one can eliminate those constraints such as orthogonality to obtain an unconstrained optimisation problem that, by construction, will only use feasible points. This allows one to incorporate Riemannian geometry in the resulting optimisation problems, thus producing far more accurate numerical results. The Riemannian optimisation have been successfully applied in

---

X. Hong (✉)
Department of Computer Science, School of Mathematical and Physical Sciences, University of Reading, Reading, UK
e-mail: x.hong@reading.ac.uk

J. Gao
Discipline of Business Analytics, The University of Sydney Business School, The University of Sydney, New South Wales, NSW 2006, Australia
e-mail: junbin.gao@sydney.edu.au

M. Bramer and M. Petridis (eds.), *Research and Development in Intelligent Systems XXXIII*, DOI 10.1007/978-3-319-47175-4_11

machine learning, computer vision and data mining tasks, including fixed low rank optimisation [2], Riemannian dictionary learning [3], and computer vision [4].

The probability density function (PDF) estimation, e.g., the Parzen window (PW) and finite mixture model, is of fundamental importance to many data analysis and pattern recognition applications [5–12]. There is a considerable interest into research on sparse PDF estimation which can be summarized into two categories. The first category is based on constrained optimization [13–16]. The second category of sparse kernel density estimators construct the PDF estimator in a forward regression manner [17–21].

The identification of the finite mixture model is usually based on the expectation-maximisation (EM) algorithm [11] which provides the maximum likelihood (ML) estimate of the mixture model's parameters, while the number of mixtures is preset. This associated ML optimisation is generally a highly nonlinear optimisation process requiring extensive computation, while the EM algorithm for Gaussian mixture model enjoys an explicit iterative form [22]. Since the constraint on the mixing coefficients of the finite mixture model is the multinomial manifold, recently we have introduced Riemannian trust-region (RTR) algorithm for sparse finite mixture model based on minimal integrated square error criterion [16].

Note that all the aforementioned algorithms are aimed at estimating the probability density function. However in this work we investigate the less studied problem of estimating the square root of the probability density function (PDF) estimation [23]. While the resultant estimator is equally useful, the constraint on parameters is different, which can lead to computational advantages.

In this paper, we introduce a fast maximum likelihood estimator based on a linear combination of Gaussian kernels which represents the square root of probability density function. Initially we apply the k-means clustering algorithm and a grid search to determine the centers and kernel width. It is shown that the underlying parameter estimation problem can be formulated as a Riemannian optimization on the sphere manifold. The first order Riemannian geometry of the sphere manifold and vector transport are explored, and the well-known Riemannian conjugate gradient algorithm is used to estimate the model parameters. Numerical examples are employed to demonstrate that the proposed approach is effective in constructing the estimate of the square root of probability density function.

2 Preliminary on Sphere Manifold

This section briefly introduces the concept of sphere manifold and the necessary ingredients used in the retraction based framework of Riemannian optimization. As a reference, the main notations on Riemannian geometry on sphere manifold in this section is summarized in Table 1. We refer the readers to [1] for the general concepts of manifolds.

The sphere manifold is the set of unit Frobenius norm vectors of size M, denoted as

Table 1 Notations for sphere manifold

$\{\mathbb{S}^{M-1}, g\}$	Sphere manifold for parameter matrix $\boldsymbol{\theta}$ and the inner product of the manifold
$T_{\boldsymbol{\theta}}\mathbb{S}^{M-1}$	Tangent space of the sphere manifold
$\boldsymbol{u}_{\boldsymbol{\theta}}, \boldsymbol{v}_{\boldsymbol{\theta}}$	Tangent vectors at $\boldsymbol{\theta}$
$\mathrm{Proj}_{\boldsymbol{\theta}}(z)$	Orthogonal projector from a vector in ambient space onto the tangent space at $\boldsymbol{\theta}$
$\mathrm{grad}\, F(\boldsymbol{\theta})$	Riemannian gradient of $F(\boldsymbol{\theta})$ on the manifold $\mathbb{S}^{M-1}$
$\mathrm{Grad}\, F(\boldsymbol{\theta})$	Classical gradient of $F(\boldsymbol{\theta})$ as seen in Euclidean space
$\mathrm{Exp}_{\boldsymbol{\theta}}$	Retraction mapping
$\mathscr{T}_{\boldsymbol{\theta}_{k+1} \leftarrow \boldsymbol{\theta}_k}(\boldsymbol{u}_{\boldsymbol{\theta}_k})$	Vector transport

$$\mathbb{S}^{M-1} = \{\boldsymbol{\theta} \in \mathbb{R}^M : \|\boldsymbol{\theta}\|_2 = 1\}. \tag{1}$$

It is endowed with a Riemannian manifold structure by considering it as a Riemannian submanifold of the embedding Euclidean space $\mathbb{R}^M$ endowed with the usual inner product

$$g(\boldsymbol{u}_{\boldsymbol{\theta}}, \boldsymbol{v}_{\boldsymbol{\theta}}) = \boldsymbol{u}_{\boldsymbol{\theta}}^{\mathrm{T}} \boldsymbol{v}_{\boldsymbol{\theta}}, \tag{2}$$

where $\boldsymbol{u}_{\boldsymbol{\theta}}, \boldsymbol{v}_{\boldsymbol{\theta}} \in T_{\boldsymbol{\theta}}\mathbb{S}^{M-1} \subset \mathbb{R}^M$ are tangent vectors to $\mathbb{S}^{M-1}$ at $\boldsymbol{\theta}$. The inner product on $\mathbb{S}^{M-1}$ determines the geometry such as distance, angle, curvature on $\mathbb{S}^{M-1}$. Note that the tangent space $T_{\boldsymbol{\theta}}\mathbb{S}^{M-1}$ at element $\boldsymbol{\theta}$ can be described by

$$T_{\boldsymbol{\theta}}\mathbb{S}^{M-1} = \{\boldsymbol{u}_{\boldsymbol{\theta}} : \boldsymbol{u}_{\boldsymbol{\theta}}^{\mathrm{T}} \boldsymbol{\theta} = 0\}. \tag{3}$$

Riemannian gradient: Let the Riemannian gradient of a scalar function $F(\boldsymbol{\theta})$ on $\mathbb{S}^{M-1}$ be denoted by $\mathrm{grad}\, F(\boldsymbol{\theta})$, and its classical gradient as seen in the Euclidean space as $\mathrm{Grad}\, F(\boldsymbol{\theta})$. Then we have

$$\mathrm{grad}\, F(\boldsymbol{\theta}) = \mathrm{Proj}_{\boldsymbol{\theta}}\big(\mathrm{Grad}\, F(\boldsymbol{\theta})\big), \tag{4}$$

where $\mathrm{Proj}_{\boldsymbol{\theta}}(z)$ is the orthogonal projection onto the tangent space, which can be computed as

$$\mathrm{Proj}_X(z) = z - (\boldsymbol{\theta}^{\mathrm{T}} z)\boldsymbol{\theta} \tag{5}$$

in which z represents a vector in the ambient space.

Retraction mapping: An important concept in the recent retraction-based framework of Riemannian optimization is the retraction mapping, see Sect. 4.1 of [1]. The exponential map Exp_X, defined by

$$\text{Exp}_{\boldsymbol{\theta}}\left(\lambda \boldsymbol{u}_{\boldsymbol{\theta}}\right) = \cos\left(\|\lambda \boldsymbol{u}_{\boldsymbol{\theta}}\|_2\right)\boldsymbol{\theta} + \frac{\sin(\|\lambda \boldsymbol{u}_{\boldsymbol{\theta}}\|_2)}{\|\boldsymbol{u}_{\boldsymbol{\theta}}\|_2}\boldsymbol{u}_{\boldsymbol{\theta}}, \tag{6}$$

is the canonical choice for the retraction mapping, where the scalar λ is a chosen step size. The retraction mapping is used to locate the next iterate on the manifold along a specified tangent vector, such as a search direction in line search in the Newton's algorithm or the suboptimal tangent direction in the trust-region algorithm, see Chap. 7 of [1]. For example, the line search algorithm is simply given by

$$\boldsymbol{\theta}_{k+1} = \text{Exp}_{\boldsymbol{\theta}_k}\left(\lambda_k \boldsymbol{u}_{\boldsymbol{\theta}_k}\right). \tag{7}$$

where the search direction $\boldsymbol{u}_{\boldsymbol{\theta}_k} \in T_{\boldsymbol{\theta}_k}\mathbb{S}^{M-1}$ and λ_k is a chosen step size at iteration step k.

Vector transport: In Riemannian optimization algorithms, the second derivatives can be approximated by comparing the first-order information (tangent vectors) at distinct points on the manifold. The notion of vector transport $\mathscr{T}_{\boldsymbol{\theta}_{k+1} \leftarrow \boldsymbol{\theta}_k}(\boldsymbol{u}_{\boldsymbol{\theta}_k})$ on a manifold, roughly speaking, specifies how to transport a tangent vector $\boldsymbol{u}_{\boldsymbol{\theta}_k}$ from a point $\boldsymbol{\theta}_k$ to another point $\boldsymbol{\theta}_{k+1}$ on the manifold. The vector transport for the sphere manifold is calculated as

$$\mathscr{T}_{\boldsymbol{\theta}_{k+1} \leftarrow \boldsymbol{\theta}_k}(\boldsymbol{u}_{\boldsymbol{\theta}_k}) = \text{Proj}_{\boldsymbol{\theta}_{k+1}}(\boldsymbol{u}_{\boldsymbol{\theta}_k}) \tag{8}$$

3 Proposed Estimator for the Square Root of Probability Density Function

In this work we investigate the less studied problem of estimating the square root of the probability density function (PDF) estimation [23]. The reason that the root square of the probability density function is estimated is that the problem can be formulated as a Riemannian optimisation one for computational advantage, as well as that the resultant estimator can be used as an alternative to a probability density function estimator. Specifically, given the finite data set $D_N = \{\boldsymbol{x}_j\}_{j=1}^N$ consisting of N data samples, where the data $\boldsymbol{x}_j \in \mathbb{R}^m$ follows an unknown PDF $p(\boldsymbol{x})$, the problem under study is to find a sparse approximation of the square root of $\psi(\boldsymbol{x}) = \sqrt{p(\boldsymbol{x})} > 0$ using M component Gaussian kernels, given by

$$\psi(\boldsymbol{x}) = \sum_{i=1}^{M} \omega_i K_\sigma\left(\boldsymbol{x}, \boldsymbol{c}_i\right) = \boldsymbol{\omega}^{\mathrm{T}}\boldsymbol{k}(\boldsymbol{x}) \tag{9}$$

subject to

$$\int \psi^2(\boldsymbol{x})d\boldsymbol{x} = 1 \tag{10}$$

where

$$K_\sigma\left(\boldsymbol{x},\boldsymbol{c}_i\right)=\frac{1}{\left(2\pi\sigma^2\right)^{m/2}}\exp\left(-\frac{\|\boldsymbol{x}-\boldsymbol{c}_i\|^2}{2\sigma^2}\right), \tag{11}$$

in which $\boldsymbol{c}_i=\left[c_{i,1}\ c_{i,2}\cdots c_{i,m}\right]^{\mathrm{T}}$ is the center vector of the ith kernels and $\sigma>0$ is the width parameter, and ω_i is the ith kernel weight. $\boldsymbol{\omega}=[\omega_1\cdots\omega_M]^{\mathrm{T}}$, $\boldsymbol{k}(\boldsymbol{x})=[K_\sigma\left(\boldsymbol{x},\boldsymbol{c}_1\right)\cdots K_\sigma\left(\boldsymbol{x},\boldsymbol{c}_M\right)]^{\mathrm{T}}$.

Applying (9) to the constraint (10), we have

$$\begin{aligned}\int\psi^2(\boldsymbol{x})d\boldsymbol{x}&=\sum_{i=1}^{M}\sum_{j=1}^{M}\omega_i\omega_j\int K_\sigma\left(\boldsymbol{x},\boldsymbol{c}_i\right)K_\sigma\left(\boldsymbol{x},\boldsymbol{c}_j\right)d\boldsymbol{x}\\&=\boldsymbol{\omega}^{\mathrm{T}}\boldsymbol{Q}\boldsymbol{\omega}=1\end{aligned} \tag{12}$$

where $\boldsymbol{Q}=\{q_{i,j}\}$, with $q_{i,j}=K_{\sqrt{2}\sigma}\left(\boldsymbol{c}_i,\boldsymbol{c}_j\right)$, as shown in Appendix A.

The proposed algorithm consists of two consecutive steps; (i) determining M kernel centres using the k-means clustering algorithm; (ii) for a grid search of kernel width, estimate kernel parameters using Riemannian conjugate gradient algorithm $\boldsymbol{\omega}$ based on the maximum likelihood criterion as introduced below.

Provided M kernel centers (via the k-means clustering algorithm) and a preset kernel width, we initially denote the eigenvalue decomposition $\boldsymbol{Q}$ as $\boldsymbol{Q}=\boldsymbol{U}\boldsymbol{\Sigma}\boldsymbol{U}^{\mathrm{T}}$, where $\boldsymbol{U}$ is an orthogonal matrix consisting of the eigenvectors and $\boldsymbol{\Sigma}$ is a diagonal matrix of which the entries are eigenvalues $\boldsymbol{Q}$.

Let $\boldsymbol{\theta}=\sqrt{\boldsymbol{\Sigma}}\boldsymbol{U}^{\mathrm{T}}\boldsymbol{\omega}$, $\psi(\boldsymbol{x})$ can be written as

$$\psi(\boldsymbol{x})=\boldsymbol{\theta}^{\mathrm{T}}\bar{\boldsymbol{k}}(\boldsymbol{x}) \tag{13}$$

where $\bar{\boldsymbol{k}}(\boldsymbol{x})=\sqrt{\boldsymbol{\Sigma}}^{-1}\boldsymbol{U}^{\mathrm{T}}\boldsymbol{k}(\boldsymbol{x})$.

In order to satisfy $\psi(\boldsymbol{x})>0$, we initialize all ω_i as the same positive number ξ, and $\boldsymbol{\omega}_{Ini}=\xi\boldsymbol{1}_M$, $\boldsymbol{1}_M$ is a length M vector with all elements as ones. The corresponding $\boldsymbol{\theta}_0$ can be calculated as

$$\boldsymbol{\theta}_0=\sqrt{\boldsymbol{\Sigma}}\boldsymbol{U}^{\mathrm{T}}\boldsymbol{1}_M \tag{14}$$

followed by a normalization step $\boldsymbol{\theta}_0=\boldsymbol{\theta}_0/\|\boldsymbol{\theta}_0\|$ so that $\boldsymbol{\theta}_0$ is on the sphere manifold. We are now ready to formulate the maximum likelihood estimator as the following Riemannian optimization problem, given as

$$\boldsymbol{\theta}^{\mathrm{opt}}=\min_{\boldsymbol{\theta}\in\mathbb{S}^{M-1}}\left\{F(\boldsymbol{\theta})-\sum_{j=1}^{N}\log(\boldsymbol{\theta}^{\mathrm{T}}\bar{\boldsymbol{k}}(\boldsymbol{x}_j))\right\} \tag{15}$$

followed by setting $\boldsymbol{\theta}^{\mathrm{opt}}=\sqrt{\boldsymbol{\Sigma}}\boldsymbol{U}^{\mathrm{T}}\boldsymbol{\omega}^{\mathrm{opt}}$. For our objective function $F(\boldsymbol{\theta})$, it is easy to check that Euclidean gradient Grad $F(\boldsymbol{\theta})$ can be calculated respectively as

$$\mathrm{Grad}\,F(\boldsymbol{\theta}) = -\sum_{j=1}^{N} \frac{\bar{\boldsymbol{k}}(\boldsymbol{x}_j)}{\boldsymbol{\theta}^{\mathrm{T}}\bar{\boldsymbol{k}}(\boldsymbol{x}_j)} \tag{16}$$

Based on Grad $F(\boldsymbol{\theta})$, Riemannian gradient of the objective function $F(\boldsymbol{\theta})$ on the sphere manifold can be calculated according to (4), (5). We opt to Riemannian conjugate gradient algorithm to solve (15), which generalizes the classical conjugate gradient algorithm [24] to optimization problems over Riemannian manifolds [25]. Since the logarithm function acts as a natural barrier at zero and the Riemannian conjugate gradient algorithm is a local minimization algorithm, the constraint $\psi(\boldsymbol{x}) > 0$ can be met.

With all the ingredients available, we form the algorithm for solving (15) in Algorithm 1, which is well implemented in the Manifold Optimization Toolbox Manopt http://www.manopt.org, see [25]. We used the default parameter settings in Manopt, so that in Step 4 of Algorithm 1 α_k is based on line search backtracking procedure as described in Chap. 4, p63 of Sect. 4 of [1]. In Step 5 of Algorithm 1 ω_{k+1} is based on the default option "Hestenes-Stiefel's modified rule". Specifically, we have

$$\beta_{k+1} = \max\left\{0, \frac{<\mathrm{grad}\,F(\boldsymbol{\theta}_{k+1}), \gamma_{k+1}>}{<\mathscr{T}_{\boldsymbol{\theta}_{k+1}\leftarrow\boldsymbol{\theta}_k}(\boldsymbol{\eta}_k), \gamma_{k+1}>}\right\} \tag{17}$$

where $< \cdot, \cdot >$ denotes inner product, and

$$\gamma_{k+1} = \mathrm{grad}\,F(\boldsymbol{\theta}_{k+1}) - \mathscr{T}_{\boldsymbol{\theta}_{k+1}\leftarrow\boldsymbol{\theta}_k}(\mathrm{grad}\,F(\boldsymbol{\theta}_k)) \tag{18}$$

Algorithm 1 Riemannian conjugate gradient Algorithm for solving (15)

Require: $\bar{\boldsymbol{k}}(\boldsymbol{x}_j)$, $j = 1, \cdots, N$. Initial point $\boldsymbol{\theta}_0$ which is on the sphere manifold $\mathbb{S}^{M-1}$, and default threshold ϖ, e.g., $\varpi = 10^{-6}$ or $\varpi = 10^{-5}$;

Ensure: $\boldsymbol{\theta}$ that yields the minimum $F(\boldsymbol{\theta})$.

1: Set $\boldsymbol{\eta}_0 = -\mathrm{grad}\,F(\boldsymbol{\theta}_0)$ and $k = 0$;

2: **while** $\|\mathrm{grad}\,F(\boldsymbol{\theta}_k)\|_2 < \varpi$ **do**

3: $k = k + 1$;

4: Compute a step size α_k and set

$$\boldsymbol{\theta}_k = \mathrm{Exp}_{\boldsymbol{\theta}_{k-1}}(\alpha_k \boldsymbol{\eta}_{k-1}); \tag{19}$$

5: Compute β_k and set

$$\boldsymbol{\eta}_k = -\mathrm{grad}\,F(\boldsymbol{\theta}_k) + \beta_k \mathscr{T}_{\boldsymbol{\theta}_k\leftarrow\boldsymbol{\theta}_{k-1}}(\boldsymbol{\eta}_{k-1}); \tag{20}$$

6: **end while**

7: Return $\boldsymbol{\theta} = \boldsymbol{\theta}_k$, $F(\boldsymbol{\theta})$, and the associated $\boldsymbol{U}$, $\boldsymbol{\Sigma}$.

For completeness, the proposed estimator is summarized in Algorithm 2,

Algorithm 2 Proposed estimator for the square root of probability density function

Require: $\boldsymbol{x}_j, j = 1, \cdots, N$. Preset number of kernels M;
Ensure: $\psi(\boldsymbol{x})$ that maximizes the likelihood.
Use the k-means clustering algorithm to find $\boldsymbol{c}_i, i = 1, \cdots, M$;
for $i = 1, 2, \ldots, (Iter + 1)$ **do**
  $\sigma = \sigma_{min} + \frac{i-1}{n}(\sigma_{max} - \sigma_{min})$. where $(Iter + 1)$, σ_{min} and σ_{max} are preset, and they denote the minimum, maximum value and the total number of kernel width on a grid search.
  Form $\boldsymbol{Q}$ and its eigen-decomposition $\boldsymbol{Q} = \boldsymbol{U}\boldsymbol{\Sigma}\boldsymbol{U}^{\mathrm{T}}$
  Obtain $\bar{\boldsymbol{k}}(\boldsymbol{x}_j) = \sqrt{\boldsymbol{\Sigma}}^{-1}\boldsymbol{U}^{\mathrm{T}}\boldsymbol{k}(\boldsymbol{x}_j), j = 1, \cdots, N$
  Find $\boldsymbol{\theta}_0$ according to (14) and normalize it.
  Apply the Riemannian conjugate gradient Algorithm to return $F(\boldsymbol{\theta})$.
end for
Find σ^{opt} that minimizes $F(\boldsymbol{\theta})$ over the grid search. Return the associated $\boldsymbol{\theta}, \boldsymbol{U}, \boldsymbol{\Sigma}$.
Return $\boldsymbol{\omega}^{\mathrm{opt}} = \boldsymbol{U}\sqrt{\boldsymbol{\Sigma}}^{-1}\boldsymbol{\theta}^{\mathrm{opt}}$.

4 Illustrative Example

A data set of $N = 600$ points was randomly drawn from a known distribution $p(\boldsymbol{x})$ and used to construct the estimator of the square root of probability density function $\psi(\boldsymbol{x})$ using the proposed approach. $p(\boldsymbol{x})$ is given as a mixture of two Gaussians, as defined by

$$\begin{aligned} p(\boldsymbol{x}) = & \frac{1}{6\pi}\exp\left(-\frac{(x_1-1)^2+(x_2-1)^2}{2}\right) \\ & + \frac{1}{3\pi}\exp\left(-\frac{(x_1+1)^2+(x_2+1)^2}{2}\right) \end{aligned} \tag{21}$$

We preset $M = 10$, the k-means algorithm was implemented using Matlab command *kmeans.m*, producing M centers $\boldsymbol{c}_i, i = 1, \cdots, M$. Based on the centers, the kernel matrices $\boldsymbol{Q}$ are generated using any kernel width on the grid point on $[0.5, 2]$, with a grid width of 0.1. The Riemannian conjugate gradient algorithm was used to minimize the negative likelihood for the range of kernel widths. The result of $logF(\theta)$ versus the kernel width was shown in Fig. 1. The convergence of Riemannian conjugate gradient algorithm on the sphere manifold based on the optimal kernel width was shown shown in Fig. 2, showing that the algorithm can converge rapidly. The performance of proposed estimator was shown in Fig. 3(a), (b) & (c), which plots $\psi^2(\boldsymbol{x})$, the true density $p(\boldsymbol{x})$, and the estimation error $e(\boldsymbol{x}) = \psi(\boldsymbol{x})^2 - p(\boldsymbol{x})$ respectively over a 41×41 meshed data, with a grid size of 0.2, ranging from -4 to 4 for both x_1 and x_2.

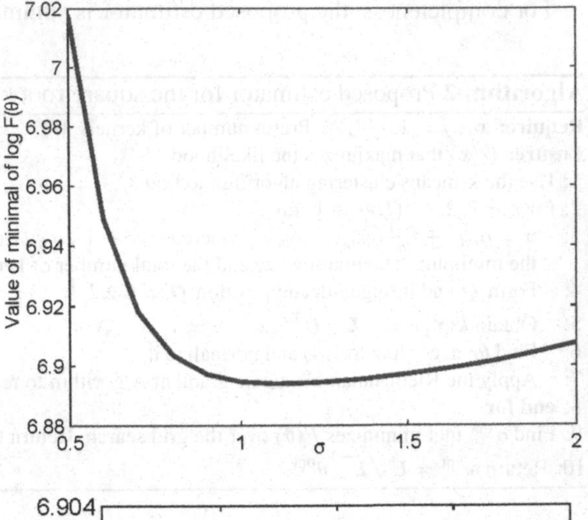

Fig. 1 $\log(F(\theta))$ versus the kernel width

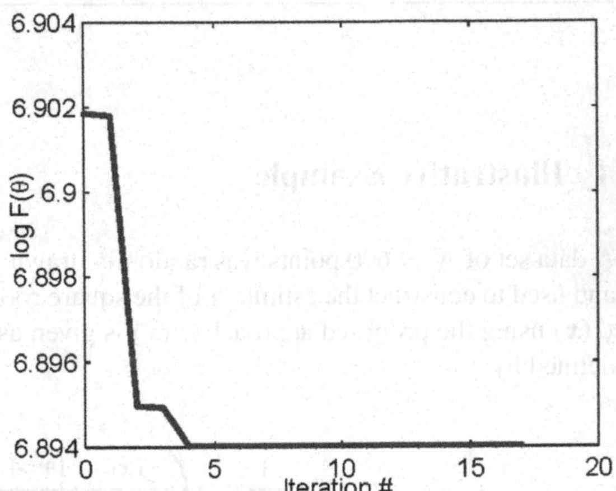

Fig. 2 Convergence of Riemannian conjugate gradient algorithm

The experiment was repeated for 100 different random runs in order to compare the performance of the proposed algorithm with the Gaussian mixture model (GMM) that is fitted using the EM algorithm [5]. A separate test data set of $N_{test} = 1681$ was generated using a 41×41 meshed data, with a grid size of 0.2, ranging from -4 to 4 for both x_1 and x_2. These points were used for evaluation according to

$$L_1 = \begin{cases} \frac{1}{N_{test}} \sum_{k=1}^{N_{test}} |p(\boldsymbol{x}_k) - \psi^2(\boldsymbol{x}_k)| & \text{Proposed} \\ \frac{1}{N_{test}} \sum_{k=1}^{N_{test}} |p(\boldsymbol{x}_k) - \hat{p}_{\text{GMM}}(\boldsymbol{x}_k)| & \text{GMM} \end{cases} \tag{22}$$

where $\hat{p}_{\text{GMM}}$ is the resultant GMM model based on two Gaussian mixtures, with the constraint that the covariance matrix is diagonal. The GMM model fitting is implemented using Matlab command *gmdistribution.fit.m*. The results are as

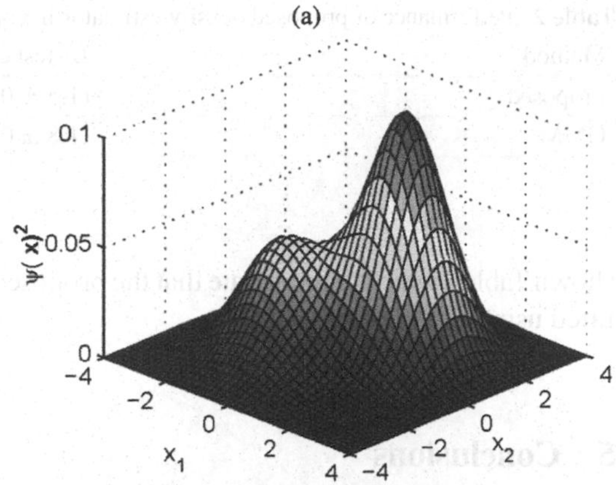

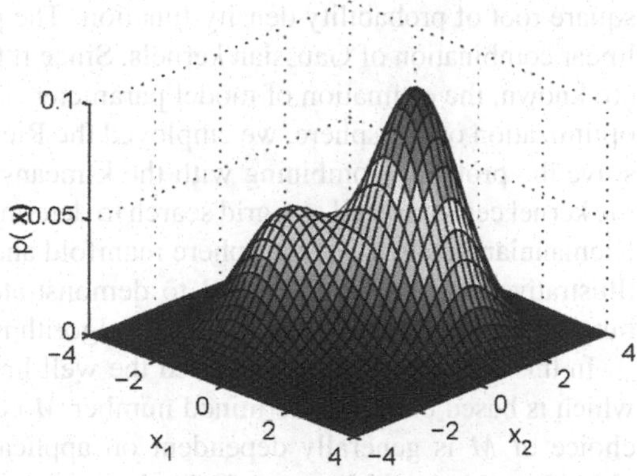

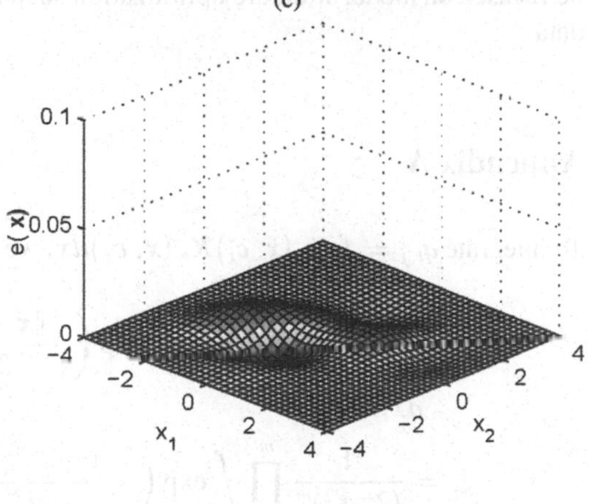

Fig. 3 The result of the proposed estimator; **a** The resultant pdf estimator $\psi(\boldsymbol{x})^2$. **b** The true pdf $p(\boldsymbol{x})$; and **c** The estimation error $e(\boldsymbol{x})$

Table 2 Performance of proposed density estimator in comparison with GMM model

Method	L_1 test error (mean ± STD)
Proposed	$(1.5 \pm 0.3) \times 10^{-3}$
GMM	$(1.8 \pm 0.6) \times 10^{-3}$

shown Table 2 which demonstrate that the proposed algorithm outperforms the GMM fitted using EM algorithm.

5 Conclusions

In this paper we have proposed a fast maximum likelihood estimator to represent the square root of probability density function. The proposed model is in the form of a linear combination of Gaussian kernels. Since if the kernel centres and kernel width are known, the estimation of model parameters can be formulated as a Riemannian optimization on the sphere, we employed the Riemannian optimization algorithm to solve the problem, combining with the k-means clustering algorithm to determine the kernel center as well as a grid search to determine the kernel width. The first order Riemannian geometry of the sphere manifold and vector transport are explored. An illustrative example is employed to demonstrate that the proposed approach can outperform the GMM model fitted EM algorithm.

In this research we have included the well-known k-means clustering algorithm which is based on a predetermined number M centers which may not optimal. The choice of M is generally dependent on applications, and for k-means clustering algorithm, M should be set sufficiently large to perform well. Future research will be focused on model structure optimization so that the model size can be learnt from data.

Appendix A

To integrate $q_{i,j} = \int K_\sigma(\boldsymbol{x}, \boldsymbol{c}_i) K_\sigma(\boldsymbol{x}, \boldsymbol{c}_j) d\boldsymbol{x}$, we let $\boldsymbol{x} = [x_1, ...x_m]^{\mathrm{T}}$, we have

$$\begin{aligned} q_{i,j} &= \frac{1}{(2\pi\sigma^2)^m} \int \cdots \int \exp\left(-\frac{\|\boldsymbol{x} - \boldsymbol{c}_i\|^2}{2\sigma^2} - \frac{\|\boldsymbol{x} - \boldsymbol{c}_j\|^2}{2\sigma^2}\right) \\ &\quad dx_1 ... dx_m \\ &= \frac{1}{(2\pi\sigma^2)^m} \prod_{l=1}^{m} \int \exp\left(-\frac{(x_l - c_{i,l})^2}{2\sigma^2} - \frac{(x_l - c_{j,l})^2}{2\sigma^2}\right) dx_l \end{aligned} \tag{23}$$

in which

$$\begin{aligned}
&\int \exp\Big(-\frac{(x_l-c_{i,l})^2}{2\sigma^2}-\frac{(x_l-c_{j,l})^2}{2\sigma^2}\Big)dx_l \\
&=\int \exp\Big(-\frac{x_l^2-(c_{i,l}+c_{j,l})x_l+(c_{i,l}^2+c_{j,l}^2)/2}{\sigma^2}\Big)dx_l \\
&=\exp\Big(-\frac{\frac{c_{j,l}^2+c_{i,l}^2}{2}-\big(\frac{c_{i,l}+c_{j,l}}{2}\big)^2}{\sigma^2}\Big) \\
&\times\int \exp\Big(-\frac{\big[x_l-(c_{i,l}+c_{j,l})\big]^2}{\sigma^2}\Big)dx_l
\end{aligned} \tag{24}$$

By making use of $\int \frac{1}{\sqrt{2\pi\sigma^2}}\exp\Big(-\frac{(x_l-c)^2}{2\sigma^2}\Big)dx_l=1$, i.e. Gaussian density integrates to one, we have

$$\begin{aligned}
&\int \exp\Big(-\frac{(x_l-c_{i,l})^2}{2\sigma^2}-\frac{(x_l-c_{j,l})^2}{2\sigma^2}\Big)dx_l \\
&=\sqrt{\pi\sigma^2}\exp\Big(-\frac{\frac{c_{j,l}^2+c_{i,l}^2}{2}-\big(\frac{c_{i,l}+c_{j,l}}{2}\big)^2}{\sigma^2}\Big) \\
&=\sqrt{\pi\sigma^2}\exp\Big(-\frac{(c_{j,l}-c_{i,l})^2}{4\sigma^2}\Big)
\end{aligned} \tag{25}$$

Hence

$$q_{i,j}=\frac{1}{(4\pi\sigma^2)^m}\exp\Bigg(-\frac{\|c_i-c_j\|^2}{4\sigma^2}\Bigg)=K_{\sqrt{2}\sigma}(c_i,c_j) \tag{26}$$

References

1. Absil, P.-A., Mahony, R., Sepulchre, R.: Optimization Algorithms on Matrix Manifolds. Princeton University Press (2008)
2. Mishra, B., Meyer, G., Bach, F., Sepulchre, R.: Low-rank optimization with trace norm penalty. SIAM J. Optim. **23**(4), 2124–2149 (2013)
3. Harandi, M., Hartley, R., Shen, C., Lovell, B., Sanderson, C.: Extrinsic methods for coding and dictionary learning on Grassmann manifolds, pp. 1–41 (2014). arXiv:1401.8126
4. Lui, Y.M.: Advances in matrix manifolds for computer vision. Image Vision Comput. **30**, 380–388 (2012)
5. McLachlan, G., Peel, D.: Finite Mixture Models. Wiley, New York (2000)
6. Silverman, B.W.: Density Estimation for Statistics and Data Analysis. Chapman and Hall, London (1986)
7. Duda, R.O., Hart, P.E.: Pattern Classification and Scene Analysis. Wiley, New York (1973)
8. Chen, S., Hong, X., Harris, C.J.: Particle swarm optimization aided orthogonal forward regression for unified data modelling. IEEE Trans. Evol. Comput. **14**(4), 477–499 (2010)
9. Rutkowski, L.: Adaptive probabilistic neural networks for pattern classification in time-varying environment. IEEE Trans. Neural Netw. **15**(4), 811–827 (2004)
10. Yin, H., Allinson, N.W.: Self-organizing mixture networks for probability density estimation. IEEE Trans. Neural Netw. **12**(2), 405–411 (2001)

11. Dempster, A.P., Laird, N.M., Rubin, D.B.: Maximum likelihood from incomplete data via the EM algorithm. J. R. Stat. Soc. B **39**(1), 1–38 (1977)
12. Parzen, E.: On estimation of a probability density function and mode. Ann. Math. Stat. **33**(3), 1066–1076 (1962)
13. Weston, J., Gammerman, A., Stitson, M.O., Vapnik, V., Vovk, V., Watkins, C.: Support vector density estimation. In: Schölkopf, B., Burges, C., Smola, A.J. (eds.) Advances in Kernel Methods—Support Vector Learning, pp. 293–306. MIT Pres, Cambridge, MA (1999)
14. Vapnik, V., Mukherjee, S.: Support vector method for multivariate density estimation. In: Solla, S., Leen, T., Müller, K.R. (eds.) Advances in Neural Information Processing Systems, pp. 659–665. MIT Press, Cambridge, MA (2000)
15. Girolami, M., He, C.: Probability density estimation from optimally condensed data samples. IEEE Trans. Pattern Anal. Mach. Intell. **25**(10), 1253–1264 (2003)
16. Hong, X., Gao, J., Chen, S., Zia, T.: Sparse density estimation on the multinomial manifold. IEEE Trans. Neural Netw. Learn. Syst. (In Press, 2015)
17. Choudhury, A.: Fast machine learning algorithms for large data. Ph.D. dissertation, School of Engineering Sciences, University of Southampton (2002)
18. Chen, S., Hong, X., Harris, C.J., Sharkey, P.M.: Sparse modeling using forward regression with PRESS statistic and regularization. IEEE Trans. Syst. Man Cybern. Part B **34**(2), 898–911 (2004)
19. Chen, S., Hong, X., Harris, C.J.: Sparse kernel density construction using orthogonal forward regression with leave-one-out test score and local regularization. IEEE Trans. Syst. Man Cybern. Part B **34**(4), 1708–1717 (2004)
20. Chen, S., Hong, X., Harris, C.J.: An orthogonal forward regression techniques for sparse kernel density estimation. Neurocomputing **71**(4–6), 931–943 (2008)
21. Hong, X., Chen, S., Qatawneh, A., Daqrouq, K., Sheikh, M., Morfeq, A.: Sparse probability density function estimation using the minimum integrated square error. Neurocomputing **115**, 122–129 (2013)
22. Bilmes, J.A.: A gentle tutorial of the EM algorithm and its application to parameter estimation for Gaussian mixture and hidden Markov models. Technical Report ICSI-TR-97-021. University of California, Berkeley (1998)
23. Pinheiro, A., Vidakovic, B.: Estimating the square root of density via compactly supported wavelets. Comput. Stat. Data Anal. **25**(4), 399–415 (1998)
24. Hager, W.W., Zhang, H.: A survey of nonlinear conjugate gradient methods. Pac. J. Optim. **2**(1), 35–58 (2006)
25. Boumal, N., Mishra, B., Absil, P.-A., Sepulchre, R.: Manopt, a Matlab toolbox for optimization on manifolds. J. Mach. Learn. Res. **15**, 1455–1459 (2014)

3Dana: Path Planning on 3D Surfaces

Pablo Muñoz, María D. R-Moreno and Bonifacio Castaño

Abstract An important issue when planning the tasks that a mobile robot has to reach is the path that it has to follow. In that sense, classical path planning algorithms focus on minimizing the total distance, generally assuming a flat terrain. Newer approaches also include traversability cost maps to define the terrain characteristics. However, this approach may generate unsafe paths in realistic environments as the terrain relief is lost in the discretisation. In this paper we will focus on the path planning problem when dealing with a Digital Terrain Model (DTM). Over such DTM we have developed 3Dana, an any-angle path planning algorithm. The objective is to obtain candidate paths that may be longer than the ones obtained with classical algorithms, but safer. Also, in 3Dana we can consider other parameters to maximize the path optimality: the maximum slope allowed by the robot and the heading changes during the path. These constraints allow discarding infeasible paths, while minimizing the heading changes of the robot. To demonstrate the effectiveness of the algorithm proposed, we present the results for the paths obtained for real Mars DTMs.

Keywords Path Planning · Slope · DTM · Rover

1 Introduction

Classical path planning algorithms used by on-ground operators are based on (i) flat terrains with free or blocked cells or (ii) terrain models that usually are produced merging different terrain characteristics (slope, rocks, quicksands, etc.) into a single layer: a traversability cost map. These maps provide the estimated effort required to

P. Muñoz (✉) · M.D. R-Moreno
Departamento de Automática, Universidad de Alcalá, Alcalá de Henares, Spain
e-mail: pmunoz@aut.uah.es

M.D. R-Moreno
e-mail: mdolores@aut.uah.es

B. Castaño
Departamento de Matemáticas y Físicas, Universidad de Alcalá, Alcalá de Henares, Spain
e-mail: bonifacio.castano@uah.es

M. Bramer and M. Petridis (eds.), *Research and Development in Intelligent Systems XXXIII*, DOI 10.1007/978-3-319-47175-4_12

cross an area of the map with a unique number. The merging process simplifies the path planning algorithm, but at the expense of lost specific information that can be useful during the path search. For instance, analysing slopes, rocks or other terrain characteristic independently will allow the path planning algorithm to obtain safer paths, avoiding certain slopes, rocks concentration or other constraints, rather than only minimize the path cost as done till now.

In this paper we present a new path planning algorithm called 3D Accurate Navigation Algorithm (3Dana). It is based on heuristic path planning algorithms such as A* [1], Theta* [2] and S-Theta* [3]. 3Dana is developed having in mind its potential application to exploration robots (e.g. rovers) operating in planetary surfaces. Then, constraints such as the relief of the terrain or the heading changes for a given path are relevant parameters to be considered. 3Dana exploits a Digital Terrain Model (DTM) that provides an accurate abstraction of the terrain relief. Then, the algorithm can avoid paths that overcomes a maximum slope defined by the human operator. It also considers the heading of the robot during the path search, generating smoother routes. Using 3Dana, human operators can select the best performance path considering different parameters, while keeping the safety of the robot.

The next section provides the description of the path planning problem and the notation used for this paper. Section 3 presents a brief revision of path planning algorithms. Following, Sect. 4 defines the DTM employed to model the surface used by our algorithm. Then, the 3Dana path planning algorithm is presented. Section 6 shows an empirical evaluation of the algorithm on real Mars maps. Finally, conclusions and future work are outlined.

2 Path Planning Notation and Representation

The most common discretisation of the environment is a 2D representation formed by a uniform regular grid with blocked and unblocked cells [4]. The edges of the cell represents the nodes over which the robot traverses, being constant the distance between adjacent nodes (except for diagonal moves in an 8-connected grid). Then, each node represents a position with coordinates (x_i, y_j) in the map. In the following, we represent nodes using a lower-case letter and for the sake of simplicity we will symbolize any node p as a coordinate pair (x_p, y_p). Also, we assume that each node can have an elevation value z_p, so we obtain a DTM discretised as a grid. To define the path planning problem, we assume that s and g are the start and goal nodes of respectively. Then, a candidate path will be a set $(p_1, p_2, ..., p_{n-1}, p_n)$ with initial node $p_1 = s$ and goal $p_n = g$. A path is valid iff it does not cross a blocked cell. Besides their geometrical values, each node has four attributes required by heuristic search algorithms:

- $p.G$: the cumulative cost to reach the node p from the initial node.
- $p.H$: the heuristic value, i.e., the estimated distance to the goal node.

- *p.F*: the node evaluation function: $p.F = p.G + p.H$.
- *p.parent*: the reference to the parent node. The parent of a node *p* must be a reachable node from the *p* node.

Finally, some algorithms can deal with traversability cost maps. A cost map is an extension of the introduced grid in which each cell has an associated cost. This cost is used to represent a terrain characteristic and, sometimes, the cost is generated combining different parameters, e.g. rocky and hazardous areas, slopes, etc. During the path search, the cell cost is used as a multiplicative factor applied to the length of the path that traverses over such cell. Using cost maps, some algorithms are able to avoid potentially hazardous areas. However, exploiting a combination of parameters into a single value, could lead to miss information that can be useful separately.

3 Related Works

The objective of path planning is to obtain paths between different points in an environment that can be partially or completely known. There are several variations to solve this problem such as Rapidly exploring Random Trees (RRT) [5] or Probabilistic Road Maps (PRM) [6]. However, we will focus on path planning methods based on heuristic search algorithms as 3Dana inherits from these algorithms.

The most representative heuristic search algorithm is A* [1]. A* has been so widely used because is simple to implement, very efficient and has lots of scope for optimization [7]. But it has an important limitation: it is based on a graph search. Typically, an eight-connected graph is used, which implies a restriction in the path headings to multiples of $\pi/4$. Thus, A* generates a sub-optimal paths with zig-zag patterns. For the heuristic computation A* uses the *Octile* distance [2].

In order to avoid the zig-zag patterns a new family of path planning algorithms called *any-angle* has appeared. These algorithms are based on A* as well, and called *any-angle* since the paths generated are not restricted to to multiples of $\pi/4$. The most representative is Theta* [2]. During the search, for a given position *p*, Theta* evaluates the line of sight between the successors of *p* and the parent of the current node, $q = p.parent$, i.e., if the straight line between *p* and *q* cross or do not cross blocked cells. Given a node $t \in successors(p)$, if there is line of sight between *t* and *q*, the parent of *t* will be *q* instead of *p* (as happens in A*). If there are obstacles between *q* and *t*, Theta* behaves as A*. This allows removing the intermediate node (*p*), smoothing the path during the search process, while maintain a free-obstacle path. As the parent of a node is no longer restricted to be an adjacent node, the path generated can have any heading change, i.e., the A* heading restriction of $\pi/4$ is overcome. However, the line of sight checking is performed frequently, which has a significant computational overhead. In any case, the paths generated are never longer than the obtained with previous approaches. As there is no angle restriction, Theta* obtains better results using the *Euclidean* distance (the straight line distance) as the heuristic function as opposed to the Octile one. Also, as a consequence of the expansion process, Theta* only performs heading changes at the corners of the blocked cells.

A recent algorithm developed from Theta* is Smooth Theta* (S-Theta*) [3]. It aims to reduce the amount of heading changes that the robot should perform to reach the goal. To do this, S-Theta* includes a new term, $\alpha(t)$. This value gives a measure of the deviation from the optimal trajectory to achieve the goal, considering the heading of the robot at each step during the path search. In an environment without obstacles, the optimal path between two points is the straight line. Therefore, applying the triangle inequality, any node that does not belong to that line will involve both, a change in the direction and a longer distance. Therefore, $\alpha(t)$ causes that nodes far away from that line will not be expanded during the search. S-Theta* reduces the number and amplitude of heading changes with respect to Theta*, but at the expense of slightly longer path lengths. Besides, S-Theta* can produce heading changes at any point, not only at the vertex of the obstacles as Theta* does.

Above algorithms try to minimize the total distance and the heading changes for a path in uniform flat environments. Although it is possible to exploit them in robotics, when applying to natural scenarios, e.g., planetary exploration robots, they do not provide enough safe routes. Path generated can cross rocky or high-hills areas that the robot will never be able to reach. Algorithms such as the D* family (the most representative one is Field D* [8]) works with cost maps aiming to provide safer paths in realistic environments. The objective of these algorithms is not to minimize the distance travelled but to minimize the path cost. In particular, the NASA rover drivers employ a variation of Field D* to plan the operations of the MER and MSL.

In a similar direction, Garcia et al. [9] presented a path planning architecture that uses a cost map that combines the elevation, slope and roughness of the terrain. These parameters are acquired using a laser scanner and processed using a fuzzy engine to generate the cell costs. Then, a path planning module generates the path based on a modified version of A*. However, this approach inherits quite unrealistic paths generation since they exploit A*.

Another possibility is to employ a DTM, as Ishigami et al. [10] do. They present a technique to evaluate the motion profiles of a rover when it has to follow the shortest path generated by the Dijkstra algorithm [11]. Based on a DTM, they are able to evaluate the path considering the rover dynamics and the wheel-soil contact model. This allows generating really safe paths, but the process is very time consuming. The algorithm also imposes a constraint that previous approaches do not have: it requires a specific model of the robot.

Following the DTM representation, Page et al. [12] exploit a 3D triangle mesh for the terrain discretisation. Using such DTM, their algorithm generates a path following either the valleys or ridges of the terrain depending on the criteria selected by the user. Although the idea is quite novel, it is not clear how the algorithm performs in real scenarios.

4 Terrain Interpolation

We have presented a terrain discretisation based on a 2D representation that can be easily enhanced to a 3D terrain or DTM considering the elevation at each node. If we have a DTM, the ground representation is defined as a set of $k = n \times m$ spatial points $(x_i, y_j, z_{i,j})$, where $1 \leq i \leq n$ and $1 \leq j \leq m$. The value $z_{i,j}$ represents the height of the terrain over the node located at the position (x_i, y_j). We assume a regular grid, i.e., the distance between two nodes (x_i, y_j) and (x_{i+1}, y_j), $1 \leq i \leq n-1$; $1 \leq j \leq m$ is constant. As well, we have the same distance for two nodes (x_i, y_j) and (x_i, y_{j+1}), $1 \leq i \leq n$; $1 \leq j \leq m-1$. A possible graphical representation can be seen in Fig. 1.

Given a DTM, the elevation of each node is known, so, algorithms restricted to move between nodes (e.g. A*) can be used without problems. If we want to traverse from $(x_i, y_j, z_{i,j})$ to $(x_{i+1}, y_j, z_{i+1,j})$, we can obtain the path length by applying the Pythagoras theorem. Instead, *any-angle* algorithms can traverse among non-adjacent nodes. Thus, it is required to compute the elevation of points that do not belong to the rectangular grid. For example, a movement between nodes $(x_i, y_j, z_{i,j})$ and $(x_{i+\gamma}, y_{j+\delta}, z_{i+\gamma,j+\delta})$ for arbitrary $\gamma > 1, \delta > 1$ and $\gamma \neq \delta$, implies to cross more than one cell and traverse in between nodes. For such coordinates we do no have the elevation: we need to interpolate the elevation for these points to obtain the most approximate distance travelled by the robot if we want to exploit an *any-angle* algorithm.

To get the elevation of a node (x_u, y_u) that not belongs to the rectangular grid (x_i, y_j), $1 \leq i \leq n, i \in \mathbb{N}$; $1 \leq j \leq m, j \in \mathbb{N}$, we will employ a lineal interpolation when $x_u = x_i$ for some value i or $y_u = y_j$ for some j. First, we need to consider the shape belonging to the cell formed by the nodes $(x_i, y_j, z_{i,j})$, $(x_{i+1}, y_j, z_{i+1,j})$, $(x_i, y_{j+1}, z_{i,j+1})$ and $(x_{i+1}, y_{j+1}, z_{i+1,j+1})$. Usually, four points in the space are not guaranteed to fit to a plane. As we see in other approaches, it is possible to model the DTM as a triangular mesh. In this regard, we assume that each cell is composed of four triangles whose base is the connection between two nodes, while the cathetus are the joint between each node and the central point of the cell, (x_c, y_c, z_c), as shown in Fig. 2. The altitude at the central point is computed as the mean value

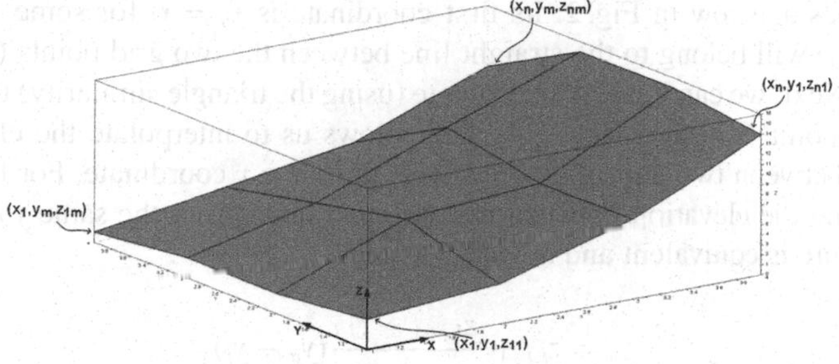

Fig. 1 DTM representation

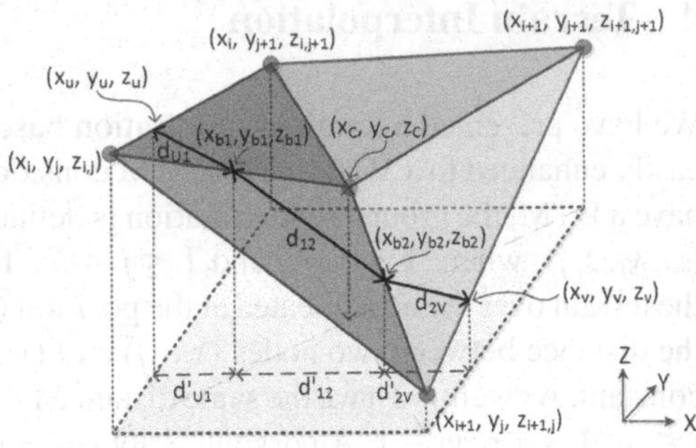

Fig. 2 Representation of a DTM cell

of the four nodes that comprise the cell $(x_i, y_j, z_{i,j})$, $(x_{i+1}, y_j, z_{i+1,j})$, $(x_i, y_{j+1}, z_{i,j+1})$ and $(x_{i+1}, y_{j+1}, z_{i+1,j+1})$. Calculating the distance travelled using this terrain representation is computationally expensive (as explained next), but unambiguous.

With this in mind, we need to compute the length for any arbitrary pair of nodes. This implies:

1. Obtaining the list of crossed cells.
2. For each cell, we need to calculate:
   (a) The coordinates in which the path enters and exits the cell.
   (b) The coordinates in which the path intersects with the segments that conforms the four triangles.
3. Obtaining the elevation for all the previous coordinates. If the coordinate is not a node, we need to interpolate its altitude.

We can determine the cells crossed by a straight line using a Cohen-Sutherland clipping algorithm [13]. As well, such algorithm can be used to compute the entry and exit point to the cell. Then, we have to calculate the intersection points between the path and the four triangles that conform the cell. In this point there are different possibilities depending on the entry and the exit points.

Assuming that the movement does not start and end in a node, we need to interpolate the altitude at different points. First, consider a point (x_u, y_u, z_u) that belongs to the x axis as show in Fig. 2. Its first coordinate is $x_u = x_i$ for some $1 \leq i \leq n$, then (x_u, y_u) will belong to the straight line between the two grid points (x_i, y_j) and (x_i, y_{j+1}). Next, we can linearly interpolate (using the triangle similarity) the altitude z_u for the point (x_u, y_u) as in Eq. 1. This allows us to interpolate the elevation of any point between two adjacent nodes with the same x coordinate. For the case of interpolating the elevation between two adjacent nodes with the same y coordinate the procedure is equivalent and not shown here.

$$z_u = z_{i,j} + \frac{z_{i,j+1} - z_{i,j}}{y_{j+1} - y_j}(y_u - y_j) \tag{1}$$

So, let (x_u, y_u, z_u) be the entry point to the cell and let $\overline{(x_v, y_v, z_v)}$ be the exit point that are now known. Considering that (x_u, y_u, z_u) belongs to the line $\overline{(x_i, y_j), (x_i, y_{j+1})}$ we can exit through one of the other three sides of the cell. Depending on the exit point, it is possible that we need to cross two or three planes of the cell. The possibilities are the following:

(a) Exit at the side defined by the line $\overline{(x_i, y_j), (x_{i+1}, y_j)}$. Two planes are crossed.
(b) Exit at the opposite side of a point that belongs to $\overline{(x_{i+1}, y_j), (x_{i+1}, y_{j+1})}$ constrained to $y_v < z_c$. Then three planes are crossed, and the one not crossed is the one formed by the nodes $(x_i, y_{j+1}), (x_c, y_c), (x_{i+1}, y_{j+1})$. This is the example presented in Fig. 2.
(c) Exit at the opposite side of a point belong to $\overline{(x_{i+1}, y_j), (x_{i+1}, y_{j+1})}$ constrained to $y_v \geq z_c$. Then three planes are crossed, and the one not crossed is formed by the nodes $(x_i, y_j), (x_c, y_c), (x_{i+1}, y_j)$.
(d) Exit at the side defined by the line $\overline{(x_i, y_{j+1}), (x_{i+1}, y_{j+1})}$. Two planes are crossed.

We can also consider entering the cell at a point within $(x_i, y_j), (x_{i+1}, y_j)$. This case is equivalent to the previously presented and can be solved using symmetry. Next, we briefly present how to compute the traversed length in a cell for case (b). Cases (a) and (d) are easier to compute and case (c) is symmetric to (b). We do not provide the mathematical model as it is excessively large to be presented here.

We need to compute the points in which the straight line that the path follows, $\overline{(x_u, y_u, z_u), (x_v, y_v, z_v)}$, changes from one plane to another in the cell. For (b) case three planes are cut, so we need to compute two points. Let (x_{b1}, y_{b1}, z_{b1}) and (x_{b2}, y_{b2}, z_{b2}) be these points as Fig. 2 shows. To obtain these points we need the coordinates (x, y) and then interpolate the elevation at such point. The way to obtain the intersect points is to employ the equation of the line and obtain the points in which the lines $\overline{(x_u, y_u), (x_v, y_v)}$ and $\overline{(x_i, y_j), (x_{i+1}, y_{j+1})}$ and/or $\overline{(x_i, y_{j+1}), (x_{i+1}, y_j)}$ (diagonals) intersect.

Once we have (x_{b1}, y_{b1}, z_{b1}) and (x_{b2}, y_{b2}, z_{b2}) we can obtain their altitude using triangle similarity. We can use Eq. 1 to do this, using the correct triangles to interpolate the altitude. If we want to obtain the elevation of the point (x_{b1}, y_{b1}, z_{b1}), we need to evaluate the triangle formed by such point, the node with coordinates $(x_i, y_j, z_{i,j})$ and the center of the cell, (x_c, y_c, z_c) Finally, we can compute the distance travelled through the cell using the Pythagoras Theorem for each path segment.

Using this representation we can also obtain the slope of the terrain. To compute the slope, we need to obtain the normal vector of each plane that comprises the cell. Considering the point (x_u, y_u, z_u) that belongs to the plane formed by the points $(x_i, y_j, z_{i,j}), (x_c, y_c, z_c)$ and $(x_{i+1}, y_j, z_{i+1,j})$, we can obtain the normal vector, $\overrightarrow{n_\pi}$, of the plane as in Eq. 2. This normal vector forms an angle, α_z, with the Z axis that gives us the slope of the terrain. Thus, the slope for such plane can be obtained as in Eq. 3. For each cell, we have four α_z, i.e., we have four slopes. Computing the slope during the path search enables the algorithm to avoid dangerous path.

$$\begin{aligned} \overrightarrow{n_\pi} &= (A, B, C)\ with: \\ A &= (y_c - y_j)\cdot(z_c - z_{i+1,j}) - (z_c - z_{i,j})\cdot(y_c - y_j) \\ B &= (z_c - z_{i,j})\cdot(x_c - x_{i+1}) - (z_c - z_{i+1,j})\cdot(x_c - x_i) \\ C &= (x_c - x_i)\cdot(y_c - y_j) - (y_c - y_j)\cdot(x_c - x_{i+1}) \end{aligned} \tag{2}$$

$$\alpha_z = \arccos\frac{C}{\sqrt{A^2 + B^2 + C^2}} \tag{3}$$

5 3Dana Algorithm

The 3D Accurate Navigation Algorithm, abbreviated as 3Dana, is a path planning algorithm developed to obtain safer routes based on heuristic search over a DTM. 3Dana is an evolution of the A* search algorithm, and it takes advantage of the newest *any-angle* path planning algorithms such as Theta* or S-Theta*. Its application scope is those mobile robots in which the elevation of the surface can affect their mobility. The main features of 3Dana are:

- Evaluation of the path cost using the terrain altitude. 3Dana performs path planning over realistic surface models, using the DTM as explained in the previous section. The length of a movement is a function of the distance between two points given their altitudes.
- Evaluation of heading changes during the search. Just like the S-Theta* algorithm [3], 3Dana calculates the necessary turns needed to reach the next position taking into consideration the current heading of the robot and the position of the goal. This allows obtaining smoother routes.
- Evaluation of the terrain slope. 3Dana avoids paths that exceed the maximum slope allowed by the robot. This allows obtaining safer and feasible paths.

During the path search 3Dana maintains a list of reachable nodes, i.e., the *open* list. Such list is ordered by the F value of the nodes. If that list is empty, all the reachable nodes from the start position have been evaluated and none is the goal position. Then, there is no feasible path between the desired points. Otherwise, the first node from the *open* list, that is, the most promising one, is extracted. If that node is the objective, the algorithm returns the path between the start and the goal through backtracking of the parents nodes from the goal to the start. Otherwise, a successor function returns a set with the visible adjacent nodes for the current node.

Instead of the previous algorithms, 3Dana uses node re-expansion, which means that all nodes will be analysed instead if they are previously expanded. This may leads to better paths, but increasing the runtime due to the possibility of expanding the same node several times. This is required when dealing with elevation maps. For example, the algorithm can reach a node by climbing a hill, that can be more expensive that surrounding it. This usually implies to expand more nodes, and thus, that path is discovered later during node re-expansion.

For the heuristic function, 3Dana uses a variation of the *Euclidean* distance to take into consideration the altitude difference between any two nodes p and t. We call this heuristic *EuclideanZ*, and is computed as in Eq. 4. The objective of this heuristic is to prioritize nodes without (or with lower) elevation changes. The *EuclideanZ* heuristic in 3D scenarios is admissible and consistent.

$$EuclideanZ(p, t) = \sqrt{(x_t - x_p)^2 + (y_t - y_p)^2 + (z_t - z_p)^2} \quad (4)$$

Besides the heuristic described, 3Dana also uses the value provided by the $\alpha(t)$ function, inherited from S-Theta*. This value measures the heading changes necessary to reach the next node as function of the current node's parent and the goal position. Alpha is computed as in Eq. 5. Using $\alpha(t)$ in the node's heuristic allows us to consider the heading changes during the search process, delaying the expansion of nodes that require a high turn to be reached. The function gives a value in the interval [0°, 180°]. If we apply a weight factor to this heuristic we can determine the relevance of the heading changes in our path. Small weights implies a soft restriction, whereas larger weights tend to make the algorithm trends to follow smoother paths with low number of heading changes, in spite of the distance travelled. Generally the weight used, called α_w, takes values in the range 0 (heading changes are not considered during search) and 1. Experimentally, we have realised that values higher than 1 usually do not reduce the heading changes.

$$\alpha(p, t, g) = \arccos \frac{\text{dist(p, t)}^2 + \text{dist(p, g)}^2 - \text{dist(t, g)}^2}{2 \cdot \text{dist(p, t)} \cdot \text{dist(p, g)}} \cdot \alpha_w \quad (5)$$

$$\text{with dist(p, t)} = \sqrt{(x_t - x_p)^2 + (y_t - y_p)^2}$$

As 3Dana is an *any-angle* algorithm, we need to perform the line of sight checking during the search. When evaluating the line of sight, we perform the path length computation and the slope analysis. This procedure is divided in two phases: first, we compute the points in which the line that connects the two nodes intersects with the horizontal or vertical axis; and second, we compute the length of each segment formed by two consecutive points.

For the first step, we have implemented an algorithm to compute the axes intersection points. For two given nodes, p_0 and p_n, the algorithm returns a list of points that intersect the axes and belong to the line that connects p_0 and p_n.

Once the points list has been calculated, the second step is to treat the segments formed by each pair of consecutive points obtained in the previous phase. So, to compute the length traversed for two non-adjacent nodes (e.g., p_0 and p_n) and check if there is a line of sight between them, we need to evaluate the intermediate positions. For all pairs of consecutive points of the list, p_i and p_{i+1}, we need to perform the process described below:

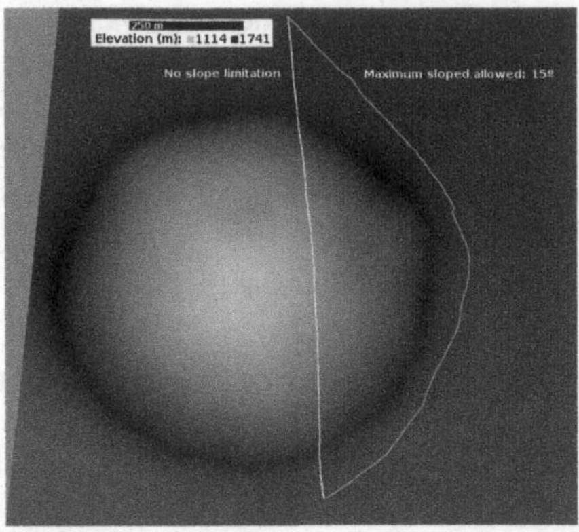

Fig. 3 No slope limited path (*center path*) versus limited slope path (*right path*) over a DTM

1. Evaluate if the cell that contains the segment $\overline{p_i p_{i+1}}$ is an obstacle. If it is, the algorithm returns that the path between p_0 and p_n is blocked.
2. If we have defined a maximum slope allowed, we compute the terrain slope of the cell using the normal vector (as in Eqs. 2 and 3). If the slope is higher than the maximum defined, such path is not considered.
3. Compute the length of the segment $\overline{p_i p_{i+1}}$ using the DTM data. For this, we use the formulation shown in Sect. 4. The path length consider the elevation change using the four planes that describes the cell crossed by the segment under consideration.

When considering a maximum slope during the path search, 3Dana can generate safer routes. Consider the paths presented in Fig. 3, both paths are obtained using 3Dana over the same DTM, but varying the maximum slope allowed. The left path does not consider slope limitations so then, it is highly undesirable since it crosses a crater. Meanwhile, when the maximum slope is setted to 15°, the algorithm avoids the crater and surrounds it.

The implementation of 3Dana allows safely generating candidate paths for mobile robots. Providing a DTM of the environment, we can select the best paths based not only in the distance travelled (considering also the elevation changes), but also reducing the heading changes and avoiding terrains with excessive slopes.

6 Experimental Results

In this section, we test the behaviour of the 3Dana algorithm using high resolution DTMs from Mars. The elevation data is obtained by the Mars Reconnaissance Observer (MRO) spacecraft, which provides elevation data with a vertical accuracy of 25 cm [14].[1] The maps used here have a resolution of 2 m, i.e., we have a uniform grid with elevation points every two meters. The objective of the experiments

[1]The DTMs are publicly accessible in http://uahirise.org/dtm.

Table 1 Data for DTEED_017147_1535. In bold: best path length + total turns for each slope

Alg.	Max. slope	α_w	Length (m)	Turn (°)	Time (s)	Expanded nodes
A*	–	–	11010	31770	1120	2497903
3Dana	–	–	**10189**	**751**	776	2503970
A*		–	11427	39915	1023	3032953
	20°	0.0	10822	4918	1842	4087662
3Dana		0.5	**10887**	**3742**	4112	3870873
		1.0	11022	3720	5670	3990782
A*		–	15434	73260	833	5213506
	10°	0.0	14703	10815	1617	5514326
3Dana		0.5	**14979**	**10355**	2485	5384046
		1.0	15193	10923	3147	5391994

is to generate different paths when varying the constraints, i.e., we evaluate paths with different heading changes consideration (modifying the α_w factor) and various maximum slopes allowed. In order to provide a comparative, we have adjusted the A* algorithm to work with the DTMs following the same model presented in Sect. 4 and including the altitude difference in the *Octile* heuristic.

For the experiments we have taken into consideration the following parameters: (1) the path length, (2) the total number of accumulated degrees by the heading changes (total turn), (3) the CPU time or search run-time, and, (4) the number of expanded nodes during the search. The execution is done on a 2.5 GHz Intel Core i7 with 8 GB of RAM under Ubuntu 14.04.

The first map considered presents a central structure and layered bedrock in a 25-km diameter crater.[2] The total area covered is near 40 km^2. The dimension in nodes is 3270 × 6636. In this map we set the initial point to the coordinates (700,500) and the goal in (2800,6000). Then, we have run A* and 3Dana with different configurations. These configuration entails no heading changes consideration ($\alpha_w = 0$), and two more values $\alpha_w = 0.5$ and $\alpha_w = 1$. Also, we have considered three possible values for the maximum slope allowed in the path: no limitation, 10° and 20°. The results obtained for the different paths are presented in Table 1, while some paths are represented in Fig. 4.

Given the data, we can observe that the path length and the heading changes obtained by 3Dana are better than the values obtained by A* when we do not consider maximum slope neither heading changes. As we consider maximum slopes for the path, we can see that as higher is the restriction imposed (i.e., smaller sloped allowed), both, the path length and the total turns increase. This is specially notorious when we restrict the slope to 10°, in which the path length is 1.35 times longer that the path restricted to 20° (without considering heading changes). Regarding to the heading

[2]http://www.uahirise.org/dtm/dtm.php?ID=ESP_017147_1535.

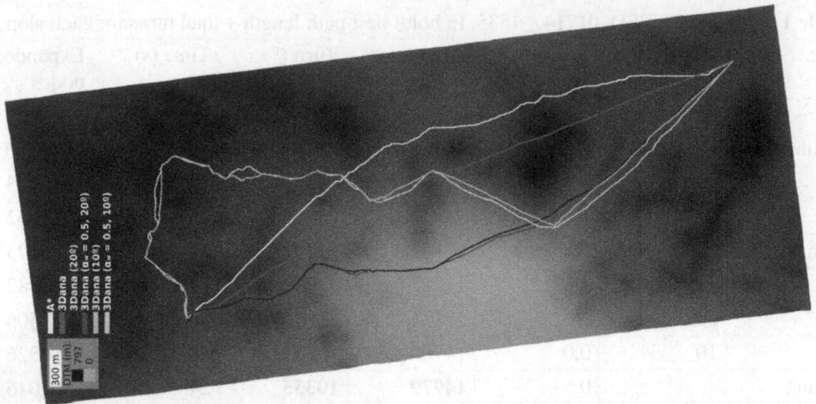

Fig. 4 Paths obtained for the DTEEC_017147_1535 using A* and different configurations of 3Dana. Colour and full size image available online: http://goo.gl/Bb4OIb

changes, we can appreciate that $\alpha_w = 0.5$ effectively reduces the total turn parameter. However, $\alpha_w = 1.0$ increases the turns respect to its previous configuration. While this parameter works fine in flat environments (see the S-Theta* evaluation [3]), considering the elevation seems to affect negatively. Particularly, attempting to avoid heading changes can lead to follow longer paths, as we can discard preferable paths (as function of the slope) in spite of reducing the turns. Also, we can observe that the runtime increases with higher α_w values. 3Dana requires some time to find paths as a consequence of both, the re-expansion process of the nodes and the computational cost related to the management of the DTM.

The second map considered presents an uplift of a 30-km diameter crater in Noachis Terra.[3] This map coves near 15 km^2. The dimension in nodes is 2960×2561. In this map we set the initial point to the coordinates (800,600) and the goal in (1800,2500). We have run the same experiments, i.e., $\alpha_w \in 0, 0.5, 1.0$ and maximum slopes 10° and 20°. However, in this map, all paths stay above of the 10° slope limitation. Thus, we have evaluated paths with 15° of maximum slope. Table 2 provides the results of these executions and Fig. 5 the paths representation.

As for the first case, 3Dana outperforms A* in both path length and total turns when not considering maximum slope and heading changes. As well, we can appreciate that the path length increases from the case with maximum slope of 20° to the case with 15°. If we analyse the heading changes, in this map considering $\alpha_w = 0.5$ slightly decreases the total turns only when the maximum slope is set to 20°. However, in any case, using $\alpha_w = 1.0$ not only increases the path length, but also the total turns. Then, seems that α_w is not well suitable for complex maps.

---

[3]http://www.uahirise.org/dtm/dtm.php?ID=ESP_030808_1535.

Table 2 Data for DTEED_030808_1535. In bold: best path length + total turns for each slope

Alg.	Max. slope	α_w	Length (m)	Turn (°)	Time (s)	Expanded nodes
A*	–	–	4800	13995	49	407544
3Dana	–	–	**4493**	**478**	77	720744
A*		–	5990	28080	343	2129666
	20°	0.0	5665	4442	433	2230932
3Dana		0.5	**5776**	**4241**	777	2150387
		1.0	5786	4428	983	2153923
A*		–	7387	41850	244	2712912
	15°	0.0	**7010**	**8569**	409	2789358
3Dana		0.5	7175	9066	749	2721862
		1.0	7450	9061	901	2774165

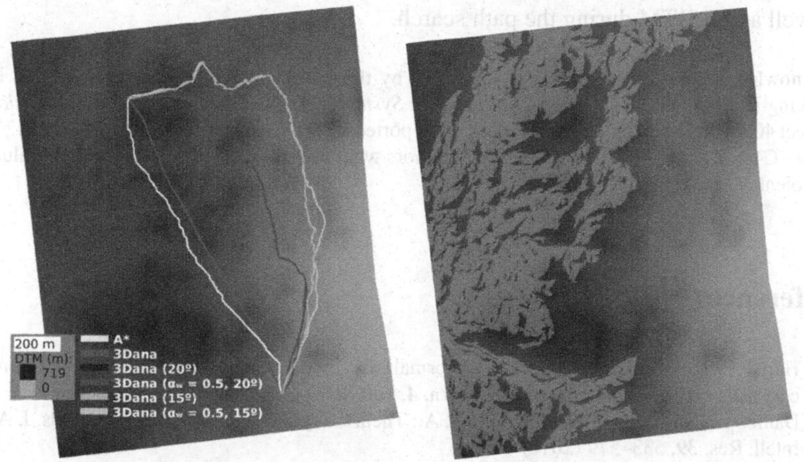

Fig. 5 Paths obtained for the DTEED_030808_1535 using A* and different configurations of 3Dana (*left*). Colour and full size image available online: http://goo.gl/wPRtgm. *Right* area of the map explored (marked in *light gray*) by 3Dana considering a maximum slope of 10° (no path found in this case)

In Fig. 5 (right) we can appreciate the nodes expanded during the search for a maximum slope of 10°. This case is remarkable because 3Dana (as well as A*) is not able to find a path with such a constraint. We can see that the algorithm expands several nodes, but there is no path that allows safely reaching the desired goal. Then, Fig. 5 (right) provides a vision of the reachable areas of the map when we restrict the maximum slope to 10°. This could provide an insight of the terrain that can be useful for human operators during the mission planning.

7 Conclusions

Heuristic search path planning algorithms such as A* or S-Theta* try to minimize the total distance that the robot should travel. Although this criteria has been widely used to compare algorithms, it is not enough if the robot cannot cross certain rocky or cumbersome areas. In this paper we presented a new *any-angle* algorithm named 3Dana. It is designed with the purpose of considering the terrain model and minimizing the heading changes of a path. 3Dana integrates a DTM during the search, which enables to avoid potentially dangerous areas, and then generate safer routes. This is done discarding paths that exceed slopes restrictions imposed by the user. Moreover, 3Dana computes the necessary turns for a path, providing smoother routes.

Experiments performed with Mars DTMs show that 3Dana can generate routes restricted by the slope, avoiding dangerous terrains. Also, some configurations allows properly reducing the heading changes performed during the path. As a future work, we will work to improve the capabilities of the algorithm, e.g., exploiting cost maps as well as the DTM during the path search.

Acknowledgments Pablo Muñoz is supported by the European Space Agency under the Networking and Partnering Initiative *"Cooperative Systems for Autonomous Exploration Missions"* project 4000106544/ 12/NL/PA. The work is supported by MINECO project EphemeCH TIN2014-56494-C4-4-P and UAH 2015/00297/001. Authors want to thanks the reviewers for their valuable comments.

References

1. Hart, P.E., Nilsson, N.J., Raphael, B.: A formal basis for the heuristic determination of minimum cost paths. IEEE Trans. Syst. Sci. Cybern. **4**, 100–107 (1968)
2. Daniel, K., Nash, A., Koenig, S., Felner, A.: Theta*: any-angle path planning on grids. J. Artif. Intell. Res. **39**, 533–579 (2010)
3. Muñoz, P., R-Moreno, M.D.: S-Theta*: low steering path-planning algorithm. In: Proceedings of the 32nd SGAI International Conference, Cambridge, UK, Dec 2012, pp. 109–121
4. Yap, P.: Grid-based path-finding. In: Advances in Artificial Intelligence, Lecture Notes in Computer Science, vol. 2338. Springer, Berlin/Heidelberg (2002)
5. LaValle, S.M., Kuffner Jr., J.J.: Randomized kinodynamic planning. Int. J. Robot. Res. **20**(5), 378–400 (2001)
6. Akinc, M., Bekris, K.E., Chen, B.Y., Ladd, A.M., Plaku, E., Kavraki, L.E.: Probabilistic roadmaps of trees for parallel computation of multiple query roadmaps. In: Robotics Research. The Eleventh International Symposium, Springer Tracts in Advanced Robotics, vol. 15, pp. 80–89. Springer, Berlin, Heidelberg (2005)
7. Millington, I., Funge, J.: Artificial Intelligence for Games, 2nd edn. Morgan Kaufmann Publishers (2009)
8. Ferguson, D., Stentz, A.: Field D*: an interpolation-based path planner and replanner. Robot. Res. **28**, 239–253 (2007)
9. Garcia, A., Barrientos, A., Medina, A., Colmenarejo, P., Mollinedo, L., Rossi, C.: 3D path planning using a fuzzy logic navigational map for planetary surface rovers. In: Proceedings of the 11th Symposium on Advanced Space Technologies in Robotics and Automation (2011)

10. Ishigami, G., Nagatani, K., Yoshida, K.: Path planning for planetary exploration rovers and its evaluation based on wheel slip dynamics. In: Proceedings of the IEEE International Conference on Robotics and Automation, Roma, Italy, Apr 2007, pp. 2361–2366
11. Dijkstra, E.: A note on two problems in connexion with graphs. Numerische Mathematik **1**, 269–271 (1959)
12. Page, D.L., Koschan, A.F., Abidi, M.A.: Ridge-valley path planning for 3D terrains. In: Proceedings of the 2006 IEEE International Conference on Robotics and Automation, Orlando, Florida, USA, May 2006, pp. 119–124
13. Foley, J., van Dam, A., Feiner, S., Hughes, J.: Computer Graphics: Principles and Practice. Addison-Wesley (1992)
14. Kirk, R.L., Howington-Kraus, E., Rosiek, M.R., Anderson, J.A., Archinal, B.A., Becker, K.J., Cook, D.A., Galuszka, D., Geissler, P.E., Hare, T.M., Holmberg, I.M., Keszthelyi, L.P., Redding, B.L., Delamere, W.A., Gallagher, D., Chapel, J., Eliason, E.M., King, R., McEwen, A.S.: Ultrahigh resolution topographic mapping of Mars with MRO HiRISE stereo images: meter-scale slopes of candidate Phoenix landing sites. J. Geophys. Res.: Planets **113**(E3), n/a–n/a (2008)

Natural Language Processing

Covert Implementations of the Turing Test: A More Level Playing Field?

D.J.H. Burden, M. Savin-Baden and R. Bhakta

Abstract It has been suggested that a covert Turing Test, possibly in a virtual world, provides a more level playing field for a chatbot, and hence an earlier opportunity to pass the Turing Test (or equivalent) in its overt, declared form. This paper looks at two recent covert Turing Tests in order to test this hypothesis. In one test (at Loyola Marymount) run as a covert-singleton test, of 50 subjects who talked to the chatbot avatar 39 (78 % deception) did not identify that the avatar was being driven by a chatbot. In a more recent experiment at the University of Worcester groups of students took part in a set of problem-based learning chat sessions, each group having an undeclared chatbot. Not one participant volunteered the fact that a chatbot was present (a 100 % deception rate). However the chatbot character was generally seen as being the least engaged participant—highlighting that a chatbot needs to concentrate on achieving legitimacy once it can successfully escape detection.

Keywords Turing Test · Natural Language · Chatbot · Virtual worlds · Second Life

1 Introduction

Burden [1] described how the Turing Test as commonly implemented provides the computer (in the guise of a chatbot) with a significant challenge, since both the judges and hidden humans are aware that they are taking part in a Turing Test, and so the dialogues which take place during the test are rarely "normal" [2]. In the same paper Burden described how a more level playing field could be created by conducting

D.J.H. Burden (✉)
Daden Limited, Birmingham B7 4BB, UK
e-mail: david.burden@daden.co.uk

M. Savin-Baden · R. Bhakta
University of Worcester, Worcester WR2 6AJ, UK
e-mail: m.savinbaden@worc.ac.uk

R. Bhakta
e-mail: r.bhakta@worc.ac.uk

M. Bramer and M. Petridis (eds.), *Research and Development in Intelligent Systems XXXIII*, DOI 10.1007/978-3-319-47175-4_13

a covert Turing Test, where neither judge nor hidden human know that the test is taking place. Burden also suggested that virtual environments could provide the ideal location for such a covert Turing Test.

Burden defined four possible Turing Test situations:

- The robotar (computer/chatbot driven avatar) is undeclared and part of a group conversation (the Covert Group test).
- A robotar is declared present (but unidentified) as part of a group conversation (the Overt Group test).
- The robotar is undeclared and part of a set of one-on-one conversations (the Covert Singleton test).
- A robotar is declared present (but unidentified) as part of a set of one-on-one conversations (the Overt Singleton test, the original Turing Test, the Imitation Game [3] as typically implemented by competitions such as the Loebner Prize [4]).

Burden identified a potential area of future research as *"to what extent do the Covert/Overt and Singleton/Group options present earlier opportunities to pass the Turing Test (or equivalent) in a virtual world?"*

This paper will review two covert Turing Tests which have been inspired by Burden's paper, one in a virtual world and one in an on-line chat room. For the second test the paper will also analyse some of the previously unpublished metrics obtained during the test with regards to the performance of the chatbot. This paper will then examine the extent to which these tests have borne out the predictions and hypothesis of the original paper, and consider what further work could be done in this area in order to further the creation of "Turing-capable" chatbots.

Cooper and Van Leeuwen [5] provide a useful survey of current thought on the Turing Test, in particular Sloman's observations that the test is a poor measure of intelligence, and that the "average" interrogator envisaged by Turing is now far more sophisticated and aware of computer capabilities—and so potentially far harder to fool. Little appears to have been written around the concept of a Covert or Group test. It is interesting that in his analysis of the problems of the first Loebner Prize and suggestions for improvements Shieber [6] does not consider the potential for a Covert or Group test to overcome some of the perceived issues. However Mauldin [7] does describe an "*unsuspecting Turing Test*" using a text based Multi-User Dungeon as the setting, but reported that the chatbot only "*fooled 'average' questioners in the TINYMUD domain for a few minutes.*" Colby's experiments with PARRY [8] appear to be the first significant use of a covert Turing approach—albeit in a non-interactive setting (a passive cf. active detection test). Kurzweil [9] does describe the sort of covert Turing Test considered here in the Worcester experiment, "*each system logs on to various human chat rooms and tries to pass for a human, basically a covert Turing test. If one of the humans in a chat room says something like 'What are you, a chatterbot?' .... the evaluation is over*".

2 The Loyola Marymount Experiment

Gilbert and Forney [10] describe how they ran a covert Turing Test in the virtual world of Second Life. A summary of that experiment is provided below for comparison with the Worcester experiment, and Gilbert and Forney otherwise provide a full description of their experiment.

The experiment was based around the premise/deception of a new shop in Second Life, with subjects being invited into the shop to discuss and comment on its layout through interacting with the shop assistant. Both subject and assistant were represented by Second Life avatars, could see each other's avatars, and were able to converse by text-chat—the standard interaction form in Second Life. The key features of the experiment were that:

- In 50 % of the sessions the shop assistant was controlled by a computer, which included both a chatbot and the ability to move through the environment and sense the subject's position
- In the other 50 % of the sessions the shop assistant was controlled by a human. Crucially the human was not told about the hidden Turing element of the experiment, they believed they were just there to show the subjects around the new shop and handle any questions
- It was only in the final question of the post-evaluation debrief at the subjects were asked whether they thought the avatar (which always presented as female) was being controlled by a man, woman or computer.

The chatbot software used for the experiment was Discourse, an Artificial Intelligence Markup Language (AIML) derived chatbot engine developed by Daden Limited (Disclosure: The primary author is the owner and Managing Director of Daden Limited).

The key finding was that 78 % of subjects (39 out of 50) did not think that the avatar, when controlled by a computer, was being controlled by a computer. The only idea that many of them had that it could have been controlled by a computer was when the question was asked.

It was evident that the natural setting, the use of avatars, and the human-ness of the language all played a part in deceiving the participants. Typical quotes included: *"She seemed human to me in looks, body language, etc.,"* and *"I assumed she was human because I expected her to be"*.

Gilbert concluded *"as the current study demonstrates, it is possible to achieve deception rates approaching 80 % using only a moderately capable chat engine when all of [the] psychological and contextual factors are favorably represented"* and that *"The current study suggests that 3D virtual environments, a platform that wasn't even contemplated when Turing first proposed his test, may offer the most favorable context to achieve these challenging outcomes because of their unique ability to activate the anthropomorphic tendency in support of humanizing the computer."*

3 The University of Worcester Experiment

With the Loyola Marymount experiment there is a clear challenge that the use of human-looking 3D avatars could have biased the results - participants saw something human (even if only in digital form) and so may have assumed that the controlling agency was also human. The Loyola Marymount experiment was also a one-on-one test, and it was possible that an even higher level of deception could be possible in a group test (Covert-Group).

Daden, working with the University of Worcester, conducted an on-line Problem-Based Learning (PBL) experiment to assess the capability and value of chatbots (both covert and overt) in an educational context. Savin-Baden et al. [11] presented the initial results from this experiment but focused on its implications for pedagogical agents. The analysis below considers the experiment within the context of a covert Turing Test.

3.1 Experimental Design

The experiment was based around the premise/deception of an on-line Problem Based Learning (PBL) exercise for Health & Care students at the University of Worcester and at Walsall College. PBL involves students being presented with a scenario which they then need to work on in groups in order to identify what information they need to address the scenario, and then carry out their own research, before further discussion and the presentation of an answer to a facilitator. The scenarios often have no "right" answer and decisions/solutions may be purely down to value judgements.

For the experiment 42 students (and 3 staff) were recruited from the two institutions—principally by requesting volunteers after lectures. The experiment was run in two phases, Phase 1 being a covert Turing Test, and Phase 2 being an overt Turing Test. In Phase 1 the students were divided into groups of 3–5 (depending on availability), and undertook 3 one-hour, online discussion sessions. Ideally:

- The first session was used to present the scenario and have an initial discussion.
- The second session was held 2 days later for further discussion after the students had done some research.
- The third session was held another 2 days later to finish the discussion and come up with an answer/recommendation.

In practice only 4 of the Phase 1 groups did 3 sessions, with 11 groups doing 2 sessions and 2 groups (both staff) doing only 1. A facilitator was present and active in each session. The scenario for Phase 1 was that of a flatmate, Carolyn, being suspected of having drink and money problems.

Phase 2 followed a similar structure with 4 groups doing 2 sessions and 3 groups doing only 1 session. The Phase 2 scenario was one of advising a professional relocating her family in Europe. Seventeen students were involved, all but 2 having also completed Phase 1. In this Phase each group was briefed that one of its number was a chatbot, and that once a participant had a good idea of who they thought the chatbot was they should either declare it to the group or in private chat to the facilitator. Due to a number of constraints it was not possible to run groups with no chatbot as a control. Since the task was more straightforward, and given that its main focus was on the un-masking of the chatbot Phase 2 typically only ran for 1 or 2 sessions.

Since the sessions were held in later afternoon/early evening the students dialled in remotely for the sessions, often from home or even from mobile devices.

A chatbot was added to each group. Its presence was not declared at all in Phase 1, and in Phase 2 its presence, but not identity, was declared. The chatbot was programmed to participate in the text chat discussion just as though it was an ordinary student. The chatbot was always in the group even if, due to failure of some participants to attend, there was only one human student in the group. The facilitators were aware of both the presence and the identity of the chatbot since the intention was to perform a "hot-debrief" when the presence of the chatbot was detected.

Given that the experiment required a deception to be maintained from first advertising the experiment to the completion of Phase 1, and would require students to be giving up their own time to participate, the team was careful to ensure that the experiment gained all the necessary ethical clearances from the University. In addition, since the sessions were straight-forward PBL sessions which were designed to contribute to the students learning, although not as part of their formal course (and completely optional), it was agreed that even if the chatbot were unmasked in the first minutes of the first session then the students would still be given the opportunity to continue with the whole 3 × 1 h exercise, with or without the chatbot, so as to get the benefits of the exposure to the on-line PBL method.

3.2 Technical Design

Technically the chatbot was again implemented using Daden's Discourse system. It is notable that the Discourse system is nothing special. It is a commercially proven chatbot—having been used in training applications and as a virtual librarian and student support agent—but it is based fundamentally on the AIML model, although extended to make it easier to add keyword spotting, to use synonyms and to define and track context. It still uses a pattern matching/response template approach with no element of machine learning or grammatical analysis.

Discourse was implemented using C# and ASP.NET with a SQLLite database for the Worcester Experiment (a Perl version using text files had been used for the Loyola Experiment). A simple forms based editor was provided to the chatbot author so that they did not need to know the XML markup of AIML.

The chatroom environment was based around industry standard software—so that it appeared no different to the participants than a "real" chat room. A locally hosted server running chat room software (in this case Prosody) controlled the chatrooms, and the students accessed the chat rooms using standard chat room clients (principally Psi client, although some other clients were also used). The Prosody software used the Extensible Messaging and Presence Protocol (XMPP) to communicate with the clients. The chatbot was interfaced with the Prosody using the same XMPP protocol—so it technically appeared no different to a human user.

The chatbot's interface to the XMPP protocol was implemented in a bespoke software element, termed the "humaniser". This took an XMPP message from a human participant (via the server) and passed it to the Discourse chatbot as input, and could optionally filter the input so that the chatbot only saw 1/N messages—it not being human to respond to every single message. The humaniser would, though, always pass the messages from the facilitator to the chatbot.

Once the chatbot had created a response to the message this was passed back to the humaniser where a number of functions were performed:

- Spelling errors were randomly introduced, but using typical error models
- Sentences were split onto multiple lines
- The message was delayed based on both length of question and of reply
- Tags for system items such as "Speaker's name" were replaced by their values for the current session/message

The message was then sent using the XMPP protocol to the Prosody server to appear in the chat room.

3.3 Principal Results: Phase 1—Covert Turing

During Phase 1 a total of 42 subjects took part in 17 groups, between them logging a total of around 88 participant-hours of chat with the chatbot (e.g. 4 humans in one group with a Chabot talking for 1 h = 4 participant-hours).

At no point during Phase 1 did any participant raise with the group or facilitator the fact that they thought that one of the participants was a computer, or even a suspicion that all was not as it seemed.

At the end of the last session for each group all participants were sent a survey to complete (which included a peer feedback survey), and 8 participants were randomly selected for a semi-structured interview. In addition to the survey and interview the chat logs from the session were also analysed to provide additional data.

Sample responses from the interviews after the presence of the bot had been revealed as to why the bot was not detected included:

"I wouldn't have expected there to be a bot present"

"There are people who like to repeat themselves and ignore others, so it's not so easy being able to tell. Especially through an online chat"

"Joined in conversation very well, responding appropriately and so on, as well as seeming to have done the homework research."

"because some of his questions were quite interesting and got us onto some very good talking points"

"the person contributed a lot, with varied responses"

"Nikos (the chatbot) was probably one of the other better people to actually get involved with and we had some good discussions. I think about it and from what we said he had quite a similar opinion to me"

From the interviews, surveys and analysis of the logs for Phase 1 there were 3 key findings of significance:

- Whilst the responses given by the computer were logically correct, and often relevant, the language used betrayed the fact that they did not come from someone from the same cohort as the human participants (see below). Whilst this would not always be necessary, in situations where the chatbot is trying to portray a particular character then how it says things is as important as what it says. As one student said, it sounded more like a tutor masquerading as a student.
- The chatbot was typically identified as being the least engaged, helpful or co-operative member of the group. However, its behaviour was seen as being within the bounds of "normal" human behaviour, some participants assuming (or rationalizing) that the chatbot was either a foreign student with poor English or someone with language difficulties. Whilst such a judgement may not be much for a human it does perhaps represent a significant milestone for a computer.
- Following on from this although the bot was successful at hiding its presence (or at least not giving away its true nature), it was generally unsuccessful at establishing any credibility or legitimacy within the group. As one student said "*I just thought it was a git... They knew what they were talking about and didn't bother talking to us.*". Something is legitimate if it is in accord with the norms, values, and practices accepted by a group [12]. Presenting legitimating behavior can facilitate community responsiveness [13] and enable newcomer integration. This suggests that if we now want to move beyond simply "hiding" the presence of chatbot we need to consider the research around legitimacy in the online community space in order that the chatbot can gain acceptance and legitimacy within the group.

It was notable that none of the participants felt aggrieved by the deception. Indeed, most had a very positive attitude to the experience and found the idea of a covert Turing Test fascinating. As one participant said:

"It was a surprise. I knew there was something not quite right with this person but I didn't really think much of it ... and it wasn't until later that, of course, I realised it was the AI and I thought... I didn't feel deceived. It was more like a jigsaw piece and you've just handed me the last piece in the puzzle. I think, ah, now everything makes sense."

3.4 Principal Results: Phase 2—Overt Turing

During Phase 2 a total of 17 subjects took part in 7 groups, between them logging a total of 28 participant-hours of chat with the chatbot. Each group was told that a chatbot was present, but not its identity (an overt Turing Test). They were asked to identify the chatbot once they thought they knew which participant it was, either publicly in the group discussion or by private message to the facilitator.

Only three participants identified the chatbot and informed the facilitator proactively during the Phase 2 sessions (one incorrectly). The remaining 14 participants correctly identified the chatbot when completing the questionnaire after the Phase 2 sessions ended.

At the end of the last session for each group all participants were sent a survey to complete, and 10 participants were randomly selected for a semi-structured interview. In addition to the survey and interview the chat logs from the session were also analysed to provide additional data.

Sample responses during the Phase 2 debrief when asked what the factors were that made then uncertain about the chatbot's identity included:

"It responded directly to my point about dyslexia and wasn't particularly off topic at any point. I thought it was at the start, but then got more believable as it went on"

"Seemed to make more relevant points than last time I did a PBL session with the bot."

"Misspelling "cyslexia" and referring back to something I had said."

Many of the participants claimed to have been deceived several times by the chatbot because of the kinds of statements it gave, i.e. perceptions of it probing earlier comments or referring to a past comment made by a human.

The analysis for Phase 2 focused on the 'tells' that the students used to positively identify the chatbot. These included:

- Excessive repetition
- No opinions relating to the topic being discussed
- Not effectively directing questions at individuals
- Not saying things related to what the previous person said
- Not referring much to what other people were saying
- Not picking up changes in conversational moves
- Delivering confused utterances
- Providing inconsistent responses.

However, responses were viewed as being factually correct and not particularly evasive. It should also be noted that there were also many criteria that were important to some participants whilst being less important for others. For instance, "spelling mistakes" and "use of names" appear on both lists—suggesting that sensitivity to specific tells may be a very personal attribute.

3.5 Technical Analysis of the Worcester Experiment

Whilst Savin-Baden [11] expands on much of the analysis above it does not provide the technical analysis of the performance of the chatbot which was used in the experiment.

3.5.1 Cases

Knowledge within an AIML chatbot is defined as a set of pattern-response pairs called cases. Chatbots entered into the Loebner Prize (an overt Turing Test) can have tens or even hundreds of thousands of cases in their database—ALICE has 120,000 [14].

To develop the cases for the Worcester experiment the research team completed the same on-line PBL scenarios as the students, with the same facilitators, and over the same time period (3 × 1 h sessions over a week). From these sessions and general discussion a mind-map was created to define all the areas in which the chatbot was likely to require responses.

The cases were then written either using replies developed by the 2 person authoring team, or manually extracted from the chat logs of the internal session. As noted above one of the downsides of this was that the chatbot had the linguistic style and vocabulary of its primary author (white English, male, 50s) and the internal session participants (mainly white English, mainly male, 20–50 s), rather than that of the study participants (college age, predominantly female, mix of races).

Table 1 shows the distribution of cases created by topic areas for Phase 1. No general knowledge cases were loaded, so the chatbot had only 410 cases available in Phase 1—encoded as 748 Discourse patterns, and only 309 cases in Phase 2. 187 cases were common between the two phases. It should be noted that the flexible

Table 1 Distribution of cases (15 topics with less than 5 cases omitted)

Content/Discussion topic	Number of cases	Content/Discussion topic	Number of cases
Default responses to generic questions	50*	Scenario needs	8
Ethical frameworks	14	Other discussions	9*
General factors	15	How to solve the problem	12
Drinking	6	Information about the bot's course	7*
Family	8	Information about the bot (its legend)	8*
Friends	7	The solution	9
Money	11	Default responses common openings (e.g. open/closed questions)	79*
Subject's personality	6	Who's doing which task	12
Police	5	Responses to key generic phrases	5
Rent	7	Test cases	5*
Trust	9	Redirects to bot common responses	11
University	7	Time sensitive statements	11*
Forming bad habits	6	Bad peer groups	5

* = common in Phase 1 and 2

structure of a Discourse pattern means that it can achieve in one pattern what may require multiple cases of pure AIML to achieve, but the equivalence in reality is unlikely to be more than 1:3 or so.

The usual approach with commercial chatbots is to go through a period of iterative development and improvement (so called convologging) both prior to and after live. However, in order to have a consistent set of cases for the experiment we undertook only some minor improvements to the cases after the first student group of each Phase, and left the cases alone for the remainder of the Phase.

3.5.2 Case Usage

Given the relatively small number of cases available to the chatbot it is interesting that an even smaller number were actually used.

In Phase 1 with 410 cases available only 166 cases were used (40 %) during the total of 3442 exchanges between a participant (student or facilitator) and the chatbot. Of these exchanges:

- 9 % were handled by problem specific patterns
- 7 % were handled by generic "stub patterns" (e.g. "do you think….")
- 32 % handled by simple defaults (responses to "yes", "no" etc.)
- 50 % were handled by the catch all wildcard (*) responses—which usually results in the bot trying to restart the discussion around a new topic, or make some non-committal utterance.

In Phase 2 with 309 cases available only 83 cases were used (27 %) during the 937 exchanges.

- 19 % handled by problem specific patterns
- 10 % handled by generic "stub patterns"
- 25 % handled by defaults (yes, no etc.)
- 44 % handled by the catch-all wildcard responses

Despite the significantly smaller pool of cases used the general feeling of the participants was that the Phase 2 chatbot performed better than the Phase 1 chatbot.

3.5.3 Technical Failings

Whilst the chatbot suffered no technical glitches during any session there were elements of poor authoring that rapidly became apparent, during Phase 1 in particular. In a normal deployment such authoring would be rapidly corrected as part of the convologging process, but since a consistent case database was needed for the experiments this was not feasible. The errors included:

- Allowing some default responses to be triggered a large number of times (e.g. "Sorry") rather than giving more specific or varied responses

- Handling facilitator responses directed specifically to the chatbot with a narrower range of cases than the student conversation (since it was assumed that a direct comment from the facilitator would be of a different nature—but it wasn't, so it often got handled by the wildcard response)

These errors featured significantly in the participants' comments about poor chatbot performance, and so an even better performance could be expected had these errors been fixed on the fly, or the chatbot system enhanced to reduce repetition more automatically.

3.5.4 Technical Conclusions

Whilst accepting that the chatbot was operating within a fairly constrained environment/dialog space, and that this was a covert Turing Test, the fact that the chatbot managed to avoid detection using only a "standard" chatbot engine and only a hundred or so cases suggests that creating an effective chatbot is probably more a task of finesse than of brute force.

The results suggest that to be successful at escaping detection then the bot must:

- Not give itself away through technical faults/authoring errors that generate "computer error" type responses
- Be able to stay reasonably on topic and make statements that are not too evasive
- Be able to secure enough "lucky hits" where what it says sounds so human that any doubts that a user has that it may be a bot are (at least temporarily) allayed.

As has been noted this is not enough to give the chatbot legitimacy, but it may be enough to stop it from being identified. It is interesting to compare this analysis to that of Shah [15] where the strategy of silence for a chatbot is considered, and certainly in creating wildcard responses for the bot (and the sampling of exchanges that was taking place) then having a silence response rather than risking making an erroneous responses was certainly one of the strategies employed.

Our conclusion is therefore that a relatively small number of carefully crafted responses tuned to the situation may be a more effective strategy for chatbot creation than creating ever greater databases of responses and more complex dialogue engines.

4 Revisiting the Original Hypothesis

One of the key research questions presented at the end of Burden [1] was *"To what extent do the Covert/Overt and Singleton/Group options present earlier opportunities to pass the Turing Test (or equivalent) in a virtual world?"*. Although the Worcester experiment was conducted within an on-line discussion group rather than in a virtual world, its results, along with that of the Loyola Marymount test that was conducted in a virtual world do allow us to start to answer this question.

Analysing the results of a conventional Turing Test (Singleton Overt) in the form of the Loebner Prize [4] there has only been one instance in the 13 year history of the prize were a chatbot has fooled at least one (and indeed only one) judge. This was Suzette by Bruce Wilcox in 2012 The chatbot achieved a deception rate of 25 % on the basis that there were 4 judges for that one experiment and one was deceived (but given that Wilcox has entered 5 more times since with Angela, Rose and Rosette but failed to fool any judges it could also be taken as a $1/(7 \times 4) = 3.6\,\%$ deception rate across all his Loebner Prize attempts). By contrast the Loyola Marymount experiment (Singleton Covert) achieved a 78 % deception rate. The Worcester Phase 1 experiment (Group Covert) achieved a 100 % deception rate, although the Phase 2 experiment (Group Overt) achieved only a 6 % deception rate (though the reticence of most students to identify the chatbot during the session, and their comments about its improved quality, suggests that a high level of doubt may have existed).

The results would therefore appear to bear out the original hypothesis that the Singleton Covert and Group Covert conditions can indeed provide an earlier opportunity to pass the Turing Test than the more traditional Singleton Overt case. Further covert singleton and covert group tests (and indeed overt group tests) should be conducted to confirm this.

5 Further Work

There are a number of areas were further work would be useful to improve chatbot performance and to produce a more solid set of results.

In the case of experimental design a significant improvement to the Worcester experiment would be to not have the facilitator aware of the chatbot identity. In the experiment the two principal investigators from Worcester were the facilitators since they best understood how to facilitate on-line PBL sessions, as well as understanding what the experiment was trying to achieve. Whilst the facilitator did not try to hide the identity of the bot, their interactions with it may have been unintentionally biased.

It would be interesting to compare the performance of a machine-learning based chatbot with that of the Discourse chatbot. The conversation logs could provide some training material for such a bot.

Both the experiments were conducted in quite constrained environments (in time, topic and technology), and with a correspondingly constrained expected dialogue. Relaxing those constraints whilst still trying to maintain the high deception rates would provide a useful challenge. The most beneficial constraint for the chatbot is likely to be that of topic. Technology interfaces can be readily added. Longer individual sessions are likely to result in all participants getting stale, and more sessions would perforce need new topics, or deeper analysis of the current ones—and broadening the chatbot to cope with more topics is the real challenge.

Given that the chatbot in the Worcester Experiment was successful in the deception it suggests that as well as looking at how to maintain that deception in a less constrained environment we should also look at how to move beyond deception and focus as well on how to establish and maintain the legitimacy of the chatbot within the group (and indeed in the Singleton test). A good grounding for this would be a firm understanding of how humans achieve this legitimacy, and then apply that to the chatbot.

A chatbot which could achieve legitimacy whilst maintaining the deception, and operate within a relatively unconstrained environment would be a significant achievement.

References

1. Burden, D.J.: Deploying embodied AI into virtual worlds. Knowl.-Based Syst. **22**(7), 540–544 (2009)
2. Wakefield, J.: Intelligent Machines: chatting with the bots. BBC Web Site. http://www.bbc.co.uk/news/technology-33825358 (2015). Accessed 30 May 2016
3. Turing, A.M.: Computing machinery and intelligence. Mind **59**, 433–460 (1950)
4. Bradeško, L., Mladenić, D.: A survey of chatbot systems through a loebner prize competition. In: Proceedings of Slovenian Language Technologies Society Eighth Conference of Language Technologies, pp. 34–37 (2012)
5. Cooper, S.B., Van Leeuwen, J. (eds.) Alan Turing: His Work and Impact. Elsevier (2013)
6. Shieber, S.M.: Lessons from a restricted Turing test (1994). arXiv:cmp-lg/9404002
7. Mauldin, M.L.: Chatterbots, tinymuds, and the turing test: entering the Loebner prize competition. AAAI **94**, 16–21 (1994)
8. Heiser, J.F., Colby, K.M., Faught, W.S., Parkison, R.C.: Can psychiatrists distinguish a computer simulation of paranoia from the real thing?: The limitations of turing-like tests as measures of the adequacy of simulations. J. Psychiatr. Res. **15**(3), 149–162 (1979)
9. Kurzweil, R.: Why we can be confident of Turing test capability within a quarter century. In: The Dartmouth Artificial Intelligence Conference: The next 50 Years, Hanover, NH (2006)
10. Gilbert, R.L., Forney, A.: Can avatars pass the Turing test? Intelligent agent perception in a 3D virtual environment. Int. J. Hum.-Comput. Stud. **73**, 30–36 (2015)
11. Savin-Baden, M., Bhakta, R., Burden, D.: Cyber Enigmas? passive detection and pedagogical agents: can students spot the fake? In: Proceedings of Networked Learning Conference (2012)
12. Zelditch, M.: 2 theories of legitimacy in the psychology of legitimacy: emerging perspectives on ideology, justice, and intergroup relations, pp. 33–53. Cambridge University Press (2001)
13. Burke, M., Joyce, E., Kim, T., Anand, V., Kraut, R.: Introductions and requests: rhetorical strategies that elicit response in online communities. In: Communities and Technologies 2007, pp. 21–39. Springer, London (2007)
14. Wilcox, B., Suzette, W.S.: The most human computer. http://chatscript.sourceforge.net/Documentation/Suzette_The_Most_Human_Computer.pdf (2016). Accessed 30 May 2016
15. Warwick, K., Shah, H.: Taking the fifth amendment in Turing's imitation game. J. Exp. Theor. Artif. Intell. 1–11 (2016)

Context-Dependent Pattern Simplification by Extracting Context-Free Floating Qualifiers

M.J. Wheatman

Abstract Qualification may occur anywhere within a temporal utterance. To reduce the ensuing pattern complexity for context-dependent systems such as *Enguage*™, it is necessary to remove the qualified value from the utterance; rendering the utterance atemporal and presenting the value as the contextual variable *when*. This is possible because a qualifier—*at* 7:30 or *until* today—is immediately recognisable as such if preceding a time value: *when* is context-free. This appropriation gives insight into the nature of the context-dependent processing of habitual natural language. While the difference between the resultant concepts—*how many coffees do I have* and *how old am I*—is perhaps not that great despite their differing origins, this work ensures the mediation system remains practical and effective. This research is informed by a prototype for the health-tech app Memrica Prompt in support of independent living for people with early stage dementia.

1 Introduction

Enguage™ [1] is an effective and practical open source Java class [2] for utterance understanding—or *mediation*—on mobile devices. Broadly, it follows Austin's approach [3]: natural language is context-dependent and is understood through outcome, both felicitousness and its perlocutionary effect. While there is a surfeit of chatbots maintaining conversations, fuelled by the Loebner Prize [4], Enguage is directed at utility—it is used as the user interface for the showcase app iNeed [5]. Whereas chatbots, such as ELIZA [6], are simply keyword based, or key-phrase based in the case of A.L.I.C.E. [7], Enguage models context dependency. Natural language has to be understood in natural language: it is autopoietic—self-describing; its

M.J. Wheatman (✉)
Withinreap Barn, Moss Side Lane, Thornley PR3 2ND, UK
e-mail: martin@wheatman.net

M. Bramer and M. Petridis (eds.), *Research and Development in Intelligent Systems XXXIII*, DOI 10.1007/978-3-319-47175-4_14

analysis is achieved through further utterances; and, most importantly to this paper, the whole utterance must be matched *exactly*. Context-dependency is further discussed in Sect. 4.

Enguage is effective in that it uses micro-interpreters—*signs*—as units of interpretation to effect a deep and subtle understanding [8]. Each sign contains a pattern with which to match an utterance, e.g. *i need X*, and a monad of utterances—a *train-of-thought*—to analyse a match. A disambiguation mechanism [9] helps to distinguish between trains-of-thought. Therefore, mediation is the process of determining the most appropriate sign with which to interpret an utterance, supported by an array of contextual attributes—*variables*—from the matched pattern.

Enguage is practical in that it is based around concepts. Enguage automatically loads each *repertoire*—the set of utterances which address a concept—which matches any word in an utterance, and keeps it in memory while in use. This allows, for example, *and another* to be in the *need* repertoire [10], although the word *need* is not in this utterance, because it requires the prior use of *need* to effectively prime its understanding e.g. *I need a coffee* must be uttered before *and another* can be meaningfully uttered. Thus, autoloading means that an unlimited number of repertoires can co-exist.

However, the introduction of time qualification variables in the *meeting* repertoire [10], developed as a prototype for the Memrica Prompt app [11], presents a problem: the proliferation of patterns.

I am meeting my brother at 7:30. *At 7:30 I am meeting my brother.*

These utterances effectively have the same meaning, however, each requires a different pattern. The number of patterns required grows quickly: this *floating* time qualifier can appear anywhere in the utterance (other than perhaps within the phrase *my brother).* Further, they could be replaced by, or combined with, a date qualifier, also floating, or any combination of the two: *tomorrow I am meeting my brother at 7:30.*

The key is that the qualifier preceding the time value is immediately recognisable as such wherever it occurs: time, date and duration qualifiers have a context-free meaning. This paper describes pre-processing utterances for time and date qualifiers, in appropriate situations. Section 2 describes the *when* class, with Sect. 3 describing the syntax used. While context-free interpreters are a standard mechanism, it also illustrates the context-dependent nature of Enguage, described in Sect. 4. Removing floating qualifiers, which will also includes spatiality, from the pattern returns Enguage to being a practical mechanism.

2 The When Class

The solution to pattern proliferation due to floating qualifiers is to remove them, leaving an atemporal utterance. This section describes the class *when*, which models temporal points, boundaries and durations. It has *fromText* and *toText* methods to interface with the user, and *toValue* and *fromValue* methods for internal representation, and reasoning, as a named value—an attribute.

A *moment* value is a decimally coded timestamp represented as a primitive long, e.g. *10:30 last Christmas* may be represented as 20151225103000. This allows simple numeric comparison, and easy recognition when printed. The Moment class supports this representation, with methods supporting the absolute form, as above, and the relative forms, such as *last Thursday*. Where a portion of the date is not given, the out-of-band component values of 88 are inserted to maintain its place-value nature. This gives some flexibility in support of odd dates such as the 35th *of May* [12], while keeping it as a primitive long.

A *duration* is composed of two moments: *from* and *to*. It is thus represented as an absolute values, for example 20151225103000-20160101103000. The Duration class supports this, and also a relative value, again represented as decimally coded long, which is the difference between the two. An instant is *from* with an identical *to*. For simplicity, examples of negative durations, e.g. *I lived at home until 2002 and then moved back in 2006,* are not detailed.

The class *when* therefore supports a duration, with a factory method to extract floating qualifiers from an utterance: modifying the text and returning the When object. For this to occur, for such sense to be felicitously made of an utterance, mediation has to be configured to declare the repertoire as temporal. How this is achieved is as follows.

A *temporal* repertoire is presented [10], which is in its entirety:

On "X is temporal", perform "temporal add X".

This contextualisation occurs while loading the repertoire; the first utterance in the *meeting* repertoire is *meeting is temporal.* The temporal property is implemented per sign. The user can then utter: *I am meeting my bother at 7:30*; which, when matched in the meeting repertoire, becomes the atemporal utterance and variable: *I am meeting my brother* and WHEN='88888888073000' This then matches the pattern *i am meeting PHRASE-X.* As with other repertoires, first person personal pronoun phrases are translated into third person phrases, using the default person _user.

Thus, the objective pattern *SUBJECT is meeting PHRASE-WHOM* is matched, with the contextual values [SUBJECT="_user", WHOM='my brother', WHEN= '88888888073000']. The identification of a time qualifier is possible because the concept (repertoire) has been deemed temporal, and the value is self-evident (context-free) through a keyword, in this case *at*, and a time value. The supported syntax is as follows.

3 A Context-Free Representation of the *When* Qualifier

The identification of the floating time qualifier is deemed to be context-free because, unlike matching with patterns as in the rest of Enguage (see below), the temporal asset can be represented by a fixed syntax. This basically means that there are enough keywords in the utterance to provide a framework for a numeric, or textual, time. An approximation to the meta-syntax used to interpret temporal phrases is represented here:

WHEN ::= {WHENCOMPONENTS}
WHENCOMPONENTS ::= 'at' TIME
| 'on' ABSDAY
| RELDAY
| 'in' [MONTH] YEAR
| FROM | UNTIL
ABSDATE ::= 'the' NUMBER ['st' | 'nd' | 'rd' | 'th']
[| 'of' MONTH [| NUMBER]]
FROM ::= from MOMENT
UNTIL ::= [until | to] MOMENT
MOMENT ::= ABSDATE | RELDAY
RELDAY ::= [| SHIFT] [
'yesterday' ['morning' | 'afternoon' | 'evening'] | 'last night'
| 'today' | 'this' ['morning' | 'afternoon' | 'evening'] | 'tonight'
| 'tomorrow' ['morning' | 'afternoon' | 'evening' | 'night']
| ['on' | 'this' | 'last' | 'next'] WEEKDAY]
WEEKDAY ::= 'Monday' | … | 'Sunday'
MONTH ::= 'January' | … | 'December'
SHIFT ::= NUMBER ['day' | 'week' | 'month' | 'year' | 'century' | 'millennium']
['ago']
TIME ::= HOUR [MINUTES] ['am' | 'pm']
HOUR ::= NUMBER
MINUTES ::= NUMBER

This is straightforward, but it presents Enguage with a problem: *at 7:30* does have an objective meaning in a temporal sense, but until now Enguage has only been presented itself as a context-dependent approach. Should floating qualifiers always be removed—is the temporal repertoire superfluous? Can we be certain that Enguage really is a context-dependent system?

4 Engauge: Context Dependent System

The difficulty in understanding context dependent utterances can be illustrated by considering the phrases: *the Jumbo has landed*—an everyday occurrence concerning flight; and, *the Eagle has landed*—one of the most iconic phrases of the C20th. Syntax cannot determine meaning—both sentences have the same structure, but their context is very different. The first sentence has a variable meaning in that *Jumbo* could be exchanged with any other aircraft name to be useful; whereas, the second only has a meaning in its entirety—it has a certain wholeness, but is still open to ambiguity. In fact the two sentences overlap: the second has two meanings.

A second example is found in the *need* repertoire. The sign *i need X* can be countered by the sign *i have X* in registering and relinquishing a need. However, the sentence *I need to go to town* has the same meaning—intentional thrust—as *I have to go to town*. Again, we see that syntax is inadequate to capture meaning unambiguously, because intention is dependent upon the words used. A context dependent system, such as Enguage, must match the entire utterance, rather than simply types of utterances which is the case with traditional language analysis (e.g. syntax trees); or, partial matches in the case of keywords or phrases with chatbots. This is achieved by a simple pattern matching algorithm, which matches against fixed boilerplate and variable hotspots, and is described below.

In comparison with the context-free description in Sect. 3, the pattern matching algorithm of Enguage is relatively simple: it has a flat structure consisting of a list of a constant (boilerplate) and variable (hotspot) pairs, which is the same for all repertoires. By convention, constant is given in lower case and variables in upper case, e.g. *i need X*. The meta-syntax of which is:

```
PATTERN ::= {CONST VAR}
CONST ::= { STRING }
VAR ::= [ | [ 'NUMERIC-'
          | 'PHRASE-'
          | 'ABSTRACT-'
          | 'QUOTED-'
          | 'SINGULAR-'
          | 'PLURAL-' ] STRING [ CONST ]]
```

This is the primary stage of applying an interpretation to an utterance—the mediation of a sign. There are no keywords, Enguage maps an utterance—an arbitrary list of strings, given here as "..."—onto another utterance by enacting a list of utterances and calls to a persistent memory management layer.

"..." → {"..." | perform <class><method><object>[<attr>]...} → "..."

The construction of such signs, two examples from the *meeting* repertoire which are given below, are supported by the autopoietic repertoire [13].

On "SUBJECT is meeting PHRASE-WHOM":
set output format to ",at WHERE,WHEN";
perform "list exists SUBJECT meeting WHOM";
reply "I know";
if not, perform "list add SUBJECT meeting WHOM";
then, reply "ok, SUBJECT is meeting ...".
On "I am meeting PHRASE-WHOM", _user is meeting WHOM.

This mapping suggests Austin's triadic model, developed by Searle [14], of: locution—utterance; illocution—action; and, perlocution—reaction. It also reflects an underlying triadic framework of Informatics [15–17]. In brief, the micro-interpreter in its role within a repertoire is, as indicated, a Peircean Sign—that which *stands to somebody for something in some respect or capacity* [18, 2.228]; the train-of-thought is *Interpretant* that which is created in the mind of the interpreter [*ibid*]; meaning, as reference to objects, is: *Iconic*—by resemblance; *Indexical*—as a pointer; or, *Symbolic*—by convention. In this case, it is Symbolic, where phrases are merely stored and replayed to effect meaning in the user [19]. This analysis supports many other analogies such as *representamen,* and *immediate* and *dynamic* objects. The reduction to a context-free analysis is removing the Symbolic nature of utterances to their underlying Iconic and Indexical states. This is similar to the obfuscation of source code—we can still see much in it, such as that it is source code (an iconic reference), and its apparent complexity (an indexical reference); even without any apparent Symbolic meaning, it still acts as source code. The values, such as 7:30, would be apparent.

Boilerplate defines a context upon which to frame variables, such as *i need PHRASE-X*. Typically, boilerplate is composed of stop-list words—normally removed from internet search queries—and instances of the repertoire name, e.g. *need* or *meeting*. The matching of boilerplate and correct association of prefixes (e.g. *NUMERIC-QUANTITY* which must be numeric, either numerals or implicit numerical values such as *a* to mean 1), determines what variables are presented to the train-of-thought: if this utterance and pattern match, then these are the corresponding values. Thus, while there are no fixed keywords, boilerplate acts as a context-dependent key-phrases. Variables are initialised from the matched pattern, either directly as *i need X*, or indirectly through the train-of-thought; they can also be set explicitly by the train-of-thought, by performing *variable set OBJECT coffee*. An example of indirect setting is as follows: *i need a coffee* matches *i need X*, triggering the objective thought with the default person: *_user needs a coffee;* which matches *SUBJECT needs NUMERIC-QUANTITY OBJECT,* resulting in quantity being set: [OBJECT="coffee", SUBJECT="_user", QUANTITY="1"].

Variables, in particular the phrase variable—with its deferred meaning—was a key driving force behind this project. One criticism of a relativistic approach, of deferred meaning, is that if something can mean anything, it means nothing [20]: once the chatbot ELIZA [6] has found a keyword, for example, it will simply output a given reply regurgitating any input required, simply translating *my* to *your*, etc. However, in a context-dependent system, such as Enguage, this proves highly useful: the meaning of phrases remain with the user, which is orthogonal to the meaning of the concept. When applying the *i need X* interpretation to the *I need a coffee* utterance, the notion of need is orthogonal to the items needed. Thus, the system of repertoires represent context-dependency. The meaning is in the *need* at the heart of the repertoire in boilerplate. The meaning is in the concept supported by the repertoire.

The intentional meaning of a concept is completed by the analysis in the train-of-thought. This is susceptible to disambiguation if it uniquely identifies the corresponding action. A built-in mechanism in Enguage to allow the action to be rolled back and another performed, represented by *no, PHRASE-X*. If *X* is equal to the previous utterance, then the disambiguation mechanism is invoked [9]. Thus the appropriate concept can be reached.

5 Discussion

A floating qualifier must be extracted as a context-free value to minimise the number of patterns. However, other values are already supported in Enguage the form of the *numeric* variable, but these do not float. For example, *I need coffee*, *I need a coffee* and *I need a cup of coffee* merely requires three signs in the repertoire, respectively: *i need PHRASE-X*, *i need NUMERIC-QUANTITY PHRASE-X*; and *i need NUMERIC-QUANTITY UNIT of PHRASE-X*. The *numeric* class, itself, has the same architecture as *when*: an objective value which is implemented with a from and to *text* methods, and a to and from *value* method to be represented as a variable internally, in this case as: an absolute or relative—so as to model *ten* or *another;* signed—so as to model more or fewer; fraction—typically with a denominator of 1. While these three signs sufficiently cover the numeric quantities in *need*, numerics are not context-free because they are not accompanied by a keyword.

Further, the objective value of these variables enable the train-of-thought to reason with variables. The numeric has allowed a simple spoken calculator to be created, e.g. *what is 1 plus 2 all squared* [1]. This also allows utterances such as i *need 1 plus 2 coffees* to be correctly evaluated. Thus, with *when* given a date of birth we can appropriate age. We could rightly expect our system to respond to meaningful when based utterances: *i was born on the 23rd of August, 1968*; *how old am i*; and simply *how many days to Christmas*.

Used in concert with the extraction of floating qualifiers, there is at least a potential issue with the unintended extraction from non-temporal utterances extraction of non-temporal utterance. This is dealt with by the use of the temporal repertoire, as described. However, there may be other issues. For this argument it is assumed *i am meeting my brother at 7 11* is a simple ambiguous example where the temporal qualifier might indeed be referring to a place—the convenience shop—not a time. In practice, this might not fail because the speech-to-text layer upon which the Enguage library depends, does make some remarkable conversions into valid text—times, for example, are given in *hh:mm* format. This hypothetical example is therefore platform dependent. Another potential problem may occur where there is more than one temporal qualifier *i am meeting my brother at 7 11 at 7:30*. This would, however, be dealt with if the spatial floating qualifier is implemented. There are other existing strategies to deal with this particular case: by recognising 711 as a sign in itself; there is a preprocessing repertoire *colloquia* [10] which supports the sign *when user says X user means Y*, which may be used to contextualise 711 as *the convenience store 7 11*, i.e. which would break the match as a temporal qualifier.

Finally, the many issues raised by the contextualisation of mediation are all dealt with by the disambiguation mechanism [9], leading to the claim that Enguage remains effective. Initially, it was thought that disambiguation would need to refute the temporal claim, and re-searching the sign-base from the current sign, reasoning that this sign may have matched in the presence of the extracted value. However, this has not been pursued. The disambiguation mechanism has been found to work well in practice: a different utterance is often used. It is simply not natural to repeat an utterance, if it has already been misunderstood.

6 Conclusion

Enguage, as a mediation system, follows a naïve cognitive model of accepting everything that is successfully understood, with the proviso that disambiguation will allow the user to access the appropriate interpretation. Meaning can be seen at two levels—user references and shared concepts. The programming of Engauge is in repertoires composed, themselves, of written natural language statements—the autopoietic repertoire. Matching patterns and post-match interpretation is informed by the discrete values extracted from the match, which now includes those pre-processed as floating qualifiers.

This paper documents the extraction of context-free values from arbitrary arrays of strings. Without the need to map floating qualifiers into the sign pattern, Enguage remains practical. It also argues that Engauge is a context-dependent system and describes the nature of that dependency—whole utterance matches and structure defined per sign. The *when* example also shows that mediation contextualisation is also achieved through the temporal concept. The contextualisation of a repertoire as temporal is not yet suspended by the disambiguation process, because the user often resolves the ambiguous situations themselves through the disambiguation process, see [9]. With such disambiguation, Enguage can also be deemed to be remaining effective.

Enguage is a context-dependent system with context-free attributes. Support for concepts such as the context-free *how old am i* and the context-dependent *what do i need* or indeed *how many coffees do i need*, are of similar worth, but are provided by very different mechanisms. One analogy suggested is that the context-free route is like acquiring a primitive type, where as the context-dependent is supporting a class—it models a concept. This mediation process supports the pan-semiotic view that everything is a sign. A triadic information model has been alluded to, but not relied on as a main argument.

The class *when* has been implemented in version 1.7 of Enguage, which is in the showcase iNeed app [5] and the GitHub repository [2]. Further work includes investigating if the temporal repertoire is necessary; the development of spatiality—a further class—to support *I am meeting my brother at the pub*. This necessitates the redevelopment of the object model as outlined in [8], which is not yet implemented in the mobile version, but is due in Autumn 2016. Finally, modelling context-dependency creates a vocal user interface, but does not approach full textual understanding, because it is based on habitual utterances—one of the last great issues for computing [21].

References

1. Wheatman, M.J.: iNeed, Version 1.5—context and arithmetic demo (2015). https://www.youtube.com/watch?v=HsJyrdtk0GM, Accessed 25 May 2016
2. Wheatman, M.J.: GitHub Enguage repository. https://github.com/martinwheatman/Enguage, Accessed 25 May 2016

3. Austin, J.L.: In: Urmson, J.O., Sbisa, M. (eds.) How to Do Things with Words. Oxford University Press (1962)
4. Loebner, H.G.: In Response. Commun. ACM **37**(6) (1994)
5. Wheatman, M.J.: I Need: a Vocal Shopping List. http://bit.ly/1fIrEdZ (2014). Accessed April 2014
6. Weizenbaum, J.: ELIZA–a computer program for the study of natural language communication between man and machine. Commun. ACM **9**(1), 36–45 (1966)
7. Wallace, R.S.: The anatomy of A.L.I.C.E. http://www.alicebot.org/anatomy.html (2016). Accessed 16 July 2016
8. Wheatman, M.J.: A semiotic analysis of if we are holding hands, whose hand am I holding. J. Comput. Inf. Technol. **22**, LIS 2013. http://cit.srce.unizg.hr/index.php/CIT/article/view/2278/1658 (2014)
9. Wheatman, M.J.: A pragmatic approach to disambiguation. In: Proceedings of the 17th International Conference on Informatics and Semiotics in Organisations, Campinas, 1–3 Aug 2016, Brazil (2016)
10. Wheatman, M.J.: All repertoires including need, meeting, temporal and colloquia (2016). https://github.com/martinwheatman/Enguage/tree/master/iNeed/assets. Accessed 25 May 2016
11. Memrica Ltd. Memrica Prompt. http://www.memricaprompt.com (2016)
12. Wikpedia The 35th of May Disambiguation Page. https://en.wikipedia.org/wiki/May_35 (2016). Accessed 25 May 2016
13. Wheatman, M.J.: An autopoietic repertoire. In: AI-2014: Proceedings of the Thirty-fourth SGAI International Conference Cambridge, UK, 9–11 Dec 2014
14. Searle, J.R.: Speech Acts: An Essay in the Philosophy of Language. Cambridge University Press (1969)
15. Stamper, R.K.: Towards a theory of information; Information: mystical fluid or a subject for scientific enquiry. Comput. J. **28**(3), 195–199 (1985)
16. Andersen, P.B.A.: Theory of Computer Semiotics. Cambridge University Press (1997)
17. Tanaka-Ishii, K.: Semiotics of Programming, Cambridge University Press (2010)
18. Peirce, C.S.: Collected papers of C.S. Peirce. In: Hartshorne, C., Weiss, P. (Eds) vol. 2, 2.2270–2.306. Harvard University Press, Cambridge, MA (1935–57)
19. Wheatman, M.J.: A semiotic model of information systems. In: Proceedings of the 13th International Conference on Informatics and Semiotics in Organisations: Problems and Possibilities of Computational Humanities, Leeuwarden, The Netherlands, 4–6 July 2011
20. Nirenberg, S., Raskin, V.: Ontological Semantics. Speech and Communication) MIT Press, Language (2004)
21. Vardi, M.Y.: Would turing have passed the turing test. Commun. ACM **57**(9) (2014)

Short Papers

Experiments with High Performance Genetic Programming for Classification Problems

Darren M. Chitty

Abstract In recent years there have been many papers concerned with significantly improving the computational speed of Genetic Programming (GP) through exploitation of parallel hardware. The benefits of timeliness or being able to consider larger datasets are obvious. However, a question remains in whether there are wider benefits of this high performance GP approach. Consequently, this paper will investigate leveraging this performance by using a higher degree of evolution and ensemble approaches in order to discern if any improvement in classification accuracies can be achieved from high performance GP thereby advancing the technique itself.

Keywords Genetic Programming · Classification · Parallel Processing

1 Introduction

Genetic Programming (GP) [1] is recognised as a computationally intensive technique arising from the use of an interpreter to execute candidate programs generated by the evolutionary process. However, recently there has been considerable work on improving the speed of GP through the exploitation of highly parallel hardware known as Graphics Processing Units (GPUs) [2, 3] or multi-core CPUs [4] enabling the evolution of candidate programs thousands of times faster. The obvious benefits are improved timeliness and consideration of larger datasets. However, a third option exists in increasing the computational complexity of GP itself such as evolving solutions for thousands of generations of evolution in short timescales rather than the typical hundred or so generations. Alternatively, an ensemble of classifiers can now be generated with ease. Thus, this paper seeks to answer the question does high performance GP fundamentally improve the technique itself when used for classification tasks or does it merely enable experimentation in a more timely manner?

D.M. Chitty (✉)
Department of Computer Science, University of Bristol, Merchant Venturers Bldg, Woodland Road, Bristol BS8 1UB, UK
e-mail: darrenchitty@googlemail.com

M. Bramer and M. Petridis (eds.), *Research and Development in Intelligent Systems XXXIII*, DOI 10.1007/978-3-319-47175-4_15

2 GP Evolution for Thousands of Generations

GP uses the principles of evolution to generate programs as potential solutions to a problem such as classification. Candidate programs consist of simple functions like addition or subtraction and inputs such as dataset features or constants. For classification, programs are tested for their ability to correctly classify a set of training samples. In traditional GP, programs are re-interpreted on every training case making GP computationally intensive resulting in low degrees of evolution. However, recent high performance implementations have significantly reduced this computational cost enabling many thousands of generations of evolution.

By evolving for many more generations it can be expected that improved accuracy on the training data will be achieved. However, the evolved classifiers could overfit the training data leading to reduced accuracy on the test data. In order to test the hypothesis that GP will merely overfit the data if allowed to evolve programs for thousands of generations, experiments will be conducted on six differing binary classification problems from the UCI Machine Learning database [5]. The parameter settings used for the GP implementation are shown in Table 1.

Experiments are conducted with tree-based GP and a fast CPU based parallel implementation of GP using a two-dimensional stack such that each GP program need only be interpreted a single time [4]. Lexicographic parsimony pressure is used to reduce *bloat* whereby if two solutions have the same classification error, the shorter will be considered the fitter. Each dataset is split into ten training and test data sets using ten-fold cross-validation. Ten random runs are used for each set resulting in 100 experiments in total. Figure 1 shows the classification error for the training and test data as evolution progresses averaged across the experiments. From these results it can be observed that as evolution continues the error on the training data reduces demonstrating that significant improvements in the accuracy on the training data can be achieved by evolving for many thousands of generations. However, Fig. 1 illustrates that as training data accuracy improves there is a slight loss of accuracy on the test data indicating overfitting although the loss is minimal.

3 GP Ensemble Approaches

A slight degradation in the accuracy on the test data has been observed by evolving GP classifiers for thousands of generations indicating a small degree of overfitting. Therefore, it could be argued that there is no benefit to continuing to evolve candidate

Table 1 GP parameters used throughout this paper unless otherwise stated

Population size: 1,000	Crossover probability: 0.9
Max generations: 10,000	Mutation probability: 0.1
Tournament size: 10	Set of GP Functions: + − % * < > && || == if

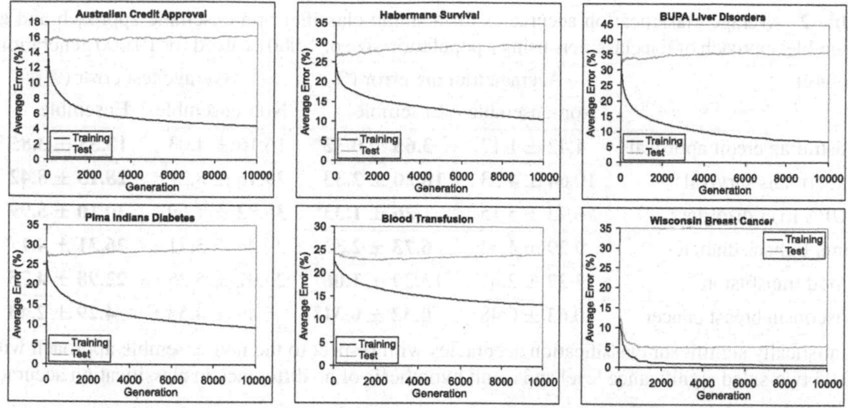

Fig. 1 The average best found training data accuracy and corresponding test data accuracy at each generation for each of the six classification problem instances

GP programs for many generations even though a considerable gain in accuracy is observed on the training set. Consequently, an alternative methodology of exploiting high performance GP needs to be considered in creating an ensemble of classifiers. Instead of generating a single highly evolved classifier which potentially overfits the training data, a number of separate base classifiers can be generated and combined to provide a team classifier. The differing classifiers should reflect differing patterns in the data and when combined into an ensemble, provide increased accuracy. Bagging [6] is a well known ensemble approach. With high performance GP, generating an ensemble of base classifiers is now much easier than previously.

To test the effectiveness of a GP ensemble approach to classification, the results from earlier are repeated but now ten separate classifiers are generated and combined using a majority voting approach. These results are shown in Table 2 whereby it can be observed that for all problems there has been an improvement in the accuracy on both training and test data. Indeed, there has been an average 14.5 % improvement in test data accuracy from using an ensemble approach indicating that the true benefit of high performance GP for classification lies in an ensemble approach rather than just increasing the degree of evolution.

An ensemble approach has demonstrated improved accuracy and high performance GP has facilitated the timeliness of generating ensemble classifiers. However, a question remains as to generate either a small number of highly evolved classifiers in the ensemble that are potentially over-trained or generate a larger set of less evolved classifiers. To answer this issue, experiments using increasing degrees of evolution and ensemble size are conducted. These results are shown in Tables 3 and 4 whereby the best accuracy on the training data occurs when using an ensemble of a few highly evolved classifiers. However, with regards the test data, only the Australian Credit Approval problem instance achieves the best accuracy using a small ensemble of highly evolved classifiers indicating perhaps that highly evolved

Table 2 Average classification accuracies for a single classifier non-ensemble approach and an ensemble approach of ten classifiers using a population size of 1,000 evolved for 10,000 generations

Dataset	Average training error (%)		Average test error (%)	
	Non-ensemble	Ensemble	Non-ensemble	Ensemble
Australian credit approval	4.42 ± 1.17	**3.65 ± 0.32***	16.16 ± 4.03	**13.33 ± 3.85***
Habermans survival	10.64 ± 4.33	**10.20 ± 2.33**	29.16 ± 8.21	**28.15 ± 8.42**
BUPA liver disorders	6.73 ± 5.15	**1.26 ± 1.33***	35.82 ± 7.69	**30.10 ± 5.99***
Pima Indians diabetes	9.29 ± 4.34	**6.73 ± 2.32***	29.24 ± 5.71	**26.31 ± 4.17***
Blood transfusion	13.27 ± 2.67	**13.29 ± 1.06**	23.92 ± 5.35	**22.98 ± 4.75**
Wisconsin breast cancer	0.63 ± 0.48	**0.32 ± 0.31***	5.49 ± 4.53	**4.29 ± 2.43***

*Statistically significant classification accuracies with respect to the non-ensemble approach with a 0.05 two-sided significance level and a null hypothesis of no difference in classification accuracy

Table 3 Average classification accuracies for ensembles of differing sized sets of classifiers and differing numbers of generations such that the computational load is equalised

		Australian Credit Approv.		Habermans Survival		BUPA Liver Disorders	
Num. Gens	Set. Size	Av. Training Error (%)	Av. Test Error (%)	Av. Training Error (%)	Av. Test Error (%)	Av. Training Error (%)	Av. Test Error (%)
1000	100	5.74 ± 0.46	15.80 ± 3.95	16.16 ± 1.39	28.20 ± 8.64	10.77 ± 1.47	28.14 ± 6.03
2000	50	4.97 ± 0.45	15.22 ± 3.63	14.63 ± 1.80	27.52 ± 8.06	5.74 ± 2.60	28.33 ± 6.62
3000	33	4.61 ± 0.43	14.78 ± 4.31	13.23 ± 2.06	27.51 ± 8.46	3.76 ± 2.33	28.38 ± 6.35
4000	25	4.29 ± 0.37	15.22 ± 4.49	12.36 ± 2.51	27.51 ± 8.18	2.98 ± 1.95	28.67 ± 5.67
5000	20	4.13 ± 0.27	14.06 ± 3.87	12.21 ± 2.56	**27.18 ± 8.32**	2.40 ± 2.13	**26.90 ± 7.12**
10000	10	**3.65 ± 0.32**	**13.33 ± 3.85**	**10.20 ± 2.33**	28.15 ± 8.42	**1.26 ± 1.33**	30.10 ± 5.99

Table 4 Average classification accuracies for ensembles of differing sized sets of classifiers and differing numbers of generations such that the computational load is equalised

		Pima Indians Diabetes		Blood Transfusion		Wisconsin Breast Cancer	
Num. Gens	Set Size	Av. Training Error (%)	Av. Test Error (%)	Av. Training Error (%)	Av. Test Error (%)	Av. Training Error (%)	Av. Test Error (%)
1000	100	17.33 ± 0.91	**24.88 ± 3.90**	17.46 ± 1.38	22.98 ± 4.63	1.23 ± 0.17	4.15 ± 2.17
2000	50	15.53 ± 1.33	25.40 ± 4.19	15.76 ± 1.12	21.78 ± 5.21	1.02 ± 0.20	**4.00 ± 2.21**
3000	33	12.99 ± 2.40	25.40 ± 4.15	14.42 ± 0.75	21.64 ± 5.64	0.97 ± 0.24	4.15 ± 2.17
4000	25	10.84 ± 1.92	25.40 ± 4.48	14.15 ± 0.78	**21.51 ± 5.38**	0.81 ± 0.20	4.29 ± 2.13
5000	20	9.49 ± 2.17	25.01 ± 4.48	13.94 ± 0.94	21.64 ± 4.72	0.70 ± 0.30	4.29 ± 2.13
10000	10	**6.73 ± 2.32**	26.31 ± 4.17	**13.29 ± 1.06**	22.98 ± 4.75	**0.32 ± 0.31**	4.29 ± 2.43

classifiers are indeed over-trained. However, the number of classifiers and the degree of evolution that provide the best test data accuracies varies considerably although a majority of the best accuracies occur for 4,000 generations or greater.

It could be reasoned that ten highly evolved classifiers results in a reduction in test data accuracy as this is too few. Consequently, the experiments using 10,000 generations to create classifiers are repeated but instead, a hundred classifiers are generated in the ensemble with the results shown in Table 5 whereby it can be observed that

Table 5 The average classification accuracies for each problem instance for an ensemble of 100 GP classifiers evolved using a population size of 1,000 for 10,000 generations

Dataset	Av. training error (%)	Av. test error (%)
Australian credit approval	3.74 ± 0.35	15.07 ± 3.56
Habermans survival	9.60 ± 1.71	29.78 ± 9.79
BUPA liver disorders	0.26 ± 0.30	27.86 ± 5.44
Pima Indians diabetes	5.68 ± 1.56	25.54 ± 5.06
Blood transfusion	12.63 ± 0.95	22.04 ± 5.82
Wisconsin breast cancer	0.46 ± 0.09	4.00 ± 2.31

an improvement in the training data accuracy is achieved for all but one of the problem instances. Moreover, when comparing to ten classifiers within the ensemble, an improvement in the test data accuracies has been observed for four of the problem instances although remaining less accurate than those achieved with less highly evolved classifiers. For further analysis, the training data and test accuracy is plotted in Fig. 2 as each new classifier is generated and added to the ensemble. These plots show that for three of the problems, test data accuracy degrades once the ensemble of classifiers exceeds twenty whereas for the others, accuracy continues to improve. Notably, for five of the problems, the peak test data accuracy is better than any of the results observed in Tables 3 and 4.

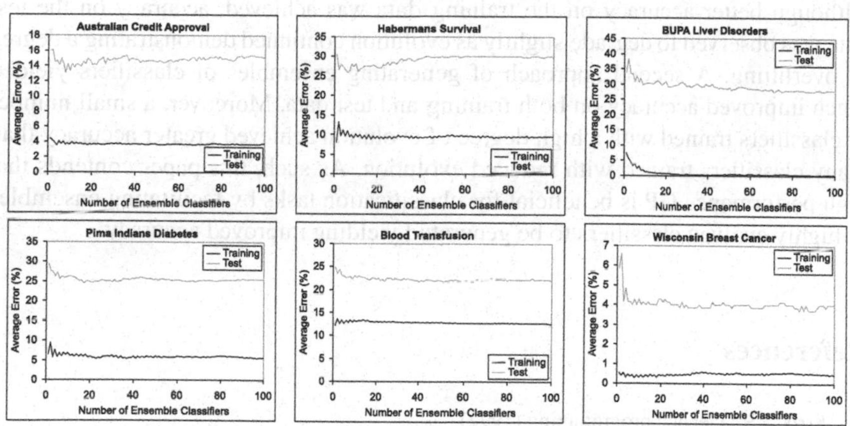

Fig. 2 The accuracies of the ensembles on both the training data and the test data as classifiers are added for each problem instance for the experiments from Table 5

4 Related Work

GP has frequently been applied to classification problems and a recent literature review can be found in [7]. With regards GP ensemble classification, Zhang and Bhattacharyya [8] demonstrated that an ensemble of GP classifiers can outperform an ensemble of decision tree and logistic regression classifiers. Bhowan et al. [9] generated an ensemble of diverse GP classifiers using a multi-objective approach and the objectives of majority and minority class accuracy. Folino et al. [10] generated a diverse set of GP classifiers by training on differing subsets of the training data. Brameier and Banzhaf [11] used Linear GP to evolve a set of classifiers and neural networks or a weighted vector to assign their relative importance. Liu and Xu [12] evolved sets of GP trees within a population whereby each solves a two class aspect of a multi-class problem. Keijzer and Babovic [13] investigated the use of subsets and variance and bias measures to the generation of ensembles of GP classifiers.

5 Conclusions

High performance GP provides improved timeliness and/or the ability to tackle larger problem sizes. However, there is the potential for further benefits through increased complexity within the technique itself. This paper firstly investigated substantially increasing the degree of GP evolution when applied to classification problems. Although better accuracy on the training data was achieved, accuracy on the test data was observed to degrade slightly as evolution continued demonstrating a degree of overfitting. A second approach of generating ensembles of classifiers yielded much improved accuracy on both training and test data. Moreover, a small number of classifiers trained with a high degree of evolution achieved greater accuracy than many classifiers trained with reduced evolution. As such, this paper contends that high performance GP is beneficial for classification tasks by facilitating ensembles of highly evolved classifiers to be generated yielding improved accuracy.

References

1. Koza, J.R.: Genetic programming (1992)
2. Chitty, D.M.: A data parallel approach to genetic programming using programmable graphics hardware. In: Proceedings of the 9th Annual Conference on Genetic and Evolutionary Computation GECCO '07, pp. 1566–1573 (2007)
3. Chitty, D.M.: Faster GPU-based genetic programming using a two-dimensional stack. Soft Comput. 1–20 (2016)
4. Chitty, D.: Fast parallel genetic programming: multi-core CPU versus many-core GPU. Soft Comput. **16**(10), 1795–1814 (2012)
5. Frank, A., Asuncion, A.: UCI machine learning repository (2010)
6. Breiman, L.: Bagging predictors. Mach. Learn. **24**(2), 123–140 (1996)

7. Espejo, P.G., Ventura, S., Herrera, F.: A survey on the application of genetic programming to classification. IEEE Trans. Syst. Man Cybern. Part C **40**(2), 121–144 (2010)
8. Zhang, Y., Bhattacharyya, S.: Genetic programming in classifying large-scale data: an ensemble method. Inf. Sci. **163**(1), 85–101 (2004)
9. Bhowan, U., Johnston, M., Zhang, M., Yao, X.: Evolving diverse ensembles using genetic programming for classification with unbalanced data. Evol. Comput. IEEE Trans. **17**(3), 368–386 (2013)
10. Folino, G., Pizzuti, C., Spezzano, G.: Ensemble techniques for parallel genetic programming based classifiers. In: Genetic Programming, pp. 59–69. Springer (2003)
11. Brameier, M., Banzhaf, W.: Evolving teams of predictors with linear genetic programming. Genet. Program Evolvable Mach. **2**(4), 381–407 (2001)
12. Liu, K.H., Xu, C.G.: A genetic programming-based approach to the classification of multiclass microarray datasets. Bioinformatics **25**(3), 331–337 (2009)
13. Keijzer, M., Babovic, V.: Genetic programming, ensemble methods and the bias/variance tradeoff–introductory investigations. In: Genetic Programming, pp. 76–90. Springer (2000)

Towards Expressive Modular Rule Induction for Numerical Attributes

Manal Almutairi, Frederic Stahl, Mathew Jennings, Thien Le and Max Bramer

Abstract The Prism family is an alternative set of predictive data mining algorithms to the more established decision tree data mining algorithms. Prism classifiers are more expressive and user friendly compared with decision trees and achieve a similar accuracy compared with that of decision trees and even outperform decision trees in some cases. This is especially the case where there is noise and clashes in the training data. However, Prism algorithms still tend to overfit on noisy data; this has led to the development of pruning methods which have allowed the Prism algorithms to generalise better over the dataset. The work presented in this paper aims to address the problem of overfitting at rule induction stage for numerical attributes by proposing a new numerical rule term structure based on the Gauss Probability Density Distribution. This new rule term structure is not only expected to lead to a more robust classifier, but also lowers the computational requirements as it needs to induce fewer rule terms.

Keywords Modular Rule Induction · Expressive Numerical Rule Terms · Computationally Efficient Classification

M. Almutairi · F. Stahl (✉) · M. Jennings · T. Le
Department of Computer Science, University of Reading, PO Box 225,
Whiteknights, Reading RG6 6AY, UK
e-mail: F.T.Stahl@reading.ac.uk

M. Almutairi
e-mail: manal.almutairi@pgr.reading.ac.uk

M. Jennings
e-mail: m.jennings@student.reading.ac.uk

T. Le
e-mail: t.d.le@pgr.reading.ac.uk

M. Bramer
School of Computing, University of Portsmouth, Buckingham Building,
Lion Terrace, Portsmouth PO1 3HE, UK
e-mail: Max.Bramer@port.ac.uk

M. Bramer and M. Petridis (eds.), *Research and Development in Intelligent Systems XXXIII*, DOI 10.1007/978-3-319-47175-4_16

1 Introduction

The classification of unseen data instances (Predictive Analysis) is an important data mining task. Most classification techniques are based on the 'divide-and-conquer' approach to generate decision trees as detailed in [6]. Many variations exist despite the inherent weakness in that decision trees often require irrelevant tests to be included in order to perform classification tasks [3]. Cendrowska's Prism algorithm addresses this issue through a 'separate-and-conquer' approach, which generates if-then rules directly from training instances [3]. Cendrowska's Prism algorithm sparked work on a range of different Prism variations that aim to improve upon Cendrowska's original algorithm. The original Prism algorithm was not designed to work with numerical attributes but an extended version of the algorithm with the ability to deal with numerical attributes was described in [1] by incorporating local discretisation. PrismTCS which is a computationally more efficient version of Prism aims to introduce an order in the rules it induces [2]. PMCRI is a parallel implementation of PrismTCS and Prism [7] etc. In addition there are various pruning methods for the Prism family of algorithms that aim to reduce the induced ruleset from overfitting, i.e. J-pruning [2] and Jmax-pruning [8]. Another advantage of the Prism family of algorithms compared with decision trees is that Prism algorithms are more expressive. Prism rules avoid unnecessary rule terms and thus are less confusing and more expressive compared with decision trees. Algorithms of the Prism family may not overfit as much as decision trees on the training data, nevertheless, they still tend to overfit if there is noise in the data, especially for numerical values. A recent development of one of the co-authors of this paper uses Gauss Probability Density Distribution in order to generate more expressive rule terms for numerical attributes. This development was part of a real-time rule based data stream classifier [4] and resulted in more expressive rulesets and faster real-time rule induction. This paper presents the ongoing work on a new member of the Prism family of algorithms that makes use of the rule term structure introduced in [4]. The new version of Prism based on Gauss Probability Density Distribution is expected to produce more expressive rulesets, produce them faster and with less overfitting compared with its predecessor N-Prism, especially on noisy training data.

This paper is organised as follows: Sect. 2 introduces Prism, Sect. 3 describes and positions our ongoing development of a new version of Prism based on Gauss Probability Density Distribution. Section 4 provides an initial empirical evaluation of G-Prism in comparison with Prism. Section 5 provides concluding remarks including ongoing developments of this new version of Prism.

2 Modular Rule Induction with Prism

One of the major criticisms of decision trees is the replicated subtree problem. The problem has been first highlighted in [3] and been given the name replicated subtrees in [9].

Assume a fictitious example training data set comprises four attributes *a, b, c* and *d* and two possible classifications *go* and *stop*. Each attribute can take 2 possible values *True* (T) and *False* (F). And the pattern encoded are the two following rules leading to classification *go*, whereas the remaining cases would lead to classification *stop*:

$$\text{IF (a) AND (b)} \rightarrow \text{go}$$
$$\text{IF (c) AND (d)} \rightarrow \text{go}$$

Using a tree induction approach to classify cases *go* and *stop* would lead to the replicated subtree problem illustrated in Fig. 1, whereas Prism can induce these two rules directly without adding any unnecessary rule terms.

Cendrowska's original Prism algorithm is described in Algorithm 1. Essentially Prism tries to maximise the conditional probability with which a rule covers a particular target class in order to specialise a classification rule. The specialisation stops once the rule only covers instances of the target class. Once a rule is complete Prism deletes all instances from the data that are covered by the rules induced so far and starts inducing the next rule. This process is repeated for every possible target class. Note that Cendrowska's original Prism algorithm [3] does not deal with numerical attributes, however, implementations of algorithms of the Prism family handle numerical attributes as described in lines 7–12 in Algorithm 1. This is very inefficient because there is a large number of probability calculations, in particular $N \cdot m \cdot 2$, where N is the number of training instances and m the number of numerical attributes. Section 3 introduces G-Prism which handles the induction of rule terms from numerical attributes in a more efficient way.

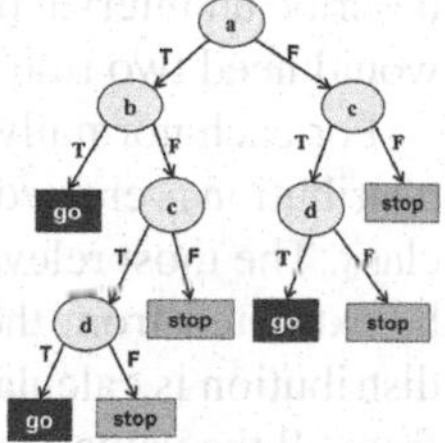

Fig. 1 Replicated subtree problem

Algorithm 1: Learning classification rules from labelled data using Prism.

```
1  for i = 1 → C do
2      D ← Dataset;
3      while D does not contain only instances of class ω_i do
4          forall the attributes α_j ∈ D do
5              if attribute α_j is categorical then
6                  Calculate the conditional probability, P(ω_i | α_j = v) for all possible attribute-value (α_j = v) from attribute α_j;
7              else if attribute α_j is numerical then
8                  sort D according to v;
9                  forall the values v of α_j do
10                     for each v of α_j calculate P(ω_i | α_j ≤ v) and P(ω_i | α_j > v);
                   end
12             end
           end
15         Select the (α_j = v), (α_j > v), or (α_j ≤ v) with the maximum conditional probability as a rule term ;
16         D ← S, create a subset S from D containing all the instances covered by selected rule term at line 15;
       end
18     The induced rule R is a conjunction of all selected (α_j = v), (α_j > v), or (α_j ≤ v) at line 15;
19     Remove all instances covered by rule R from original Dataset;
20     repeat
21         lines 3 to 9;
       until all instances of class ω_i have been removed;
   end
```

3 Inducing Expressive Module Classification Rules

As described in Sect. 2, the way Prism deals with numerical attributes requires many cut-point calculations to work out the conditional probabilities for each attribute value, which is computationally expensive. The same critique has been made in [4] for a real-time rule based data stream classifier and a new kind of rule term for numerical streaming data has been proposed. This new rule term is computationally less demanding and also more expressive. This paper proposes to incorporate the in [4] presented rule term structure into the Prism family of algorithms in order to create a more expressive, computationally efficient and robust (to noisy data) Prism classifier.

Instead of creating two rule terms for every possible value in an attribute, only one rule term of the form $(v_i < \alpha_j \leq v_k)$ is created where v_i and v_k are two attribute values. Thus frequent cut-point calculations can be avoided. Additionally, this form of rule term is more expressive than the representation of a rule term for a numerical attribute of $(\alpha_j \leq v)$ or $(\alpha_j > v)$. A rule term in the form of $(v_i < \alpha_j \leq v_k)$, can describe an interval of data whereas the previous approach described in Algorithm 1 would need two rule terms and thus result in longer less readable rules and rulesets.

For each normally distributed numerical attribute in the training data, a Gaussian distribution is created to represent all values of a numerical attribute for a given target class. The most relevant value of a numerical attribute for a given target class ω_i can be extracted from the generated Gaussian distribution of that class. The Gaussian distribution is calculated for the numerical attribute α_j with mean μ and variance σ^2 from all the values associated with classification, ω_i. The conditional probability for class ω_i is calculated as in Eq. 1.

$$\mathbb{P}(\alpha_j = v|\omega_i) = \mathbb{P}(\alpha_j = v|\sigma^2) = \frac{1}{\sqrt{2\pi\sigma^2}} exp(-\frac{((\alpha_j = v) - \mu)^2}{2\sigma^2}) \quad (1)$$

Hence, a value based on $\mathbb{P}(\omega_i|\alpha_j = v)$, or equivalently $log(\mathbb{P}(\omega_i|\alpha_j = v))$ can be calculated as shown in the Eq. 2. This value can be used to determine the probability of a given class label, ω_i for a valid value, v of a numerical attribute, α_j.

$$log(\mathbb{P}(\omega_i|\alpha_j = v)) = log(\mathbb{P}(\alpha_j = v|\omega_i)) + log(\mathbb{P}(\omega_i)) - log(\mathbb{P}(\alpha_j = v)) \quad (2)$$

Based on the created Gaussian distribution for each class label, the probability between a lower bound and an upper bound, Ω_i can be calculated for such that if $v \in \Omega_i$, then v belongs to class ω_i. Practically, from a generated Gaussian distribution for class ω_i, a range of values μ in the centre is expected to represent the most common values of the numerical attribute for class ω_i. However, if an algorithm is based on 'separate-and-conquer' strategy then the training data examples are different after each iteration and the bounds should be selected from the available values of the numerical attribute from a current iteration.

As shown in Fig. 2, the shaded area between lower bound, v_i and upper bound, v_k represents the most common values of the numerical attribute α_j for class ω_i from a given subset of training data examples. The selection of an appropriate range of the numerical attribute is simply to identify a possible rule term in the form of $(v_i < \alpha_j \leq v_k)$, which is highly relevant to a range of values from the numerical attribute α_j for a target class ω_i from the training data examples. Once $\mathbb{P}(\omega_i|v_i < \alpha_j \leq v_k)$ is calculated for each numerical attribute then the Prism learning algorithm selects the rule term with highest conditional probability from both categorical and numerical attributes. The resulting algorithm is termed G-Prism where G stands using Gaussian distribution. Please note that data stream classifiers deal with an infinite amount of numerical values per attribute and hence typically assume a normal distribution, however, this is not the case for batch algorithms like Prism. A test of normal distribution will be applied prior to applying this method. Attributes that are not normal distributed would be dealt with as described Sect. 2.

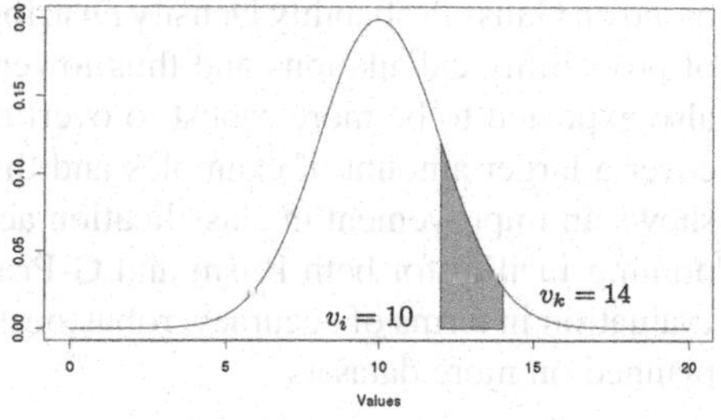

Fig. 2 The *shaded area* represents a range of values, Ω_i, of continuous attribute α_j for class ω_i

Table 1 Comparison of G-Prism's accuracy with Prism's accuracy

	Datasets	
	Glass	Iris
Accuracy G-Prism	56	93
Accuracy prism	48	76

As has been shown in [4], the new rule term structure is fitting the target class less exact and covers more training data, whilst still achieving high accuracy. This resulted in less overfitting [4] compared rule terms induced through binary splits. Thus this new rule term structure incorporated in Prism is expected to be also less prone to overfitting in comparison with binary splits on numerical attributes.

4 Preliminary Results

The first prototype version of the G-Prism algorithm has been compared in terms of classification accuracy against its Prism predecessor as described in Algorithm 1. For the comparison two datasets have been used, the Glass and the Iris dataset from the UCI repository [5]. The datasets have been randomly split into a train and test datasets, whereas the testset comprises 30 % of the data.

Both algorithms seem to struggle on the Glass dataset, however, G-Prism shows a higher accuracy on both datasets. Please note that both algorithms are expected to perform better when pruning is incorporated, which is subject to ongoing work (Table 1).

5 Conclusions and Future Work

The paper positioned the ongoing work on a new Prism classifier using a different more expressive way of inducing rule terms. The new rule term induction method is based on Gauss Probability Density Distribution and is expected to reduce the number of probability calculations and thus lower computational cost. The new method is also expected to be more robust to overfitting as the new form of rule term should cover a larger amount of examples and thus generalise better. An initial evaluation shows an improvement in classification accuracy. Ongoing work is implementing a pruning facility for both Prism and G-Prism. Thus a more exhaustive comparative evaluation in terms of accuracy, robustness to noise and computational scalability is planned on more datasets.

References

1. Bramer, M.: Principles of Data Mining. Undergraduate Topics in Computer Science. Springer International Publishing (2013)
2. Bramer, M.A.: An information-theoretic approach to the pre-pruning of classification rules. In: Neumann, B., Musen, M., Studer, R. (eds) Intelligent Information Processing, pp. 201–212. Kluwer (2002)
3. Cendrowska, J.: PRISM: an algorithm for inducing modular rules (1987)
4. Le, T., Stahl, F., Gomes, J., Gaber, M.M. Di Fatta, G.: Computationally efficient rule-based classification for continuous streaming data. In: Research and Development in Intelligent Systems XXXI, pp. 21–34. Springer (2014)
5. Lichman, M.: UCI machine learning repository (2013)
6. Ross, J.: Quinlan induction of decision trees. Mach. Learn. **1**(1), 81–106 (1986)
7. Stahl, F., Bramer, M.: Computationally efficient induction of classification rules with the PMCRI and j-pmcri frameworks. Knowl.-Based Syst. **35**, 49–63 (2012)
8. Stahl, F., Bramer, M.: Jmax-pruning: a facility for the information theoretic pruning of modular classification rules. Knowl.-Based Syst. **29**, 12–19 (2012)
9. Witten, I.H., Frank, E., Hall, M.A.: Data Mining: Practical Machine Learning Tools and Techniques: Practical Machine Learning Tools and Techniques. Elsevier Science, The Morgan Kaufmann Series in Data Management Systems (2011)

OPEN: New Path-Planning Algorithm for Real-World Complex Environment

J.I. Olszewska and J. Toman

Abstract This paper tackles with the single-source, shortest-path problem in the challenging context of navigation through real-world, natural environment like a ski area, where traditional on-site sign posts could be limited or not available. For this purpose, we propose a novel approach for planning the shortest path in a directed, acyclical graph (DAG) built on geo-location data mapped from available web databases through Google Map and/or Google Earth. Our new path-planning algorithm we called OPEN is run against this resulting graph and provides the optimal path in a computationally efficient way. Our approach was demonstrated on real-world cases, and it outperforms state-of-art, path-planning algorithms.

Keywords Path planning · Graph · Algorithms · Real-world navigation · Google Map · Google Earth

1 Introduction

In natural environment, finding the suitable/shortest path is a difficult task. Indeed, in ski resorts, traditional resources, e.g. ski piste maps or sign posts, could be limited [1] or even non-existent as illustrated in Fig. 1, while current mobile and web applications such as Navionics Ski [2] or [3] could lack of accuracy, flexibility or adaptability to environmental changes. On the other hand, many path-planning algorithms and their derivations exist [4–7], but usually restricted to artificial environments or optimised only for situations such as network traffic flow or social media. This reveals the need to deploy new path-planning approaches.

In this paper, we focus thus on the study and the development of an appropriate approach for path planning in a natural domain such as a ski area, using web available geo-location data, in order to help the growing number of winter-sport users. This

J.I. Olszewska (✉) · J. Toman
University of Gloucestershire, Cheltenham, UK
e-mail: joanna.olszewska@ieee.org

M. Bramer and M. Petridis (eds.), *Research and Development in Intelligent Systems XXXIII*, DOI 10.1007/978-3-319-47175-4_17

Fig. 1 Natural environment consisting of a ski area without sign posts

also implies the design of an adapted description and modelling of the application domain, i.e. the ski domain.

In particular, the contribution of this paper is a novel path-planning algorithm called *OPEN* based on the integration of multiple planning algorithms. Our approach provides an *O*ptimal *P*ath within a minimized *E*xecution time and a reduced number of explored *N*odes, i.e. with a capped memory size. It relies on the parallel computing of a set of path-finding algorithms embedded into our OPEN algorithm finding a triply optimised solution against three constraints such as the path length (P), the execution time (E), and the number of explored nodes (N).

The paper is structured as follows. In Sect. 2, we present our new OPEN path-planning approach. Results and discussion are presented in Sect. 3, while conclusions are drawn up in Sect. 4.

2 Proposed Method

Our method consists of two main steps: (i) the processing of the geo-location data and the building of the corresponding graph which is the core of the ski domain; (ii) the computing of the single-source shortest path to help skiers and surfers to evolve safely through natural environment.

Firstly, to apply a path-planning algorithm, we need to study the underlying graph data, its density, its size, and its environment, and select suitable data structures [8, 9] that make the appropriate compromise between speed and memory constraints and that allow to meet the planned requirements. Hence, our path-planning algorithm is run against a directed, acyclical graph (DAG) built on publicly available, real-time 3D geographical data representing geo-location coordinates of downhill ski runs. Access to geo-spatial databases has been made easier than in the past [10, 11] by using online databases such as Google Map [12] and/or Google Earth [13].

Indeed, the online data extracted from Google Map/Google Earth complies with the Keyhole Markup Language (KML) 2.2 OpenGIS Encoding Standard and contains information such as piste name, piste description; including its difficulty, and geo-location coordinates (longitude, latitude, altitude) describing the top-bottom path of the ski piste. However, data acquired from these online datasets must be pre-processed based on the analysis of geo-location data of ski resorts, before feeding graph-based operations such as finding the shortest path. Indeed, the online data could present problems like superfluous or inconsistent data and should be mitigated against. For this purpose, we apply the algorithm as presented in [14].

Secondly, to solve the single-source, shortest-path problem, we designed the OPEN algorithm. It consists in partial, parallel computing of a set of path-planning algorithms embedded into our OPEN algorithm (Algorithm 1), and in finding a triply optimised solution against three constraints which are the path length (P), the execution time (E), and the number of explored nodes (N) related to the memory size.

The OPEN algorithm relies thus on a portfolio of path-planning algorithms which are run in parallel and applied against the built network graph. In first instance, we chose five algorithms, namely, Dijkstra's, A*, Iterative Deepening A* (IDA*), and Anytime Repairing A* (ARA*) with two different inflation factors [5]. Indeed, many consider that Dijkstra's algorithm as a robust and appropriate algorithm for directed, acyclical graphs (DAG) [15] in opposite to the Bellman-Floyd-Moore algorithm which is the most appropriate algorithm for use in graphs with cycles. A* is the most widely used path-planning algorithm for both virtual and natural environments. IDA* algorithm has the smallest memory usage when run against test graphs. ARA* is suitable for a dynamic search algorithm and provides the optimal path when the inflation factors is equal to one; otherwise, it is sub-optimal in terms of path, but it is faster. Hence, our portfolio could contain heterogeneous path-planning algorithms; this design allowing the modularity of our approach, i.e. its 'openness'.

Next, the set of selected algorithms is computed through a competitive process, where the fastest solutions are processed within our OPEN algorithm (Algorithm 1) to find the optimal path, while keeping execution time and memory size low.

Thus, the OPEN path-planning algorithm can be applied to a graph built based on Google Map/Google Earth data to enable users to find the shortest path from one point to another, aiding the navigation across the ski pistes that make up a ski resort.

3 Experiments

We have carried out experiments to evaluate, test, and validate our novel approach. In particular, we have assessed OPEN's computational efficiency in terms of precision, speed, and memory size of our system implemented in Java using the Eclipse Java EE IDE for Web Developer and Google Earth version 7.1.2.41. The dataset used to test this system is freely available under the Creative Commons 3.0 licence and is in KML 2.2 format. It represents a contained area of Whistler Blackcomb (Fig. 2), the American largest Ski Area [16], with different difficulties of pistes and numerous intersection points between the pistes. A tolerance of 15 meters is set for the line

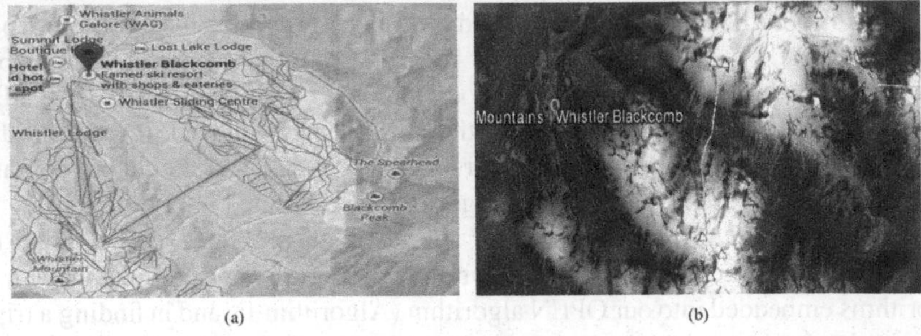

(a) (b)

Fig. 2 Data of Whistler Blackcomb ski area, available at: **a** Google Map and **b** Google Earth

intersections and all altitude coordinates are recorded as zero in the dataset. Consequently, the system will ignore in first instance the altitude when building a network graph representing this data. In a further step, the dataset could be enhanced with the altitude using services such as the Google Elevation API [17], which will return an altitude in meters above sea level for a given longitude-latitude coordinate.

In the experiments, six path-planning algorithms, i.e. Dijkstra's, A*, IDA*, ARA* with IFL = 10, ARA* with IFL = 1, and OPEN, were tested on five ski routes using the `WhistlerBlackcomb.kml` dataset. The results of the performance testing are reported in Tables 1 and 2. It is worth to note that the time measurements for the ARA* results are cumulative, whereas the node count for ARA* algorithms is not cumulative, as it shows the number of nodes expanded at each stage, each subsequent stage building on the nodes expanded in the previous stages.

As indicated in Tables 1 and 2, when searching for a non-existing path, e.g. from the start of The Saddle (TS) to the end of Lower Olympic (LO), all the algorithms find there is no path. However, IDA* occasionally results in a stack overflow exception.

For all the existing tested paths, the performance results show that Dijkstra, A*, ARA*, and OPEN algorithms have execution times within the millisecond range when applied to these graphs.

Table 1 Performance of Dijkstra, A*, and IDA* path planning algorithms, with p: the path length (in meter) computed by an algorithm, e: the execution time (in milliseconds) of an algorithm, and n: the number of explored nodes by an algorithm. The start-/end-nodes correspond to locations such as GL (Green Line), VR (Village Run), MT (Matthews Traverse), LF (Lower Franz), GR (Glacier Road), YB (Yellow Brick Road), PT (Pikas Traverse), LO (Lower Olympic), TS (The Saddle)

Start Location	End Location	Dijkstra			A*			IDA*		
		p	e	n	p	e	n	p	e	n
GL	VR	7292.3	27	1180	7292.3	12	813	7292.3	2241	8338709
MT	LF	6050.7	3	489	6050.7	2	201	6050.7	22	27259
GR	YB	8995.6	1	208	8995.6	1	208	8995.6	42	71861
PT	LO	8058.6	8	495	8058.6	8	374	8058.6	1768	6430540
TS	LO	–	–	–	–	–	–	–	–	–

Table 2 Performance of ARA* and our path planning algorithm (OPEN), with IFL: the inflation factor of the ARA* algorithm

Start Location	End Location	ARA* (IFL=10)			ARA* (IFL=1)			Our		
		p	e	n	p	e	n	p	e	n
GL	VR	7900.1	3	422	7292.3	12	743	7292.3	12	743
MT	LF	6828.2	1	70	6050.7	8	169	6050.7	2	201
GR	YB	9009.6	1	160	8995.6	10	90	8995.6	1	208
PT	LO	8119.5	1	231	8058.6	17	390	8058.6	8	374
TS	LO	–	–	–	–	–	–	–	–	–

In comparison to the other algorithms, the results from IDA* show a very slow algorithm that is very inefficient, exploring very large numbers of nodes. For example, on the route from Green Line (GL) to Village Run (VR), IDA* explores a total of 8338709 nodes, yet there are only 3349 nodes in the Whistler Blackcomb ski map. It is likely that this algorithm is not suitable for the underlying data. There are many nodes densely packed, and the edge cost between them is relatively small. For each iteration of the IDA* algorithm, the algorithm is likely only evaluating a single extra node. This is a worst-case scenario of IDA*, with the number of node expansion equal to $\Omega((N_{A*})^2)$, where N_{A*} is the number of nodes expanded by A* and Ω is the size of the subset of edges considered. Further testing looking at the memory usage profile is required to ascertain if using the IDA* algorithm is worthwhile, as well as looking at more advanced implementations of IDA* that reuse the previous iteration search results.

The use of a heuristic value in the A* algorithm provides improvements over Dijkstra's one in both execution time and number of nodes explored by the algorithm. For example, A* is far more efficient to compute the path starting at the beginning of the Green Line (GL) piste and finishing at the end of the Village Run (VR) piste than Dijkstra's. The analysis of this path shows that there are many decision points at which the algorithms need to make a decision to prioritise one route over another. More specifically, at a decision point, Dijkstra's algorithm explores additional nodes away from the destination as the cost of this path is less than the cost of the path heading toward the destination, whereas the heuristic value used in A* prioritises the path heading toward the destination and the additional nodes are not explored.

ARA* algorithm results highlight how the use of inflated heuristics produces sub-optimal results, but in less time than A* and explores fewer nodes. For example, for the path from Green Line (GL) to Village Run (VR), A* takes 12 ms and explores 813 nodes to produce the optimal route 7292.26 m long, while ARA* with inflation factor 10 takes 3 ms, explores only 422 nodes and produces a sub-optimal path. In the tests, there is no difference between using a heuristic inflation factor of 5 and 10. Further testing shows all inflation factors of 2 and greater produce the same results. As expected the inflation factor of 1 produces the optimal path.

The OPEN algorithm has computational performance at least as good as A* and in some cases even better, e.g. for the path Green Line (GL) to Village Run (VR). Moreover, the OPEN algorithm provides the optimal path unlike ARA* (IFL = 10), while OPEN outperforms the state-of-the art algorithms such as Dijkstra, IDA*, and ARA* (IFL = 1) in terms of both execution time and number of nodes explored.

4 Conclusions

In this paper, we proposed a new path-planning approach to aid people's navigation in outdoor, natural locations such as ski resorts. Indeed, automated path planning is of great utility in this novel application domain, because of the limits of traditional solutions, e.g. outdated ski resort maps or sign-post shortage in remote areas of large ski resorts, or engineering issues with current IT solutions such as native mobile apps provided by some ski resorts. Thus, we developed an original path-planning method which is focused on finding a global solution with an optimal path as well as a capped execution time and a reduced memory size, rather than on locally improving search algorithms. For this purpose, we introduced our OPEN algorithm integrating multiple path-planning algorithms. The OPEN algorithm is run on the ski domain modelled by directed, acyclic graphs based on processed online geo-location data from Google Maps/Earth. Our OPEN algorithm provides the optimal path, while it shows better computational performance than the well-established search algorithms such as A*, when tested for ski path-planning purpose. Considering the computational performance of our OPEN algorithm and the degree of generality of our proposed approach, our method could be useful for any application requiring single-source, path planning in real-world, changing 3D environments.

Appendix—OPEN Algorithm

Algorithm 1 OPEN Algorithm

```
Given a graph G, a start node S, an end (goal) node t, with s ∈ G and t ∈ G.

Considering i ∈ ℕ with i ∈ I, the algorithm set; IFL, the ARA* inflation factor;
p_i, the path length computed by an algorithm i;
e_i, the execution time of an algorithm i;
n_i, the number of node explored by an algorithm i.

Let us initialize
P, the set of computed paths by each algorithm, P = ∅;
E, the set of execution time of each algorithm, E = ∅;
N, the set of explored nodes by each algorithm, N = ∅.

Let us compute
pardo                                                  ▷ parallel computing of the algorithms
  Dijkstra(s,t); return P = P ∪ {p_1}, E = E ∪ {e_1}, N = N ∪ {n_1}
  A*(s,t); return P = P ∪ {p_2}, E = E ∪ {e_2}, N = N ∪ {n_2}
  IDA*(s,t); return P = P ∪ {p_3}, E = E ∪ {e_3}, N = N ∪ {n_3}
  ARA*(s,t,IFL=10); return P = P ∪ {p_4}, E = E ∪ {e_4}, N = N ∪ {n_4}
  ARA*(s,t,IFL=1); return P = P ∪ {p_5}, E = E ∪ {e_5}, N = N ∪ {n_5}
until # E = # I - 1                                    ▷ eliminate the slowest algorithm

if P ≠ ∅ then
    p_o = min(P)
    for all j = 1 : #P do                              ▷ find the shortest path(s)
        if p_j ≠ p_o then
            P = P \ {p_j}, E = E \ {e_j}, N = N \ {n_j}
        end if
    end for

    e_o = min(E)
    for all k = 1 : #E do                              ▷ find the fastest algorithm(s)
        if e_k ≠ e_o then
            P = P \ {p_k}, E = E \ {e_k}, N = N \ {n_k}
        end if
    end for

    n_o = min(N)
    for all l = 1 : #N do                              ▷ find the most efficient algorithm(s)
        if n_l ≠ n_o then
            P = P \ {p_l}, E = E \ {e_l}, N = N \ {n_l}
        end if
    end for

else 'no path'
end if

return head(P), head(E), head(N)                       ▷ find the optimal solution
```

References

1. Watts, D., Chris, G.: Where to Ski and Snowboard 2014. NortonWood Publishing, Bath (2014)
2. NavionicsSki: Navionics Ski (Version 3.3.2) for Android (Mobile Application Software) (2014). https://play.google.com/store
3. RTP LLC: Whistler Blackcomb Live for Android (Mobile Application Software) (2014). https://play.google.com/store
4. Baras, J., Theodorakopoulos, G.: Path Planning Problems in Networks. Morgan & Claypool Publishers, Berkley (2010)
5. Edelkamp, S., Schroedl, S.: Heuristic Search. Morgan Kaufmann, Waltham (2011)
6. Hernandez, C., Asin, R., Baier, J.A.: Reusing previously found A* paths for fast goal-directed navigation in dynamic terrain. In: Proceedings of the AAAI International Conference on Artificial Intelligence, pp. 1158–1164 (2015)
7. Uras, T., Koenig, S.: Speeding-up any-angle path-planning on grids. In: Proceedings of the AAAI International Conference on Automated Planning and Scheduling, pp. 234–238 (2015)
8. Nguyen, T.D., Schmidt, B., Kwoh, C.K.: SparseHC: a memory-efficient online hierarchical clustering algorithm. In: Proceedings of the International Conference on Computational Science, pp. 8–19 (2014)
9. Sedgewick, R., Wayne, K.: Algorithms, 4th edn. Addison-Wesley, New Jersey (2011)
10. Breunig, M., Baer, W.: Database support for mobile route planning systems. Comput. Environ. Urban Syst. **28**(6), 595–610 (2004)
11. Hulden, M., Silfverberg, M., Francom, J.: Kernel density estimation for text-based geolocation. In: Proceedings of the AAAI International Conference on Artificial Intelligence, pp. 145–150 (2015)
12. GoogleMaps: Google Maps Web Database (2015). http://maps.google.com/
13. GoogleEarth: Google Earth Web Database (2015). http://earth.google.com/
14. Toman, J., Olszewska, J.I.: Algorithm for graph building based on Google Maps and Google Earth. In: Proceedings of the IEEE International Symposium on Computational Intelligence and Informatics, pp. 80–85 (2014)
15. Cherkassky, B.V., Goldberg, A.V., Radzik, T.: Shortest paths algorithms: theory and experimental evaluation. Math. Program. **73**(2), 129–174 (1996)
16. WhistlerBlackcomb: Winter Trail Map of the WhistlerBlackcomb ski resort (2014). http://www.whistlerblackcomb.com/~/media/17fbe6c652bd4212902aa2f549a5df9f.pdf
17. Google: Google Elevation API (2016). https://developers.google.com/maps/documentation/elevation/

Encoding Medication Episodes for Adverse Drug Event Prediction

Honghan Wu, Zina M. Ibrahim, Ehtesham Iqbal and Richard J.B. Dobson

Abstract Understanding the interplay among the multiple factors leading to Adverse Drug Reactions (ADRs) is crucial to increasing drug effectiveness, individualising drug therapy and reducing incurred cost. In this paper, we propose a flexible encoding mechanism that can effectively capture the dynamics of multiple medication episodes of a patient at any given time. We enrich the encoding with a drug ontology and patient demographics data and use it as a base for an ADR prediction model. We evaluate the resulting predictive approach under different settings using real anonymised patient data obtained from the EHR of the South London and Maudsley (SLaM), the largest mental health provider in Europe. Using the profiles of 38,000 mental health patients, we identified 240,000 affirmative mentions of dry mouth, constipation and enuresis and 44,000 negative ones. Our approach achieved 93 % prediction accuracy and 93 % F-Measure.

Keywords AI Languages · Programming Techniques and Tools · Bayesian Networks and Stochastic Reasoning · Genetic Algorithms · Machine Learning

1 Introduction

Adversities associated with prescribed mediation can seriously affect the patient's wellbeing [1] and present a real financial burden on healthcare providers (estimated to lead to an annual cost of 466 million pounds in the United Kingdom alone [2]). The problem arises from the fact that trial-tested prescribed drugs are not evaluated

H. Wu (✉) · Z.M. Ibrahim · E. Iqbal · R.J.B. Dobson
King's College London, London, UK
e-mail: honghan.wu@kcl.ac.uk

Z.M. Ibrahim
e-mail: zina.ibrahim@kcl.ac.uk

E. Iqbal
e-mail: ehtesham.iqbal@kcl.ac.uk

R.J. Dobson
e-mail: richard.j.dobson@kcl.ac.uk

M. Bramer and M. Petridis (eds.), *Research and Development in Intelligent Systems XXXIII*, DOI 10.1007/978-3-319-47175-4_18

for long-term effectiveness in diverse (and possibly comorbid) patient populations, resulting in many unknowns with respect to the possibilities of the onset of adverse drug reactions (ADRs). Currently, discovering the factors affecting patients' response to treatment relies on spontaneous reporting systems that rely on patient and clinician data entry and resulting highly under-reported ADR instances [3]. However, there is a large body of unutilised knowledge embedded in the Electronic Health Records (EHRs) of hospitals, containing valuable information regarding treatment responses as well as patient profiles and disease trajectories.

Analysing the factors associated with ADRs using EHRs resolves to the task of uncovering the temporal associations connecting the different factors. Things are more complicated in long-term debilitating and comorbidity illnesses whereby ADRs are associated with multiple, overlapping and long medication episodes. From a technical perspective, a challenging task lies in modelling the time series of medication episodes in an effective way so that they can be fed with other influencing factors into predictive algorithms.

In this paper, we propose a bitwise encoding mechanism that can capture the temporal precedence and duration information of medication episodes as well as their distance from the time of inspection. We present an initial study of using the resulting model to predict the onset of three ADRs, mainly dry mouth, constipation and enuresis, from a large mental health register.

2 Temporal Encoding for Patient's Medication Episodes

Here, we present the novel vector-based encoding of dynamic and multiple medication episodes. Figure 1 illustrates an example encoding scenario. The upper part shows the sample medication episodes of a patient that are related to a certain inspecting time spot (marked as AE date in the figure). The encoding is realised through the following steps.

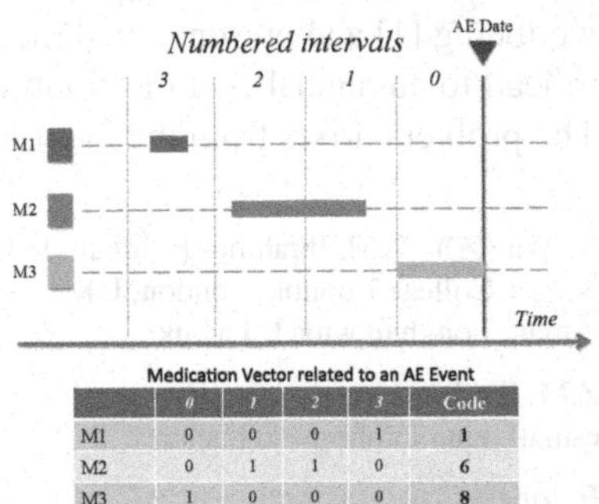

	0	1	2	3	Code
M1	0	0	0	1	1
M2	0	1	1	0	6
M3	1	0	0	0	8

Fig. 1 The bitwise encoding of dynamic medication episodes for a given adverse event

1. From the the inspecting time spot, look back to a certain period of time, e.g., 20 days. In our experiments, we selected 30 days.
2. Split the time period into intervals using a unit of time, e.g., a day or a week. Number each interval in an ascendent order (started from zero) starting from the inspecting time spot. For example, 4 intervals are identified and numbered in Fig. 1.
3. Add the patient's medication episodes to the timeline.
4. For each medication episode, allocate a sequence of bits aligned with the interval order we obtained in step 2. The interval numbered zero is aligned with the most significant bit in the sequence. For each interval, set the corresponding bit to 1 if the interval intersects with the episode, and to 0 otherwise. For example, Medication Episode 2 (M2) in Fig. 1 spans across interval number 1 and 2. Its encoding is 0110 in binary code.
5. Repeat step 4 for all medication episodes. We will get a vector representing all the relevant medication episodes of the inspecting time spot.

2.1 Predictive Model

Having illustrated the encoding of the temporal information embedded within medication episodes, we now show how this information can be combined with additional background knowledge as well as external ontologies into a feature vector for prediction.

Our predictive model considers two sets of features: (1) demographics, (2) the generated dynamic medication episodes which are further enriched by an ontology. Equation 1 shows the features of demographic information of a patient p at a given time t where $age@t$ means the patient's age at the inspecting time t.

$$F_d(p, t) = (age@t, gender, ethnicity) \tag{1}$$

Equation 2 gives the feature calculation from dynamic medication episodes given a patient p, an adverse event type a and a time t.

$$F_m(p, a, t) = (\alpha \cdot \mathrm{diag}(\mathbf{S}_a) + \mathrm{diag}(\mathbf{1})) \cdot \mathbf{M}_{p,t} \tag{2}$$

$\mathbf{S_a}$ is a binary vector describing drugs' relation with side effect a. For example, suppose there are two drugs (d_1, d_2), and d_1 has the side effect a but d_2 does not. Then, the $\mathbf{S_a}$ is $[1, 0]$. Such vector is derived from the drug ontology. $\mathbf{M}_{p,t}$ is the vector generated from the medication episodes of the patient p at a given time t using our encoding method. α is the weighting factor for adjusting side effect knowledge importance in the final prediction model.

The final feature vector for prediction is a concatenation of the above two as shown in Eq. 3. In this paper, we view the ADR prediction as a classification problem. Therefore, the feature vector is used as the inputs for training and testing classification algorithms on our datasets. Various classifiers have been tested. The experiment section will give the detail about model selection.

$$F(p, a, t) = (F_m(p, a, t), F_d(p, t)) \tag{3}$$

3 Implementation and Evaluation

Data Source and Preparation

We used data extracted from the Clinical Record Interactive Search System (CRIS) [4] to evaluate our encoding. CRIS is a database containing a de-identified replica of the EHRs used by the South London and Maudsley Foundation Trust (SLaM), the largest mental health provider in Europe. SLaM serves over 1.2 million patients and stores much of its clinical records and prescribing information in unstructured free text format. The ADR and drug episode information were extracted using an in-house developed natural language processing tools in conjunction with manual annotation [5].

In this preliminary study, we focus on three types of adverse events identified in the CRIS registry: dry mouth (#event 58,347), constipation (#event 86,602) and enuresis (#93,366), which involves 20,795 distinct patients with mental health disorders. For each identified AE, we pick up the patient's past 30 days medication episodes. It is worth mentioning that a large proportion ($\geq$62 % on average) of AEs are associated with multiple medications. To generate negative data items (none adverse event), we picked 18,038 patients who have medication episode data in CRIS registry but never had the three types of adverse events reported.

The dataset was split into a proportion of 80/20 for training and testing. 10-folds cross validation has also been used. The former setting achieved better performance across three AE types.

Demographic Versus Medication Episode Features

We have two types of feature sets—demographic features and medication episode ones. Comparing their performances will reveal some insights about what factors are more likely to be associated with adverse events. When only using demographic features, the F-Measure is around 70 %. But for AE class of dry mouth type, the performance is extremely low—12 % True Positive and 20 % F-Measure. Table 1 gives the performance of dynamic medication episode features. In all cases, about 90 % F-Measure has been achieved with the lowest at 85 % at AE class of dry mouth and the highest at 96 % at none AE class of the same case. In summary, it is quite obvious that dynamic medication episode features are much better indicators for predicting

Table 1 Medication episode-feature only results

ADE type	TP rate	FP rate	Precision	Recall	F-measure	ROC area	Class
Dry mouth	0.83	0.04	0.86	0.83	0.85	0.97	AE
	0.96	0.17	0.95	0.96	0.96	0.97	N-AE
Weighted avg.	0.93	0.14	0.93	0.93	0.93	0.97	
Constipation	0.90	0.07	0.92	0.90	0.91	0.97	AE
	0.93	0.10	0.91	0.93	0.92	0.97	N-AE
Weighted avg.	0.92	0.08	0.92	0.92	0.92	0.97	
Enuresis	0.88	0.07	0.93	0.88	0.90	0.96	AE
	0.93	0.12	0.88	0.93	0.90	0.96	N-AE
Weighted avg.	0.90	0.10	0.90	0.90	0.90	0.96	

Table 2 Medication episode and demographic combined results

ADE Type	TP rate	FP rate	Precision	Recall	F-measure	ROC area	Class
Dry mouth	0.89	0.05	0.85	0.89	0.87	0.98	AE
	0.95	0.11	0.97	0.95	0.96	0.98	N-AE
Weighted avg.	0.94	0.10	0.94	0.94	0.94	0.98	
Constipation	0.93	0.08	0.91	0.93	0.92	0.97	AE
	0.92	0.08	0.93	0.92	0.93	0.97	N-AE
Weighted avg.	0.92	0.08	0.92	0.92	0.92	0.97	
Enuresis	0.92	0.09	0.92	0.92	0.92	0.96	AE
	0.91	0.08	0.91	0.91	0.91	0.96	N-AE
Weighted avg.	0.92	0.08	0.92	0.918	0.92	0.96	

Adverse Events. The combination of the two sets of features can achieve the best results as shown in Table 2. Both average Precision and F-Measure are around 93 %.

Ontology-Based Feature Dimension Reduction

The semantics in the drug ontologies can be utilised to combine semantically similar drugs so that the dimensions of the feature vector F_m (Eq. 2) can be reduced significantly (from 226 for all drugs to 47 for using direct parent category). The performance differences between these two settings are illustrated in Fig. 2. For dry mouth and constipation cases, the performances (average classification accuracy) decreased in about 1 % in average when using drug classes, while the accuracy did improve in the enuresis case when the feature dimension was reduced. This is a very interesting observation worth further investigation to see whether it is a single special case or the drugs related to enuresis make the case special. Figure 2 also reveals that using less feature dimension can significantly increase the training speed, which is not surprising.

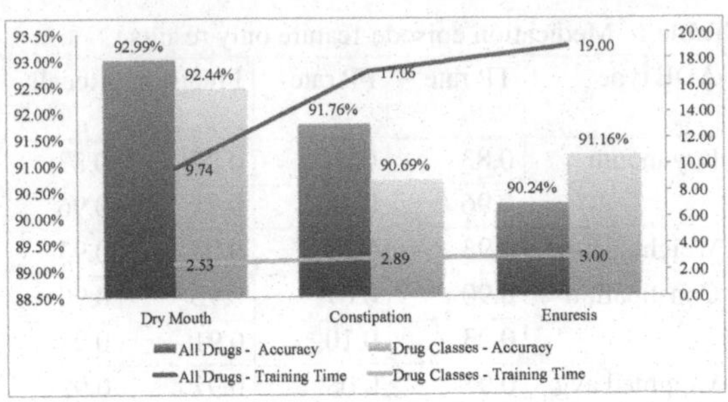

Fig. 2 Feature dimension reduction by drug ontology

4 Conclusion

We reported a retrospective study of predicting adverse events in EHRs from South London and Maudsley Foundation Trust (SLaM), the largest mental health provider in Europe. The prediction model puts a special focus on effective approaches of modelling the dynamic and multiple medication episodes, which are very common among patients with mental health disorders (observed among $\geq$62 % events in our data). Specifically, a novel bitwise encoding approach is introduced for capturing the medication trajectories, which has proved to be more effective (10 % accuracy improvement) than static medication modelling settings.

Acknowledgments This work has received support from the European Union's Horizon 2020 research and innovation programme under grant agreement No 644753 (KConnect) and National Institute for Health Research (NIHR) Biomedical Research Centre and Dementia Unit at South London and Maudsley NHS Foundation Trust and Kings College London.

References

1. Edwards, I., Aronson, J.: Adverse drug reactions: definitions, diagnosis, and management. Lancet **356**(9237), 1255–1259 (2000)
2. Pirmohamed, M., James, S., Meakin, S., Green, C., Scott, A.K., Walley, T.J., Farrar, K., Park, B.K., Breckenridge, A.M.: Adverse drug reactions as cause of admission to hospital: prospective analysis of 18 820 patients. Bmj **329**(7456), 15–19 (2004)
3. Tatonetti, N.P., Patrick, P.Y., Daneshjou, R., Altman, R.B.: Data-driven prediction of drug effects and interactions. Sci. Trans. Med. **4**(125), 125ra31–125ra31 (2012)
4. Stewart, R., Soremekun, M., Perera, G., Broadbent, M., Callard, F., Denis, M., Hotopf, M., Thornicroft, G., Lovestone, S.: The South London and Maudsley NHS foundation trust biomedical research centre (SLaM BRC) case register: development and descriptive data. BMC Psychiatry **9**(1), 1 (2009)
5. Iqbal, E., Mallah, R., Jackson, R.G., Ball, M., Ibrahim, Z.M., Broadbent, M., Dzahini, O., Stewart, R., Johnston, C., Dobson, R.J.B.: Identification of adverse drug events from free text electronic patient records and information in a large mental health case register. PloS one **10**(8), e0134,208 (2015)

Applications and Innovations in Intelligent Systems XXIV

Best Application Paper

A Genetic Algorithm Based Approach for the Simultaneous Optimisation of Workforce Skill Sets and Team Allocation

A.J. Starkey, H. Hagras, S. Shakya and G. Owusu

Abstract In large organisations with multi-skilled workforces, continued optimisation and adaptation of the skill sets of each of the engineers in the workforce is very important. However this change in skill sets can have an impact on the engineer's usefulness in any team. If an engineer has skills easily obtainable by others in the team, that particular engineer might be more useful in a neighboring team where that skill may be scarce. A typical way to handle skilling and resource movement would be to preform them in isolation. This is a sub-optimal way of optimising the workforce overall, as there would be better combinations found if the effect of upskilling some of the workforce was also evaluated against the resultant move recommendations at the time the solutions are being evaluated. This paper presents a genetic algorithm based system for the optimal selection of engineers to be upskilled and simultaneous suggestions of engineers who should swap teams. The results show that combining team moves and engineer upskilling in the same optimisation process lead to an increase in coverage across the region. The combined optimisation results produces better coverage than only moving engineers between teams, just upskilling the engineers and performing both these operations, but in isolation. The developed system has been deployed in BT's iPatch optimisation system with improvements integrated from stakeholder feedback.

Keywords Workforce · Optimisation · Genetic Algorithm · Application

---

A.J. Starkey (✉) · H. Hagras
University of Essex, Colchester CO4 3SQ, UK
e-mail: astark@essex.ac.uk

H. Hagras
e-mail: hani@essex.ac.uk

S. Shakya · G. Owusu
British Telecom, Ipswich IP5 3RE, UK
e-mail: sid.shakya@bt.com

G. Owusu
e-mail: gilbert.owusu@bt.com

M. Bramer and M. Petridis (eds.), *Research and Development in Intelligent Systems XXXIII*, DOI 10.1007/978-3-319-47175-4_19

1 Introduction

For any company with a large multi-skilled workforce, management of skills and teams poses many challenges. A multi-skilled workforce here is defined as one in which the members of the workforce are trained in multiple skill, allowing them to complete different types of tasks. The benefit is that a multi-skilled workforce is capable of completing a range of different tasks, with the aim of making the workforce more productive, more flexible to the changing demand and better at meeting customer needs [1]. This is part of the core principles of workforce optimisation and workforce management, which is about assigning the right employees with the right skills, to the right job, at the right time [2].

Additional arguments have been made for a multi-skilled workforce, such as employees with multiple skills are useful when demand is high and the company wants to maintain a high level of customer satisfaction [3, 4]. Additionally, a multi-skilled workforce can help where the labor market is scarce of the types of people you want to employ [5, 6]. Also, to get the most productivity out of a multi-skilled engineer, the skills they should be trained in should be correlated in some way [5].

The effect of the different mixture of skills in the workforce can have an impact on the utilisation of each member of the workforce and the overall performance of the company as a whole.

A study by the University of Texas in Austin [6] looked at the effects of a multi-skilled workforce in the construction industry. By conducting interviews with many large construction companies they were able to evaluate the best practice for multi-skilling on large construction projects (where more than 200 workers are needed).

They found that if all the workforce is multi-skilled, then there are no specialists, meaning more complex tasks take longer. If there are not enough multi-skilled engineers then there will be a lot of hires and fires with the changing demand as the construction project develops, multi-skilling reduces this.

It is also mentioned that as a result of a multi-skilled workforce previous studies have shown a 5–20 % reduction in labor costs and a 35 % reduction in a required workforce. Similarly, we are investigating the most optimal configurations of skills to get the maximum benefit from the multi-skilled workforce, to further increase the reduction in operating costs.

Deciding which members of the workforce will produce the most benefit when they are trained with more skills can depend on various factors, such as the location of the engineer, the type of tasks that are near to them and also the career pathway of the engineer to determine at which stage he is in terms of progression.

However when engineers are trained with more skills, other engineers in the same area will have their utilisation impacted. This may be because an engineer has low level skills that other engineer could train for and then pick up the work that engineer was doing. As a result it may be more beneficial to move the low skilled engineer to a neighboring team, that is low on resources, and could benefit from the lower skilled engineer freeing up time from the higher skilled engineers.

Due to these possibilities it may be more beneficial to evaluate the resultant effects of upskilling engineers at the time the selection of these engineers is evaluated.

A multi-skilled workforce comes with the mentioned benefits, but there is little work in the optimisation of the workforce skill sets. Hence we started our own investigation into this [7]. We built a genetic algorithm based system that would suggest engineers to be upskilled that would result in the most benefit. This work extends our previous work.

For our new system, a combined upskilling and movement is done in order to avoid a sub-optimal solution with isolated execution of these two steps. This is a form of simultaneous optimisation that uses a simulation to evaluate the solutions [2]. i.e., at the time engineers are selected for upskilling, engineers are also moved across teams and the solutions are evaluated by a simulation that feeds the solutions objective values into a fitness function.

In this paper we present a real-valued genetic algorithm for the selection of engineers to be upskilled whilst also evaluating and selecting engineers to move teams, with the hopes of producing the most benefit from the current available engineers. This application was developed for BT's actively deployed iPatch software with feedback from business stakeholders.

2 Overview of the Multi-skilled Workforce Optimisation Problem

2.1 Overview on Multi-skilled Engineers

Engineers could have varying numbers of skills based on the types of tasks they work on and how experienced they are. More experienced engineers are more likely to have more skills and more likely to have more advanced skills.

The skill sets of the engineers will differ between the different areas that the groups of engineers (teams) are assigned to. So a team with a given number of engineers and an optimised set of skills for each engineer will not necessarily be the best setup for another area.

In [8], it is noted that the skill optimisation problem is a Combinatorial Optimisation (CO) problem [9]. Algorithms designed to tackle CO problems usually aim for a metaheuristic approach [10] because the optimisation has to be completed within a reasonable amount of time. This is especially true for our problem as the environment changes on a frequent basis and needs to undergo regular optimisation. A common approach to tackling these large scale and complex optimisation problems is Genetic Algorithms (GA) [11–13].

2.2 Overview on Team Organisation Optimisation

Making sure engineers have the most optimal skill sets is just one part of the problem. This is because any change in the team's abilities can have sub-optimal implications. As such it would be necessary to reorganise the teams after the engineer skill sets have been changed.

An example of one of these implications would be that, in an area of low utilisation a few engineers may be selected to train in a specialist skill, so they can pick up more work and hence be more utilised. However for the engineer that was already a specialist, their work will be reduced. Possibly to a point where the engineer becomes grossly underutilised. As a result it may be more beneficial to move that engineer into a neighboring team. Especially if that area's team is near maximum utilisation but low completion of tasks (meaning there are more tasks than there is time available from the engineers).

This additional layer of change adds more complexity to the problem, because if the re-organisation of teams happens after the skill optimisation has taken place then the results will be sub-optimal. The team re-organisation has to be done during the solution evaluation when engineers are being selected for training.

2.3 Objective and Constraints

In this work we present a number of objectives that need to be assessed when evaluating a mobile workforce. These objectives will be present as if they are of equal weight. The objectives applicable to this problem are the following:

Task Coverage (*C*): the percentage of the tasks estimated to be completed by the engineers at the end of the simulation. This is a maximisation objective.

Travel Distance (*T*): the distance in km an engineer, on average, has to travel in the simulated area. This is a minimisation objective.

Utilisation (*U*): the average utilisation of the engineers. This is a maximisation objective

These objectives are used within the fitness function of the genetic algorithm. The fitness function is given in Eq. (1).

$$F = \frac{CU}{T} \tag{1}$$

The total coverage (C) is represented by sum total of all engineers (n) completed work Ci divided by the Total task (W). This is shown in Eq. (2).

$$C = \frac{1}{W}\sum\nolimits_{i=1}^{n} C_i \tag{2}$$

The average utilization (U) is the sum total of engineer completed work (C) divided by engineer available time (A), this is then divided by the sum total number of engineers (n) to get the average utilisation. This is shown in Eq. (3).

$$U = \frac{1}{n}\sum\nolimits_{i=1}^{n} \frac{C_i}{A_i} \tag{3}$$

The average travel (T), shown in Eq. (4), is represented as the sum total of all engineers travel distance (Ti) divided by the sum total of engineers (n). This is shown in Eq. (4).

$$T = \frac{1}{n}\sum\nolimits_{i=1}^{n} T_i \tag{4}$$

We do not include any weighting factors in the fitness function. Business objectives change on a regular basis so we will evaluate any solution to our problem with all objectives equal. This will help to determine what solutions to our problem are the best not only overall but in any particular objective.

3 The Proposed GA Based Skill and Team Allocation Optimistion System

The proposed optimisation system is a real valued genetic algorithm based solution (RVGA). The genes in each of the solutions represents an engineer ID. The solution length (number of genes) is related to the number of upskills, where an upskill is the next logical skill set for any given engineer. Figure 1 shows an example of the real valued chromosome, where an ID of an engineer is stored within each gene. This then tells the simulation that this set of engineers needs to use their upskilled skill set.

The reason why we don't use a binary valued GA here is because each gene would have to represent an engineer that could be upskilled. The GA would then switch on/off engineers to be trained, but this would be uncontrolled. The GA could select any number of engineers to train and not optimise for the number we have specified. In most situations the binary GA will switch on all engineers to be trained as this gives the most benefit. However this is not practical from a business point of view for several reasons, not least the cost of training all engineers as well as the opportunity cost of the lost time while the engineer are on training courses.

ID: 14	ID: 65	ID: 21	ID: 78	ID: 123	ID: 105	ID: 8	ID: 53

Fig. 1 Upskilling chromosome

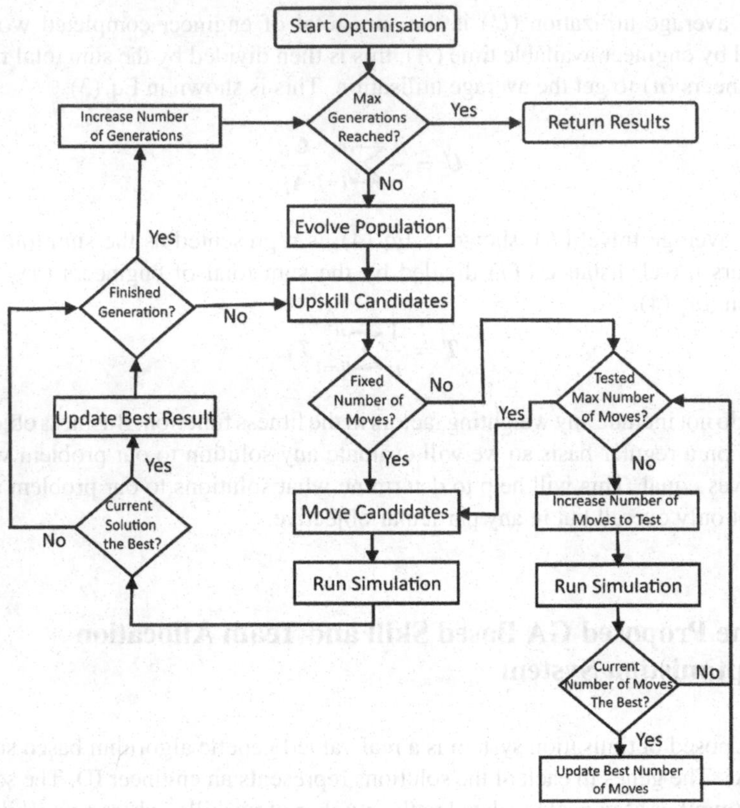

Fig. 2 Solution creation and evaluation

One problem with the RVGA that we have controlled for is the chromosome containing duplicates of the same engineer ID. As if there is a duplicate the GA will then give a result of one engineer less than we wanted. This solution will be penalised and given a zero fitness value, as it is not what we asked for.

During the solution evaluation section of the GA, shown in Fig. 2, each solution will take the workforce to be optimised and give engineers (that have been selected by the genes) their next set of skills. This is the upskill candidates step. The daily simulation will run at the end of this step and the effect of the upskills will be measured. After the upskilling we have two options for moving engineers. We either set a fixed number of engineers to move (N) or try to evaluate how many would be the best to move.

If we have a fixed number, we move onto the next step and move N number of engineers. If we want the system to decide how many engineers to move we start at 0 moves and evaluate the effect of increasing the number of moves up to the maximum number we want to test for. As a result, N number of moves will be equal to the number of moves that gave the best result.

Once we have determined N, we select N number of engineers to move to their closest alternative team. The alternative team is the one that is geographically closest to them. As we don't want engineers to travel too far to their first task.

The engineers selected here are those deemed least utilised. As the simulation runs, it collects data on the estimated jobs an engineers is likely to complete during a day. Thus it calculates the engineer's utilisation. The N number of engineers with the lowest utilisation will be moved.

The system then runs the simulation again, once the teams have been altered, and the results of this second simulation are put into the fitness function to score the solution. This way the selection of the engineers to be upskilled will affect the engineers that will move teams, with the hope that both aspects will be taken into account during the optimisation process and produces better results than upskilling or moving engineers separately.

4 Experiments and Results

Our data is collected from a real world mobile engineer workforce and allows us to evaluate current teams and their performance, then run simulations adjusting the teams and workforce skill sets and compare the affects. In our experiments we selected a region to optimise. This region contained 8 sub regions. Each sub region contains working areas. The teams are allocated to the working areas and any team reorganisation at the sub-region level involves moving engineers between the working areas.

Tables 1, 2, 3, 4, 5, 6, 7 and 8 show the optimisation results for each sub-region. The tables show the original results from the current teams with their current skill sets. Then each column shows the results from a different experiment with the aim of improving in the objectives. The results shown from these experiments are the average of 5 runs of each of the experiments. We are only showing the average of the 5 runs due to space limitations. The GA parameters were tuned empirically. Multiple runs were done and the best performing parameters were chosen for the experiments.

For the purpose of our experiments, we keep the number of upskills at 10, this is based on our previous work [7]. The number of fixed moves we set at 5. Our

Table 1 Sub-region 1

Objective	Original	Moves only	Upskill only	Upskill *then* move	Upskills and move	Upskills dynamic moves
Cov.	89.47	89.47	91.92	91.95	92.24	92.34
Trav.	9.15	9.15	8.07	8.29	8.12	8.09
Util	80.69	80.69	82.90	82.93	83.20	83.29

Table 2 Sub-region 2

Objective	Original	Moves only	Upskill only	Upskill *then* move	Upskills and move	Upskills dynamic moves
Cov.	94.18	94.96	94.18	94.18	94.82	94.18
Trav.	7.28	7.32	5.87	5.92	5.93	5.88
Util	84.30	84.99	84.30	84.30	84.87	84.30

Table 3 Sub-region 3

Objective	Original	Moves only	Upskill only	Upskill *then* move	Upskills and move	Upskills dynamic moves
Cov.	86.10	86.89	88.24	87.80	89.70	88.34
Trav.	5.81	6.24	4.96	5.59	6.32	5.33
Util	68.80	69.43	70.51	70.16	71.68	70.59

Table 4 Sub-region 4

Objective	Original	Moves only	Upskill only	Upskill *then* move	Moves and upskills	Upskills dynamic moves
Cov.	95.52	95.52	95.57	95.57	95.59	95.59
Trav.	8.15	8.15	6.85	7.07	6.83	6.84
Util	81.39	81.39	81.44	81.44	81.46	81.46

Table 5 Sub-region 5

Objective	Original	Moves only	Upskill only	Upskill *then* move	Upskills and move	Upskills dynamic moves
Cov.	89.84	89.84	90.06	90.06	89.79	90.06
Trav.	6.66	6.66	6.52	6.61	6.90	6.54
Util	80.57	80.57	80.76	80.76	80.51	80.76

Table 6 Sub-region 6

Objective	Original	Moves only	Upskill only	Upskill *then* move	Upskills and move	Upskills dynamic moves
Cov.	89.36	89.36	89.20	89.14	89.17	89.11
Trav.	9.06	9.06	7.73	7.78	7.68	7.69
Util	88.87	88.87	88.71	88.65	88.68	88.62

Table 7 Sub-region 7

Objective	Original	Moves only	Upskill only	Upskill *then* move	Upskills and move	Upskills dynamic moves
Cov.	81.56	82.30	81.87	82.08	84.47	82.06
Trav.	6.27	6.29	5.54	5.59	5.66	5.47
Util	75.86	76.55	76.15	76.35	78.57	76.30

Table 8 Sub-region 8

Objective	Original	Moves only	Upskill only	Upskill *then* move	Upskills and move	Upskills dynamic moves
Cov.	93.32	93.32	93.33	93.33	93.16	93.34
Trav.	11.65	12.00	10.54	10.67	10.71	10.53
Util	81.57	81.57	81.58	81.58	81.43	81.59

GA settings are 0.4 for crossover, 0.05 for mutation, a population of 100 and max generations is set at 30 (We get convergence of results with these settings by 30 generations).

Fixed moves without any upskilling is our first experiment. We want to know the effect of just moving the 5 least utilised workers in each sub-region.

From Tables 1, 2, 3, 4, 5, 6, 7 and 8 we can see that just moving the least utilised will increase the coverage in 3 of the 8 sub-regions with the remaining 5 having no affect. In addition travel actually increases as a result of moving engineers in 4 of the 8 sub-regions.

This is most likely for two reasons. Firstly the move only process is not part of the GA system, so it does not use the fitness function, so is not constrained by travel. Secondly, it simply looks at the least utilised engineers and assigns them to a different team. Thus the engineer will then have to travel further to their new working area, as they will now be assigned to one that is further away.

Our second set of experiments look at just running the upskilling optimisation. With 10 engineers throughout the sub-region being selected for training. This has the effect of increasing coverage in 6 of the 8 areas. Of these 6 areas, 4 of them performed better in coverage than just moving engineers. In sub-region 7 (Table 7), moving only performed better than upskill only, however both improved on the original.

Our third set of experiments looked at a step process in which we first upskill 10 engineers via the GA process, then once the GA process is complete we move the 5 least utilised workers based on the new upskills. The method produces suboptimal results as the GA has attempted to find the best solution for upskilling engineers, then the solution is impacted (usually made worse) by moving engineers between teams.

Although this method produced coverage result that were better than just moving the engineers (sub-regions 1, 3, 4, 5 and 8) in the majority of cases, 75 %, this method performed worse than upskill only when travel was also taken into account (sub-regions 2, 3, 4, 5, 6 and 8). This led us to the last two experiments.

Our fourth set of experiments looked at combining the upskilling optimisation with moving the least utilised engineers within the GA process. This experiment now has the advantage of applying the fitness function to new team configurations. This process improves in coverage in 5 of the 8 sub-regions and 6 of the 8 in travel.

The final set of experiments looks at allowing the system to run simulation tests to alter the number of moves to find the best number of engineers to move within each sub-region. The system could choose up to 10 engineers to move. This resulted in 6 of the sub-regions being improved in coverage, with only sub-region 6 (Table 6) performing worse. All sub-regions have improved travel distances when compared to upskill *then* move. The reason for sub-region 6 performing worse in coverage could be because of the fitness function. The reduction in travel of 15.12 % may be why the solutions produced for sub-region 6 has a small 0.28 % reduction in coverage given the equal weighting of these objectives.

In all cases for the combined optimisation methods, either fixed moves or dynamic moves, outperformed the step process of upskilling *then* moving engineers. With fixed number of moves being the process that outperforms the step process most often, in 75 % of cases (Sub-Regions: 1, 2, 3, 4, 6 and 7). This shows us that combining these methods of increasing coverage into a single GA process produces better results.

Table 9 gives an overview on the coverage improvements with the different experiments. On average for the 8 sub-regions we can see that only moving engineers gives a 0.29 % improvement. Only upskilling gives a 0.63 % improvement. Moving and upskilling together produce the best improvement, with the dynamic number of moves giving a 0.71 % improvement and the fixed number of moves giving a 1.20 % improvement. When this is applied to the 7571 h worth of work across the region, this 1.20 % improvement is equal to 90.85 h work.

Table 9 Coverage evaluation

Sub-region	Moves only (%)	Upskill only (%)	Upskill *then* move (%)	Upskills and move (%)	Upskills and dynamic moves (%)
1	0.00	2.45	2.48	2.77	2.87
2	0.78	0.00	0.00	0.64	0.00
3	0.79	2.14	1.70	3.60	2.24
4	0.00	0.05	0.05	0.07	0.07
5	0.00	0.22	0.22	−0.05	0.22
6	0.00	−0.16	−0.25	−0.19	−0.25
7	0.74	0.31	0.52	2.91	0.50
8	0.00	0.01	0.01	−0.16	0.02
AVG.	**0.29**	**0.63**	**0.59**	**1.20**	**0.71**

Table 10 Travel evaluation

Sub-region	Moves only (km)	Upskill only (km)	Upskill *then* Move (km)	Upskills and move (km)	Upskills and dynamic moves (km)
1	0.00	−1.08	−0.86	0.05	−0.03
2	0.04	−1.45	−1.36	0.06	−0.05
3	0.43	−1.28	−0.22	1.36	−0.99
4	0.00	−1.30	−1.08	−0.01	0.00
5	0.00	−0.14	−0.05	0.38	−0.36
6	0.00	−1.33	−1.37	−0.05	0.01
7	0.02	−0.75	−0.68	0.12	−0.18
8	0.35	−1.46	−0.98	0.17	−0.18
AVG.	**0.11**	**−1.10**	**−0.82**	**0.26**	**−0.22**

Table 10 gives an overview on the travel improvements. If coverage increases we expect travel to increase also. This is because the engineers are travelling to and completing more tasks. So these are directly conflicting objectives. If the travel is also being reduced at the same time as increased coverage, then the task allocation for the engineers has become much more efficient.

Table 10 also shows us the extra kilometers traveled per engineer as a result of the optimisation attempts. Moving only will increase travel on average for the sub-regions. Upskiling only produces the most travel benefit. This is logical as the same teams are given a wider selection of jobs to choose from, meaning they are more likely to choose jobs that are closer to them. A fixed number of moves and upskills increases travel the most on average. This makes sense as it is also the optimisation technique that increases coverage the most.

Upskilling with a dynamic number of moves actually reduces travel despite having the second highest coverage increase. The difference between fixed moves and dynamic moves is 0.48 Km per engineer. Which is significant as in this region there are 1481 engineers. Resulting in a difference of about 710 Km per day.

Given this, for the combined optimisation techniques, fixed moves increase travel 385 Km per day, while dynamic moves gives a reduction of 325 Km per day. If the regional manager is looking to reduce fuel consumption cost and CO_2 emissions, using the dynamic system looks far more attractive.

Alternatively if the regional manager has the goal of completing more tasks, which results in increased customer satisfaction and a reduced reliance on contractor work, then the fixed option looks better.

Whichever of these options is chosen it will be with the combined optimisation, as the combined optimisation techniques outperform either move only, upskill only and upskill *then* move, in coverage. If travel is of a concern the dynamic moves option may be the best, given that it is the second best at increasing coverage but also has the benefit of reducing travel.

5 Real World Application

The developed system has been deployed as part of BT's iPatch solution which is focused on organisational optimisation for two key areas, (1) geography and (2) resources. Geography optimisation looks at the geographical areas that the teams of engineers work within and how best these areas can be designed [14]. The second area of focus is the resources where it looks at how resources are organised into teams and what skills they have, the topic of this paper.

Initially a business problem with resource management was highlighted, so we looked at how we could develop the iPatch tool to help solve this problem using similar techniques we have had success with before. The problem was originally proposed by BT's Field Engineering division where they wanted to know which engineers would give the greatest benefit after they were trained in more skills. Although they were the primary stakeholder, one of our focuses was to keep the solution to the problem generic, so that it could be applied to other areas of the business. We then developed the GA to select any given number of engineers, then simulated the affect they would have with new skills.

We had to model the problem well to give a realistic view of the effect of training the selected engineers. The model to evaluate any of the proposed solutions involved simulating an average day's work. This began with setting up the engineers to be as close to reality as possible. The engineers were grouped into their current teams, placed at their know starting location, given the skills they currently have listed and giving the average amount of hours they work in on day. This data was provided by BT.

The simulation then involved allocating the closest tasks to the engineers based on the skills they have (and how much time they had left for the day). Further feedback from the stakeholders led us to reorder the task allocation so those with few skills would be allocated tasks first. A greedy logic was then implemented here so that the list will choose the engineers with the least amount of skills first and leave the engineers with the most amount of skill last. This meant that highly skilled engineers were not taking jobs from the lower skilled engineers. If this happened, those engineers would be poorly utilised and the higher skilled tasks would not be completed, because the relevant engineers would be doing something else.

The simulation was then run on areas and teams where the objective values were roughly known so that the simulation results could compare against these. The comparison of the real coverage, travel and utilisation were close enough that the stakeholders were happy with the simulation as a means of testing solutions to the resource problem.

The focus between us and the stakeholders after this point was to discuss how the optimisation of upskilling was performing and if the suggestions were logical. After the upskills were seen to be logical another problem was highlighted, with under resourced areas and under utilised resources in neighboring areas. The solution here would be to move engineers between teams, but the most optimal solution to this problem was not known. Thus we set out the discussed experiments to investigate this issue. The results presented gave the stakeholders confidence in the best methods presented for getting the most out of each engineer.

The resource optimisation software was then implemented into BT's iPatch solution and so far, the employed geography and resource optimisation system has generated an increase of productivity of 0.5 % saving an estimated £2million and cutting fuel consumption by 2.9 %, an additional saving of over £150 K.

In addition to the financial benefits, customer commitments are more effectively met, improving the service quality and due to less fuel consumption the company can promote sustainability targets with less CO_2 emitted. The system has also won a GTB award for the best business innovation of the year [15].

These outcomes shows the real world impact these AI technologies are having on a large, nationwide, mobile engineering workforce.

6 Conclusions and Future Work

In this paper we have presented a real value GA system for engineer upskilling and move recommendations. We ran several experiments to evaluate the different options of making these recommendations. The first was to only move the least utilised engineers, the second was to only upskill 10 engineers across the sub-region. The third was to combine both moves and upskilling but with a fixed number of upskills. The final experiment was to combine both the moves and the upskills but with a dynamic number of moves.

The results showed that combining team moves and engineer upskilling in the same optimisation process lead to an overall 1.20 % increase in coverage across the region with the fixed moves option and a 0.71 % increase with a dynamic number of moves chosen by the system. Both of these results produced better coverage that only moving engineers between teams, just upskilling the engineers or upskilling then moving the engineers.

Importantly we saw that combining the upskilling and move process into a single GA improved the results over the step process of upskilling then moving. This was either by a fixed number of moves or dynamic. With fixed moves combined optimisation outperforming the step process 75 % of the time.

In addition we also looked at the affect the optimisation processes had on average travel per engineer. Notably, a dynamic number of moves resulted in less travel whilst also generating the second highest increase in coverage.

The optimisation algorithms used on the mobile workforce have transferred their results well into the real world due to the communication and feedback from the stakeholders in which it was developed for. The BT iPatch tool has given a noticeable financial impact, as well as improved customer satisfaction and a reduction in CO_2 emissions, through reduced fuel consumption.

Our future work will focus on having the system suggest the best number of upskills, rather than have this as one of the systems initial parameters. Similar to how the system will dynamically choose moves number now. However this will also increase the search space and training costs have to be factored into the fitness function.

References

1. Thannimalai, P., Kadhum, M.M., Jeng Feng, C., Ramadass, S.: A glimpse of cross training models and workforce scheduling optimization. In: IEEE Symposium on Computers and Informatics, pp. 98–103 (2013)
2. Cimitile, M., Gaeta, M., Loia, V.: An ontological multi-criteria optimization system for workforce management. In: World Congress on Computational Intelligence, pp. 1–7 (2012)
3. Koole, G., Pot, A., Talim, J.: Routing heuristics for multi-skill call centers. In: Proceedings of the 2003 Simulation Conference, vol. 2, pp. 1813–1816 (2003)
4. Easton, F., Brethen, R.H.: Staffing, Cross-training, and Scheduling with Cross-trained Workers in Extended-hour Service Operations, pp. 1–28 (2011)
5. Lin, A., Ahmad, A.: SilTerra's experience in developing multi-skills technician. In: IEEE International Conference on Semiconductor, Electronics, pp. 508–511 (2004)
6. Haas, C.T., Borcherding, J.D., Glover, R.W., Tucker, R.L., Rodriguez, A., Gomar, J.: Planning and scheduling a multiskilled workforce, Center for Construction Industry Studies (1999)
7. Starkey, A., Hagras, H., Shakya, S., Owusu, G.: A Genetic Algorithm Based Approach for the Optimisation of Workforce Skill Sets, AI-2015, pp. 261–272 (2015)
8. Hu, Z., Mohd, R., Shboul, A.: The application of ant colony optimization technique (ACOT) for employees selection and training. In: First International Workshop on Database Technology and Applications, pp. 487–502 (2009)
9. Turchyn, O.: Comparative analysis of metaheuristics solving combinatorial optimization problems. In: 9th International Conference on the Experience of Designing and Applications of CAD Systems in Microelectronics, pp. 276–277 (2007)
10. Fanm, W., Gurmu, Z., Haile, E.: A bi-level metaheuristic approach to designing optimal bus transit route network. In: 3rd Annual International Conference on Cyber Technology in Automation, Control and Intelligent Systems, pp. 308–313 (2013)
11. Domberger, R., Frey, L., Hanne, T.: Single and multiobjective optimization of the train staff planning problem using genetic algorithms. In: IEEE Congress on Evolutionary Computation, pp. 970–977 (2008)
12. Liu, Y., Zhao, S., Du, X., Li, S.: Optimization of resource allocation in construction using genetic algorithms. In: Proceedings of the 2005 International Conference on Machine Learning, pp. 18–21 (2005)
13. Tanomaru, J.: Staff Scheduling by a Genetic Algorithm with Heuristic Operators International Conference on Evolutionary Computation, pp. 456–461 (1995)
14. Starkey, A., Hagras, H., Shakya, S., Owusu, G.: A Multi-objective Genetic Type-2 Fuzzy Logic Based System for Mobile Field Workforce Area optimization, Information Sciences, pp. 390–411 (2015)
15. http://www.essex.ac.uk/events/event.aspx?e_id=7695. [Last Accessed: 11/08/16]

Legal Liability, Medicine and Finance

Artificial Intelligence and Legal Liability

J.K.C. Kingston

Abstract A recent issue of a popular computing journal asked which laws would apply if a self-driving car killed a pedestrian. This paper considers the question of legal liability for artificially intelligent computer systems. It discusses whether criminal liability could ever apply; to whom it might apply; and, under civil law, whether an AI program is a product that is subject to product design legislation or a service to which the tort of negligence applies. The issue of sales warranties is also considered. A discussion of some of the practical limitations that AI systems are subject to is also included.

Keywords Social Impact · Acceptance and Implications of AI · Intelligent Decision Support Systems · Industrial Applications of Artificial Intelligence

1 Introduction

A recent issue of a popular computing journal [1] posed the following question: "It is the year 2023, and for the first time, a self-driving car navigating city streets strikes and kills a pedestrian. A lawsuit is sure to follow. But exactly which laws will apply? No-one knows."

The article goes on to suggest that the laws that are likely to apply are those that deal with products with a faulty design. However, it argues that following this legal route holds back the development of self-driving cars, as settlements for product design cases (in the USA) are typically almost ten times higher than for cases involving human negligence, and that does not include the extra costs associated with product recalls to fix the issue. It goes on to argue that such cases should instead be dealt with as cases of negligence, just as they would for a human driver; the author points out that a standard handbook of US tort law [2] states that "A bad state of mind is neither necessary nor sufficient to show negligence; conduct is everything."

J.K.C. Kingston (✉)
University of Brighton, Brighton BN2 4JG, UK
e-mail: j.k.kingston@brighton.ac.uk

M. Bramer and M. Petridis (eds.), *Research and Development in Intelligent Systems XXXIII*, DOI 10.1007/978-3-319-47175-4_20

It may be that the issue will arise even sooner than the year 2023. The author of this paper recently hired a car that included several new safety features. One of these features was that, if the car's radars detected an imminent collision while the car was travelling at between 4 and 19 mph, the car's engine would cut out to help prevent the collision.

While reversing the car out of a driveway, the author drove too close to a hedge. The car sounded its proximity alarm, and cut the engine. However, even when the steering wheel was turned so that the car would miss the hedge, the engine would not restart while the car was in reverse gear. The author had to put the car into a forward gear and travel forward slightly before he was able to continue reversing out.

All of this took place wholly within the driveway. However, if it had taken place while the rear end of the car was projecting into the road, with a heavy lorry travelling at some speed towards the car, most drivers would prefer to risk getting their paintwork scratched by a hedge than sitting in a car that refused to restart and complete the desired manoeuvre. It seems inevitable that soon, some driver will blame these 'safety features' for their involvement in a serious accident.

The purpose of this paper is to consider the current capabilities, or lack of them, of artificial intelligence, and then to re-visit the question of where legal liability might lie in the above cases.

First, it is important to establish what this paper means by the term "artificial intelligence". There are researchers in the AI field who consider anything that mimics human intelligence, by whatever method, to be "artificial intelligence"; there are others who think that the only "artificially intelligent" programs are those that mimic the way in which humans think. There are also those in the field of information systems who would classify many "artificially intelligent" programs as being complex information systems, with 'true' artificial intelligence being reserved for the meta-level decision making that is sometimes characterised as 'wisdom'.

In this paper, any computer system that is able to recognise a situation or event, and to take a decision of the form "IF this situation exists THEN recommend or take this action" is taken to be an artificially intelligent system.

2 Legal Liability

2.1 Criminal Liability

The references cited below refer primarily to US law; however, many other jurisdictions have similar legislation in the relevant areas.

In [3], Gabriel Hallevy discusses how, and whether, artificial intelligent entities might be held criminally liable. Criminal laws normally require both an *actus reus* (an action) and a *mens rea* (a mental intent), and Hallevy helpfully classifies laws as follows:

1. Those where the *actus reus* consists of an action, and those where the *actus reus* consists of a failure to act;
2. Those where the *mens rea* requires knowledge or being informed; those where the *mens rea* requires only negligence ("a reasonable person would have known"); and strict liability offences, for which no *mens rea* needs to be demonstrated.

Hallevy goes on to propose three legal models by which offences committed by AI systems might be considered:

1. Perpetrator-via-another. If an offence is committed by a mentally deficient person, a child or an animal, then the perpetrator is held to be an innocent agent because they lack the mental capacity to form a *mens rea* (this is true even for strict liability offences). However, if the innocent agent was instructed by another person (for example, if the owner of a dog instructed his dog to attack somebody), then the instructor is held criminally liable (see [4] for US case law).
   According to this model, AI programs could be held to be an innocent agent, with either the software programmer or the user being held to be the perpetrator-via-another.
2. Natural-probable-consequence. In this model, part of the AI program which was intended for good purposes is activated inappropriately and performs a criminal action. Hallevy gives an example (quoted from [5]) in which a Japanese employee of a motorcycle factory was killed by an artificially intelligent robot working near him. The robot erroneously identified the employee as a threat to its mission, and calculated that the most efficient way to eliminate this threat was by pushing him into an adjacent operating machine. Using its very powerful hydraulic arm, the robot smashed the surprised worker into the machine, killing him instantly, and then resumed its duties.
   The normal legal use of "natural or probable consequence" liability is to prosecute accomplices to a crime. If no conspiracy can be demonstrated, it is still possible (in US law) to find an accomplice legally liable if the criminal acts of the perpetrator were a natural or probable consequence (the phrase originated in [6]) of a scheme that the accomplice encouraged or aided [7], as long as the accomplice was aware that some criminal scheme was under way.
   So users or (more probably) programmers might be held legally liable if they knew that a criminal offence was a natural, probable consequence of their programs/use of an application. The application of this principle must, however, distinguish between AI programs that 'know' that a criminal scheme is under way (i.e. they have been programmed to perform a criminal scheme) and those that do not (they were programmed for another purpose). It may well be that crimes where the *mens rea* requires knowledge cannot be prosecuted for the latter group of programs (but those with a 'reasonable person' *mens rea*, or strict liability offences, can).
3. Direct liability. This model attributes both *actus reus* and *mens rea* to an AI system.
   It is relatively simple to attribute an *actus reus* to an AI system. If a system takes an action that results in a criminal act, or fails to take an action when there is a duty to act, then the *actus reus* of an offence has occurred.

Assigning a *mens rea* is much harder, and so it is here that the three levels of *mens rea* become important. For strict liability offences, where no intent to commit an offence is required, it may indeed be possible to hold AI programs criminally liable. Considering the example of self-driving cars, speeding is a strict liability offence; so according to Hallevy, if a self-driving car was found to be breaking the speed limit for the road it is on, the law may well assign criminal liability to the AI program that was driving the car at that time.
This possibility raises a number of other issues that Hallevy touches on, including defences (could a program that is malfunctioning claim a defence similar to the human defence of insanity? Or if it is affected by an electronic virus, could it claim defences similar to coercion or intoxication?); and punishment (who or what would be punished for an offence for which an AI system was directly liable?).

2.2 *The Trojan defence*

In the context of defences against liability for AI systems, it is important to mention a number of cases where a defendant accused of cybercrime offences has successfully offered the defence that his computer had been taken over by a Trojan or similar malware program, which was committing offences using the defendant's computer but without the defendant's knowledge. A review of such cases can be found in [8]; they include a case in the United Kingdom when a computer containing indecent pictures of children was also found to have eleven Trojan programs on it, and another UK case where a teenage computer hacker's defence to a charge of executing a denial of service attack was that the attack had been performed from the defendant's computer by a Trojan program, which had subsequently wiped itself from the computer before it was forensically analysed. The defendant's lawyer successfully convinced the jury that such a scenario was not beyond reasonable doubt.

2.3 *Civil Law: Torts and Breach of Warranty*

2.3.1 Negligence

When software is defective, or when a party is injured as a result of using software, the resulting legal proceedings normally allege the tort of negligence rather than criminal liability [9]. Gerstner [10] discusses the three elements that must normally be demonstrated for a negligence claim to prevail:

1. The defendant had a duty of care;
2. The defendant breached that duty;
3. That breach caused an injury to the plaintiff.

Regarding point 1, Gerstner suggests there is little question that a software vendor owes a duty of care to the customer, but it is difficult to decide what standard of care is owed. If the system involved is an "expert system", then Gerstner suggests that the appropriate standard of care is that of an expert, or at least of a professional.

On point 2, Gerstner suggests numerous ways in which an AI system could breach the duty of care: errors in the program's function that could have been detected by the developer; an incorrect or inadequate knowledge base; incorrect or inadequate documentation or warnings; not keeping the knowledge up to date; the user supplying faulty input; the user relying unduly on the output; or using the program for an incorrect purpose.

As for point 3, the question of whether an AI system can be deemed to have caused an injury is also open to debate. The key question is perhaps whether the AI system *recommends* an action in a given situation (as many expert systems do), or *takes* an action (as self-driving and safety-equipped cars do). In the former case, there must be at least one other agent involved, and so causation is hard to prove; in the latter case, it is much easier.

Gerstner also discussed an exception under US law for "strict liability negligence." This applies to products that are defective or unreasonably dangerous when used in a normal, intended or reasonably foreseeable manner, and which cause injury (as opposed to economic loss). She discusses whether software is indeed a 'product' or merely a 'service'; she quotes a case in which electricity was held to be a product [11], and therefore leans towards defining software as a product rather than a service. Assuming that software is indeed a product, it becomes incumbent on the developers of AI systems to ensure that their systems are free from design defects; manufacturing defects; or inadequate warning or instructions.

Cole [12] provides a longer discussion of the question of whether software is a product or a service. His conclusion is that treating AI systems as products is "partially applicable at best", and prefers to view AI as a service rather than a product; but he acknowledges that law in the area is ill-defined.

Cole cites some case law regarding the "duty of care" that AI systems must abide by:

1. In [13], a school district brought a negligence claim against a statistical bureau that (allegedly) provided inaccurate calculations of the value of a school that had burned down, causing the school district to suffer an underinsured loss. The duty being considered was the duty to provide information with reasonable care. The court considered factors including: the existence, if any, of a guarantee of correctness; the defendant's knowledge that the plaintiff would rely on the information; the restriction of potential liability to a small group; the absence of proof of any correction once discovered; the undesirability of requiring an innocent party to carry the burden of another's professional mistakes; and, the promotion of cautionary techniques among the informational (tool) providers.
2. Based on [14], Cole discusses the duty to provide reasonable conclusions from unreasonable inputs. He follows [15] to suggest that AI developers probably have an affirmative duty to provide relatively inexpensive, harmless, and simple, input

error-checking techniques, but notes that these rules may not apply where the AI program is performing a function in which mistakes in input may be directly life-threatening (e.g. administering medicine to a patient); in such cases, he suggests applying the rules relating to "ultra-hazardous activities and instrumentalities" instead [16].
3. Cole suggests that AI systems must be aware of their limitations, and this information must be communicated to the purchaser. It is well established that vendors have a duty to tell purchasers of any known flaws; but how can unknown weaknesses or flaws be established, and then communicated?

2.3.2 Breach of Warranty

If an AI system is indeed a product, then it must be sold with a warranty; even if there is no express warranty given by the vendor (or purchased by the user), there is an implied warranty that it is (to use the phrase from the UK Sale of Goods Act 1979), "satisfactory as described and fit for a reasonable time." Some jurisdictions permit implied warranties to be voided by clauses in the contract; however, when an AI system is purchased built into other goods (such as a car), it seems unlikely that any such contractual exclusions (e.g. between the manufacturer of the car and the supplier of the AI software) could successfully be passed on to the purchaser of the car.

2.4 *Legal Liability: Summary*

So it seems that the question of whether AI systems can be held legally liable depends on at least three factors:

- The limitations of AI systems, and whether these are known and communicated to the purchaser;
- Whether an AI system is a product or a service;
- Whether the offence requires a *mens rea* or is a strict liability offence.

If an AI system is held liable, the question arises of whether it should be held liable as an innocent agent, an accomplice, or a perpetrator.

The final section of this paper considers the first of these three factors.

3 Limitations of AI systems

The various limitations that AI systems are subject to can be divided into two categories:

- Limitations that human experts with the same knowledge are also subject to;
- Limitations of artificial intelligence technology compared with humans.

3.1 Limitations that Affect Both AI Systems and Human Experts

The limitations that affect both AI systems and human experts are connected with the knowledge that is specific to the problem.

Firstly, the knowledge may change very rapidly. This requires humans and AI systems both to know what the latest knowledge is, and also to identify which parts of their previous knowledge is out of date. Whether this is an issue depends almost entirely on the domain: in our example of automated car driving, the knowledge that is required to drive a car changes very slowly indeed. However, in the world of cyber security, knowledge of exploits and patches changes on a daily basis.

Secondly, the knowledge may be too vast for all possibilities to be considered. AI systems can actually perform better than human experts at such tasks – it is feasible to search thousands, or even hundreds of thousands of solutions – but there are still some tasks where the scope is even wider than that. This is typically the case in planning and design tasks, where the number of possible plans or designs may be close to infinite. (In contrast, scheduling and configuration, which require planning and design within a fixed framework, are less complex, though the possible options may still run into thousands). In such cases, AI systems can promise to give a good answer in most cases, but cannot guarantee that they will give the best answer in all cases.

From a legal standpoint, it could be argued that the solution to such issues is for the vendor to warn the purchaser of an AI system of these limitations. In fast-changing domains, it may also be considered legally unreasonable if the vendor does not provide a method for frequently updating the system's knowledge. This raises the question of where the boundaries of 'fast-changing' lie. As ever, the legal test is reasonableness, which is usually compared against the expected life of an AI system; so if knowledge was expected to change annually (e.g. in an AI system for calculating personal tax liability), then it would probably be judged reasonable for a vendor to warn that the knowledge was subject to change. However, it would probably not be judged 'reasonable' for the vendor to provide automatic updates to the knowledge, because the complexity of tax law is such that any updates would not merely require downloading files of data and information; they would require a newly developed and newly tested system.

In contrast, AI systems that help local councils calculate household benefits may have been built on the (apparently unshakeable) assumption that marriage was between a man and a woman. That knowledge has now changed, however, to permit marriage between any two human adults. Is it reasonable to require a vendor to warn a purchaser that those laws could change too? Such changes seem highly unlikely at present; but in the USA, there have already been attempts by a woman to marry her dog and by a man to marry his laptop, and there has been long-running lobbying from certain religious groups to legalise polygamy.

The US case of Kociemba v Searle [17] found a pharmaceutical manufacturer liable for failing to warn purchasers that use of a particular drug was associated with a pelvic inflammatory disease, even though the product had been passed as "safe and effective" by the Food and Drug Administration. It seems, therefore, that the boundary of where a warning might reasonably be required is indeed dependent on knowledge rather than on regulatory approval.

Mykytyn et al. [18] discuss issues of legal liability for AI systems that are linked to identification and selection of human experts. They quote two cases [19, 20] where hospitals were found liable for failing to select physicians with sufficient competence to provide the medical care that they were asked to provide; by analogy, AI developers could also be held liable unless they select experts with sufficient competence in the chosen domain, or warn users that the expert's competence does not extend to other domains where the system might conceivably be used.

The solution proposed by Mykytyn et al. is to use licensed and certified experts. They point out that the standards required by licensing bodies are sometimes used to determine if a professional's performance is up to the level expected [21]. They even suggest that it may be desirable to get the AI system itself licensed. The US Securities and Exchange Commission has been particularly keen on this; it required a stock market recommender system to be registered as a financial adviser [22] and also classified developers of investment advice programs as investment advisors [23].

3.2 *Limitations of AI Systems that do not Affect Human Experts*

The key limitation is that AI systems lack general knowledge. Humans carry a great deal of knowledge that is not immediately relevant to a specific task, but that could become relevant. For example, when driving a car, it is advisable to drive slowly when passing a school, especially if there is a line of parked cars outside it, or you know that the school day finishes at about the time when you are driving past. The reason is to avoid the risk of children dashing out from behind parked cars, because a human driver's general knowledge includes the fact that some children have poor road safety skills. An automated car would not know to do this unless it was programmed with a specific rule, or a set of general rules about unusually hazardous locations.

Admittedly, there are occasions when humans fail to apply their general knowledge to recognise a hazardous situation: as one commentator once said, "What is the difference between a belt-driven vacuum cleaner and a Van de Graaff generator? Very little. Never clean your laptop with that type of vacuum cleaner." However, without general knowledge, AI systems have no chance of recognising such situations.

A related issue is that AI systems are notoriously poor at degrading gracefully. This can be seen when considering edge cases (cases where one variable in the case takes an extreme value) or corner cases (multiple variables take extreme values). When human beings are faced with a situation that they previously believed to be very unlikely or impossible, they can usually choose a course of action that has some positive effect on the situation. When AI systems face a situation that they are not programmed for, they generally cannot perform at all.

For example, in the car driving example given at the start of this paper, the (hypothetical) situation where the car refuses to start while a lorry is bearing down on it is an edge case. Furthermore, the car's safety system does not seem to have been designed with city drivers in mind; the car warns drivers to see that their route is safe before making a manoeuvre, but it does not take account of the fact that in a city, a route may only be safe for a short period of time, thus making this type of 'edge' case more common than expected.

As for a corner case, September 26 1983 was the day when a Soviet early-warning satellite indicated first one, then two, then eventually that five US nuclear missiles had been launched. The USSR's standard policy at the time was to retaliate with its own missiles, and it was a time of high political tension between the USA and USSR. The officer in charge had a matter of minutes to decide what to do, and no further information; he chose to consider the message as a false alarm, reasoning that "when people start a war, they don't start it with only five missiles."

It was later discovered that the satellite had mistaken the reflection of sun from clouds as the heat signature of missile launches. The orbit of the satellite was designed to avoid such errors, but on that day (near the Equinox) the location of the satellite, the position of the sun and the location of US missile fields all combined to give five false readings.

If an AI system had been in charge of the Soviet missile launch controls that day, it may well have failed to identify any problem with the satellite, and launched the missiles. It would then have been legally liable for the destruction that followed, although it is unclear whether there would have been any lawyers left to prosecute the case.

A third issue is that AI systems may lack the information that humans use because of poorer quality inputs. This is certainly the case with the car safety system; its only input devices are relatively short range radar detectors, which cannot distinguish between a hedge and a lorry, nor can detect an object that is some distance away but is rapidly approaching. It may be that, should a case come to court regarding an accident 'caused' by these safety systems, the focus will be on how well the AI was programmed to deal with these imprecise inputs.[1]

There is also the issue of non-symbolic information. In the world of knowledge management, it is common to read assertions that human knowledge can never be

[1] After this paper was submitted for publication, the first fatality involving a self-driving car was reported from Florida. A white sided trailer had driven across the car's path; it was a bright sunny day and the car's radars failed to distinguish the trailer against the bright sky. The driver was sitting in the driver's seat and was therefore theoretically able to take avoiding action, but was allegedly watching a DVD at the time. Liability has not yet been legally established.

fully encapsulated in computer systems because it is too intuitive [24]. Kingston [25] argues that this view is largely incorrect because it is based on a poor understanding of the various types of tacit knowledge; but he does allow that non-symbolic information (information based on numbers; shapes; perceptions such as textures; or physiological information e.g. the muscle movements of a ballet dancer), and the skills or knowledge generated from such information, are beyond the scope of nearly all AI systems.

In some domains, this non-symbolic information is crucial: physicians interviewing patients, for example, draw a lot of information from a patient's body language as well as from the patient's words. Some of the criticisms aimed at the UK's current telephone-based diagnostic service, NHS Direct, can be traced back to the medical professional lacking this type of information. In the car-driving example, non-symbolic information might include headlights being flashed by other drivers to communicate messages from one car to another; such information is not crucial but it is important to being a driver who respects others.

4 Conclusion

It has been established that the legal liability of AI systems depends on at least three factors:

1. Whether AI is a product or a service. This is ill-defined in law; different commentators offer different views.
2. If a criminal offence is being considered, what *mens rea* is required. It seems unlikely that AI programs will contravene laws that require knowledge that a criminal act was being committed; but it is very possible they might contravene laws for which 'a reasonable man would have known' that a course of action could lead to an offence, and it is almost certain that they could contravene strict liability offences.
3. Whether the limitations of AI systems are communicated to a purchaser. Since AI systems have both general and specific limitations, legal cases on such issues may well be based on the specific wording of any warnings about such limitations.

There is also the question of who should be held liable. It will depend on which of Hallevy's three models apply (perpetrator-by-another; natural-probable-consequence; or direct liability):

- In a perpetrator-by-another offence, the person who instructs the AI system—either the user or the programmer—is likely to be found liable.
- In a natural-or-probable-consequence offence, liability could fall on anyone who might have foreseen the product being used in the way it was; the programmer, the vendor (of a product), or the service provider. The user is less likely to be blamed unless the instructions that came with the product/service spell out the limitations of the system and the possible consequences of misuse in unusual detail.

- AI programs may also be held liable for strict liability offences, in which case the programmer is likely to be found at fault.

However, in all cases where the programmer is deemed liable, there may be further debates whether the fault lies with the programmer; the program designer; the expert who provided the knowledge; or the manager who appointed the inadequate expert, program designer or programmer.

References

1. Greenblatt, N.A.: Self-driving Cars and the Law. IEEE Spectrum, p. 42 (16 Feb 2016)
2. Dobbs, D.B.: Law of Torts. West Academic Publishing (2008)
3. Hallevy, G.: The Criminal Liability of Artificial Intelligence Entities. http://ssrn.com/abstract=1564096 (15 Feb 2010)
4. Morrisey v. State, 620 A.2d 207 (Del.1993); Conyers v. State, 367 Md. 571, 790 A.2d 15 (2002); State v. Fuller, 346 S.C. 477, 552 S.E.2d 282 (2001); Gallimore v. Commonwealth, 246 Va. 441, 436 S.E.2d 421 (1993)
5. Weng, Y.-H., Chen, C.-H., Sun, C.-T.: Towards the human-robot co-existence society: on safety intelligence for next generation robots. Int. J. Soc. Robot. **267**, 273 (2009)
6. United States v. Powell, 929 F.2d 724 (D.C.Cir.1991)
7. Sayre, F.B.: Criminal responsibility for the acts of another, 43 Harv. L. Rev. **689** (1930)
8. Brenner, S.W., Carrier, B., Henninger, J.: The trojan horse defense in cybercrime cases, 21 Santa Clara High Tech. L.J. **1**. http://digitalcommons.law.scu.edu/chtlj/vol21/iss1/1 (2004)
9. Tuthill, G.S.: Legal Liabilities and Expert Systems, AI Expert (Mar 1991)
10. Gerstner, M.E.: Comment, liability issues with artificial intelligence software, 33 Santa Clara L. Rev. **239**. http://digitalcommons.law.scu.edu/lawreview/vol33/iss1/7 (1993)
11. Ransome v. Wisconsin Elec. Power Co., 275 N.W.2d 641, 647-48. Wis. (1979)
12. Cole, G.S.: Tort liability for artificial intelligence and expert systems, 10 Comput. L.J. **127** (1990)
13. Independent School District No. 454 v. Statistical Tabulating Corp 359 F. Supp. 1095. N.D. Ill. (1973)
14. Stanley v. Schiavi Mobile Homes Inc., 462 A.2d 1144. Me. (1983)
15. Helling v. Carey 83 Wash. 2d 514, 519 P.2d 981 (1974)
16. Restatement (Second) of Torts: Sections 520-524. op.cit
17. Kociemba v. GD Searle & Co., 683 F. Supp. 1579. D. Minn. (1988)
18. Mykytyn, K., Mykytyn, P.P., Lunce, S.: Expert identification and selection: legal liability concerns and directions. AI Soc. **7**(3), 225–237 (1993)
19. Joiner v Mitchell County Hospital Authority, 186 S.E.2d 307. Ga.Ct.App. (1971)
20. Glavin v Rhode Island Hospital, 12 R. I. 411, 435, 34 Am. Rep. **675**, 681 (1879)
21. Bloombecker, R.: Malpractice in IS? Datamation **35**, 85–86 (1989)
22. Warner, E.: Expert systems and the law. In: Boynton and Zmud (eds.) Management Information Systems, Scott Foresman/Little Brown Higher Education, Glenview Il, pp. 144–149 (1990)
23. Hagendorf, W.: Bulls and bears and bugs: computer investment advisory programs that go awry. Comput. Law J. **X**, 47–69 (1990)
24. Jarche, H.: Sharing Tacit Knowledge. http://www.jarche.com/2o1o/o1/sharing-tacit-knowledge/ (2010). Accessed April 2012
25. Kingston, J.: Tacit knowledge: capture, sharing, and unwritten assumptions. J. Knowl. Manage. Pract. **13**(3) (Sept 2012)
26. Restatement (Second) of Torts: Section 552: Information Negligently Supplied for the Guidance of Others (1977)

SELFBACK—Activity Recognition for Self-management of Low Back Pain

Sadiq Sani, Nirmalie Wiratunga, Stewart Massie and Kay Cooper

Abstract Low back pain (LBP) is the most significant contributor to years lived with disability in Europe and results in significant financial cost to European economies. Guidelines for the management of LBP have self-management at their cornerstone, where patients are advised against bed rest, and to remain active. In this paper, we introduce SELFBACK, a decision support system used by the patients themselves to improve and reinforce self-management of LBP. SELFBACK uses activity recognition from wearable sensors in order to automatically determine the type and level of activity of a user. This is used by the system to automatically determine how well users adhere to prescribed physical activity guidelines. Important parameters of an activity recognition system include windowing, feature extraction and classification. The choices of these parameters for the SELFBACK system are supported by empirical comparative analyses which are presented in this paper. In addition, two approaches are presented for detecting step counts for ambulation activities (e.g. walking and running) which help to determine activity intensity. Evaluation shows the SELFBACK system is able to distinguish between five common daily activities with 0.9 macro-averaged F1 and detect step counts with 6.4 and 5.6 root mean squared error for walking and running respectively.

Keywords Intelligent Decision Support Systems · Medical Computing and Health Informatics · Machine Learning

---

S. Sani (✉) · N. Wiratunga · S. Massie · K. Cooper
Robert Gordon University, Aberdeen, UK
e-mail: s.sani@rgu.ac.uk

N. Wiratunga
e-mail: n.wiratunga@rgu.ac.uk

S. Massie
e-mail: s.massie@rgu.ac.uk

K. Cooper
e-mail: k.cooper@rgu.ac.uk

M. Bramer and M. Petridis (eds.), *Research and Development in Intelligent Systems XXXIII*, DOI 10.1007/978-3-319-47175-4_21

1 Introduction

Low back pain (LBP) is a common, costly and disabling condition that affects all age groups. It is estimated that up to 90 % of the population will have LBP at some point in their lives, and the recent global burden of disease study demonstrated that LBP is the most significant contributor to years lived with disability in Europe [5]. Non-specific LBP (i.e. LBP not attributable to serious pathology) is the fourth most common condition seen in primary care and the most common musculoskeletal condition seen by General Practitioners [11], resulting in substantial cost implications to economies. Direct costs have been estimated in one study as 1.65–3.22 % of all health expenditure [12], and in another as 0.4–1.2 % of GDP in the European Union [7]. Indirect costs, which are largely due to work absence, have been estimated as $50 billion in the USA and $11 billion in the UK [7]. Recent published guidelines for the management of non-specific LBP [3] have self-management at their cornerstone, with patients being advised against bed rest, and advised to remain active, remain at work where possible, and to perform stretching and strengthening exercises. Some guidelines also include advice regarding avoiding long periods of inactivity.[1]

SELFBACK is a monitoring system designed to assist the patient in deciding and reinforcing the appropriate physical activities to manage LBP after consulting a health care professional in primary care. Sensor data is continuously read from a wearable device worn by the user, and the user's activities are recognised in real time. An overview of the activity recognition components of the SELFBACK system is shown in Fig. 1. Guidelines for LBP recommend that patients should not be sedentary for long periods of time. Accordingly, if the SELFBACK system detects continuous periods of sedentary behaviour, a notification is given to alert the user. At the end of the day, a daily activity profile is also generated which summarises all activities done by the user over the course of the day. The information in this daily profile also includes the durations of activities and, for ambulation activities (such as moving from one place to another e.g. walking and running), the counts of steps taken. The system then compares this activity profile to the recommended guidelines for daily activity and produces feedback to inform the user how well they have adhered to these guidelines.

The first contribution of this paper is the description of an efficient, yet effective feature representation approach based on Discrete Cosine Transforms (DCT) presented in Sect. 4. A second contribution is a comparative evaluation of the different parameters (e.g. window size, feature representation and classifier) of our activity recognition system against several state-of-the-art benchmarks in Sect. 5.[2] The insights from the evaluation are designed to inform and serve as guidance for selecting effective parameter values when developing an activity recognition system. The data collection method introduced in this paper is also unique, in that it demon-

---

[1]The SELFBACK project is funded by the European Union's Horizon 2020 research and innovation programme under grant agreement No 689043.

[2]Code and data associated with this paper are accessible from https://github.com/selfback/activity-recognition.

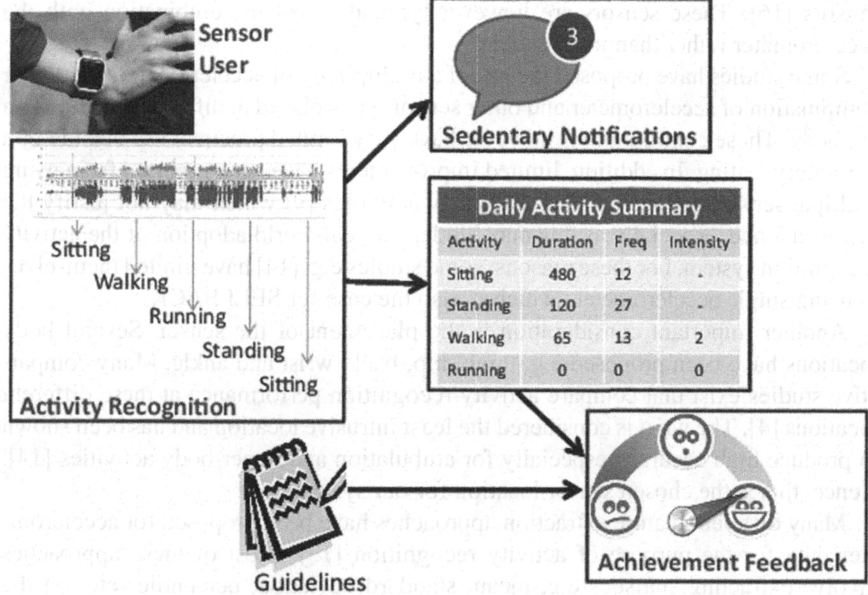

Fig. 1 Overview of SELFBACK system

strates how a script-driven method can be exploited to avoid the demand on manual transcription of sensor data streams (see Sect. 3). Related work and conclusions are also discussed and appear in Sects. 2 and 6 respectively.

2 Related Work in Activity Recognition

Physical activity recognition is receiving increasing interest in the areas of health care and fitness [13]. This is largely motivated by the need to find creative ways to encourage physical activity in order to combat the health implications of sedentary behaviour which is characteristic of today's population. Physical activity recognition is the computational discovery of human activity from sensor data. In the SELFBACK system, we focus on sensor input from a tri-axial accelerometer mounted on a person's wrist.

A tri-axial accelerometer sensor measures changes in acceleration in 3 dimensional space [13]. Other types of wearable sensors have also been proposed e.g. gyroscope. A recent study compared the use of accelerometer, gyroscope and magnetometer for activity recognition [17]. The study found the gyroscope alone was effective for activity recognition while the magnetometer alone was less useful. However, the accelerometer still produced the best activity recognition accuracy. Other sensors that have been used include heart rate monitor [18], light and temperature sensors [16]. These sensors are however typically used in combination with the accelerometer rather than independently.

Some studies have proposed the use of a multiplicity of accelerometers [4, 15] or combination of accelerometer and other sensor types placed at different locations on the body. These configurations however have very limited practical use outside of a laboratory setting. In addition, limited improvements have been reported from using multiple sensors for recognising every day activities [9] which may not justify the inconvenience, especially as this may hinder the real-world adoption of the activity recognition system. For these reasons, some studies e.g. [14] have limited themselves to using single accelerometers which is also the case for SELFBACK.

Another important consideration is the placement of the sensor. Several body locations have been proposed e.g. thigh, hip, back, wrist and ankle. Many comparative studies exist that compare activity recognition performance at these different locations [4]. The wrist is considered the least intrusive location and has been shown to produce high accuracy especially for ambulation and upper-body activities [14]. Hence, this is the chosen sensor location for our system.

Many different feature extraction approaches have been proposed for accelerometer data for the purpose of activity recognition [13]. Most of these approaches involve extracting statistics e.g. mean, standard deviation, percentiles etc. on the raw accelerometer data (time domain features). Other works have shown frequency domain features extracted from applying Fast Fourier Transforms (FFT) to the raw data to be beneficial. Typically this requires a further preprocessing step applied to the resulting FFT coefficients in order to extract features that measure characteristics such as spectral energy, spectral entropy and dominant frequency [8]. Although both these approaches have produced good results, we use a novel approach that directly uses coefficients obtained from applying Discrete Cosine Transforms (DCT) on the raw accelerometer data as features. This is particularly attractive as it avoids further preprocessing of the data to extract features to generate instances for the classifiers.

3 Data Collection

Training data is required in order to train the activity recognition system. A group of 20 volunteer participants was used for data collection. All volunteers were either students or staff of Robert Gordon University. The age range of participants is 18 54 years and the gender distribution is 52 % Female and 48 % Male. Data collection concentrated on the activities provided in Table 1.

This set of activities was chosen because it represents the range of normal daily activities typically performed by most people. In addition, three different walking speeds (slow, normal and fast) were included in order to have an accurate estimate of the intensity of the activities performed by the user. Identifying intensity of activity is important because guidelines for health and well-being include recommendations for encouraging both moderate and vigorous physical activity [1].

Table 1 Details of activities used in our data collection script

Activity name	Description
Walking slow	Walking at self-selected slow pace
Walking normal	Walking at self-selected normal pace
Walking fast	Walking at self-selected fast pace
Jogging	Jogging on a treadmill at self-selected speed
Up stairs	Walking up 4–6 flights of stairs
Down stairs	Walking down 4–6 a flights of stairs
Standing	Standing relatively still
Sitting	Sitting still with hands either on the desk or rested at the side
Lying	Lying down relatively still on a plinth

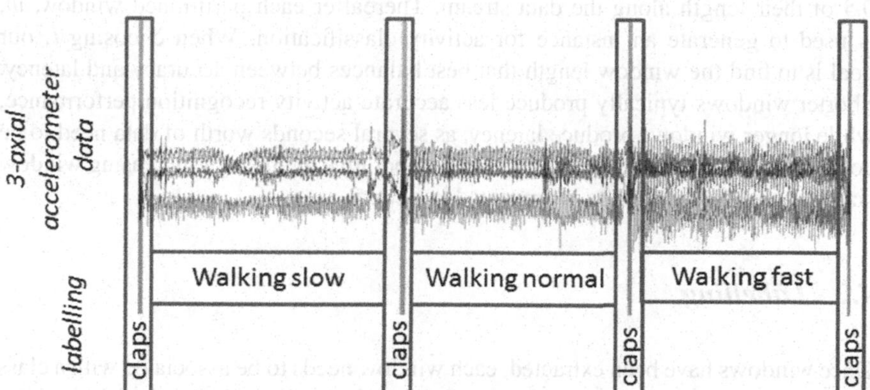

Fig. 2 Example of activity annotation with claps used to separate class transitions

Data was collected using the Axivity Ax3 tri-axial accelerometer[3] at a sampling rate of 100 Hz. Accelerometers were mounted on the wrists of participants using specially designed wristbands provided by Axivity. Participants were provided with scripts which contained related activities e.g. sitting and lying. The scripts guided participants on what activity they should do, how long they should spend on each activity (average of 3 min) and any specific details on how they should perform the activity e.g. sit with your arms on the desk.

Three claps are used to indicate the start and end of each activity. The three claps produce distinct spikes in the accelerometer signal which make it easy to detect the starts and ends of different activities in the data. This helps to simplify the annotation of the accelerometer data, by making it easy to isolate the sections of the data that correspond to specific activities. This allows the sections to be easily extracted and aligned with the correct activity labels from the script as shown in Fig. 2.

[3]http://axivity.com/product/ax3.

4 Activity Recognition Algorithm

The SELFBACK activity recognition system uses a supervised machine learning approach. This approach consists of 4 main steps which are: windowing, labelling, feature extraction and classifier training, as illustrated in Fig. 3.

4.1 Windowing

Windowing is the process of partitioning collected training data into smaller portions of length l, here specified in seconds. Figure 4 illustrates how windowing is applied to the 3-axis accelerometer data streams: x, y and z. Windows are overlapped by 0.5 of their length along the data stream. Thereafter each partitioned window, w, is used to generate an instance for activity classification. When choosing l, our goal is to find the window length that best balances between accuracy and latency. Shorter windows typically produce less accurate activity recognition performance, while longer windows produce latency, as several seconds worth of data need to be collected before a prediction is made. A comparative analysis of increasing window sizes ranging from 2 to 60 s is presented in Sect. 5.

4.2 Labelling

Once windows have been extracted, each window needs to be associated with a class label from a set of activity classes, $c \in C$. By default, this is the label of the activity

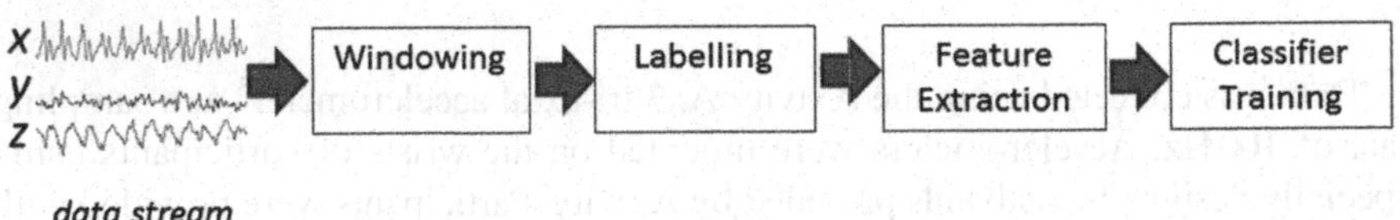

Fig. 3 SELFBACK activity recognition algorithm steps

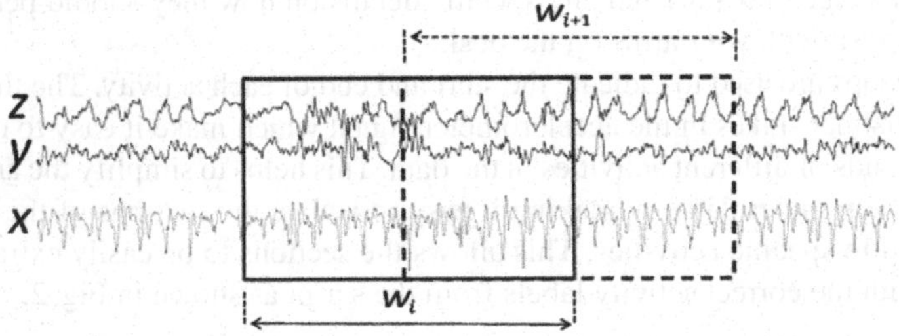

Fig. 4 Illustration of accelerometer data windowing

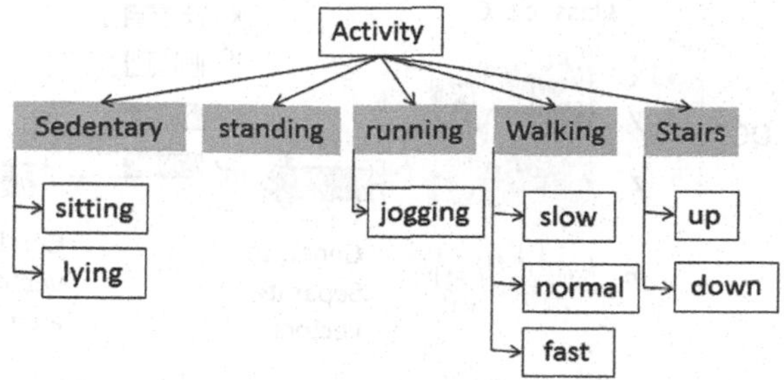

Fig. 5 Activity class hierarchy

stream from which the window was extracted. Recall from Sect. 3 that $|C|$ was 9 classes (see Table 1), and can be thought of constituting a hierarchical structure as shown in Fig. 5. However, we observed that the more granular the activity labels, the more activity recognition accuracy suffers. In the case of some closely related classes e.g. sitting and lying, it is very difficult to distinguish between these classes from accelerometer data recorded from a wearable on the wrist. This is because wrist movement tends to be similar for these activities. Also, for activity classes distinguished by intensity (i.e. walking slow, walking normal and walking fast) the speed distinction between these activity classes can be more subjective than objective. Because the pace of walking is self-selected; one participant's slow walking pace might better match another's normal walking pace. Alternatively we consider $|C|$ equal to 5 classes by using the first level of the hierarchy (shaded nodes) with sub-tree raising of leaf nodes (whereby leaf nodes are grouped under their parent node). Evaluation results for activity recognition with both $|C|$ values are presented in Sect. 5.

4.3 Feature Extraction

The 3-axis accelerometer data streams, x, y and z, when partitioned according to the sliding window method as detailed in Sect. 4.1 generates a sequence of partitions, each of length l where each partition w_i is comprised of real-valued vectors x_i, y_i and z_i, such that $\mathbf{x} = (x_{i1}, \ldots, x_{il})$ DCT is applied to each axis (in essence each windowed partition x_i, y_i and z_i) to obtain a set of DCT coefficients which are an expression of the original accelerometer data in terms of a sum of cosine functions at different frequencies [10]. Accordingly the DCT-transformed vector representations, $\mathbf{x}' = \mathrm{DCT}(\mathbf{x})$, $\mathbf{y}' = \mathrm{DCT}(\mathbf{y})$ and $\mathbf{z}' = \mathrm{DCT}(\mathbf{z})$, are obtained for each constituent in an instance. Additionally we derive a further magnitude vector, $\mathbf{m} = \{m_{i1}, \ldots, m_{il}\}$ of the accelerometer data for each instance as a separate axis, where m_{ij} is defined in Eq. 1.

$$m_{ij} = \sqrt{x_{ij}^2 + y_{ij}^2 + z_{ij}^2} \qquad (1)$$

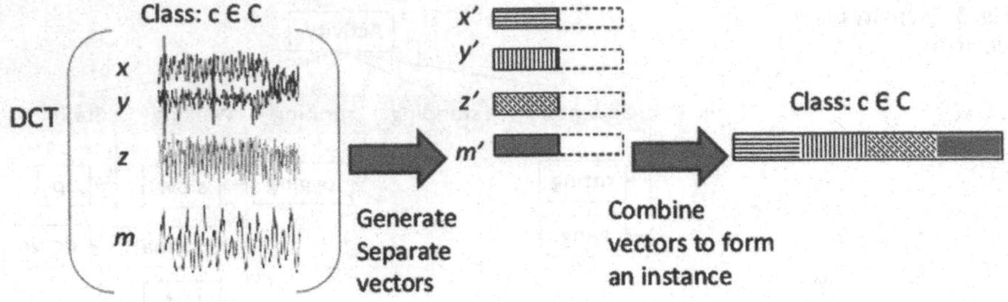

Fig. 6 Feature extraction and vector generation using DCT

As with $\mathbf{x}'$, $\mathbf{y}'$ and $\mathbf{z}'$, we also apply DCT to $\mathbf{m}$ to obtain $\mathbf{m}' = \text{DCT}(\mathbf{m})$. This means that our representation of a training instance consists of the pair $(\{\mathbf{x}', \mathbf{y}', \mathbf{z}', \mathbf{m}'\}, c)$, where c is the corresponding activity class label as detailed in Sect. 4.2. Including the magnitude in this way helps to train the classifier to be less sensitive to changes in orientation of the sensing device. Note that the coefficients returned after applying DCT are combinations of negative and positive real values. For the purpose of feature representation, we are only interested in the magnitude of the DCT coefficients, irrespective of (positive or negative) sign. Accordingly for each DCT coefficient e.g. x'_{ij}, we maintain its absolute value $|x'_{ij}|$.

DCT compresses all of the energy in the original data stream into as few coefficients as possible and returns an ordered sequence of coefficients such that the most significant information is concentrated at the lower indices of the sequence. This means that higher frequency DCT coefficients can be discarded without losing information. On the contrary, this might help to eliminate noise. Thus, in our approach we also retain a subset of the l coefficients and as proposed in [10] we retain the first 48 coefficients out of l. The final feature representation is obtained by concatenating the absolute values of the first 48 coefficients of $\mathbf{x}'$, $\mathbf{y}'$, $\mathbf{z}'$ and $\mathbf{m}'$ to produce a combined feature vector of length 192. An illustration of this feature selection and concatenation appears in Fig. 6.

4.4 *Step Counting*

An important piece of information that can be provided for ambulation activities is a count of the steps taken. This information has a number of valuable uses. Firstly, step counts provide a convenient goal for daily physical activity. Health research has suggested a daily step count of 10,000 steps for maintaining a desirable level of physical health [6]. A second benefit of step counting is that it provides an inexpensive method for estimating activity intensity. Step rate thresholds have been suggested in health literature that correspond to different activity intensities. For example, [1] identified that step counts of 94 and 125 steps per minute correspond to moderate and vigorous intensity activities respectively for men, and 99 and 135 steps per minute correspond to moderate and vigorous intensity activities for women. Accordingly, step counts are likely to provide a more objective measure for activity intensity in the SELFBACK system than classifying different walking speeds. Here, we discuss two commonly used approaches involving frequency analysis and peak counting algorithms for inferring step counts from accelerometer data specific to ambulation activity classes.

4.4.1 Frequency Analysis

The main premise of this approach is that frequency analysis of walking data should reveal the heel strike frequency (i.e. the frequency with which the foot strikes the ground when walking) which should give an idea of the number of steps present in the data [2]. For walking data collected from a wrist-worn accelerometer, one or two dominant frequencies can be observed, heel strike frequency, which should always be present, and the arm swing frequency which may sometimes be absent. Converting accelerometer data from the time domain to the frequency domain using FFT enables the detection of these frequencies. For step counting, this approach seeks to isolate the heel strike frequency. Accordingly, the step count can be computed as a function of the heel strike frequency. For example, for frequency values in Hertz (cycles per second), the step count can be obtained by multiplying the identified heel strike frequency with the duration of the input data stream in seconds.

4.4.2 Peak Counting

The second approach involves counting peaks on low-pass filtered accelerometer data where each peak corresponds to a step. This process is illustrated in Fig. 7. For filtering, we use a Butterworth low-pass filter with a frequency threshold of 2 Hz for walking and 3 Hz for running.

The low-pass filter is then applied on m, which is the magnitude axis of the accelerometer signal obtained by combining the x, y and z axes. As a result, we expect to filter out all frequencies in m that are outside of the range for walking and running respectively. In this way, any changes in acceleration left in m can be attributed to the effect of walking or running. A peak counting algorithm is then deployed to count the peaks in m where the number of peaks directly corresponds to the count of steps.

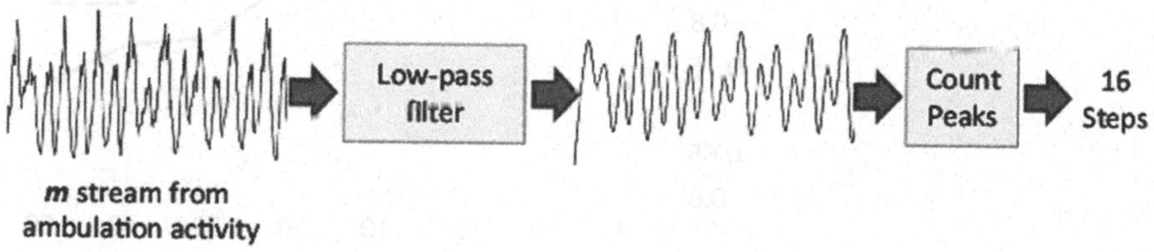

Fig. 7 Step counting using peak counting approach

5 Evaluation

In this section we present results for comparative studies that have guided the development of the SELFBACK activity recognition system. Firstly, an analysis of how window size and feature representation impact the effectiveness of human activity recognition is presented. Thereafter, we explore how classification granularity is affected by inter-class relationships and how that in turn impacts model learning. A question closely related with classification granularity is how to determine the activity intensity. For ambulation activities, step rate is a very useful heuristic for achieving this. Accordingly, we present comparative results for two step counting algorithms.

Our experiments are reported using a dataset of 20 users. Evaluations are conducted using a leave-one-person-out methodology i.e. one user is used for testing and the remaining 19 are used for training. In this way, we are testing the general applicability of the system to users whose data is not included in the trained model. Performance is reported using macro-averaged F1. SVM is used for classification after a comparative evaluation demonstrated its F1 score of 0.906 to be superior to that of kNN, decision tree, Nave Bayes and Logistic Regression; by more then 5 %, 12 %, 25 % and 3 % respectively.

5.1 Feature Representation and Window Size

For feature representation, we compare DCT, statistical time domain and FFT frequency domain features. Here time domain features are adopted from [19]. Figure 8 plots F1 scores for increasing window sizes from 2 to 60 s for each feature representation scheme.

The best F1 score is achieved with DCT features with a window size of 10 ($F1 = 0.906$). It is interesting to note that neither time or frequency domain features can match performance to that of directly using DCT coefficients for representation. Overall there is a 5 % gain in F1 scores with DCT compared to the best results of the rest.

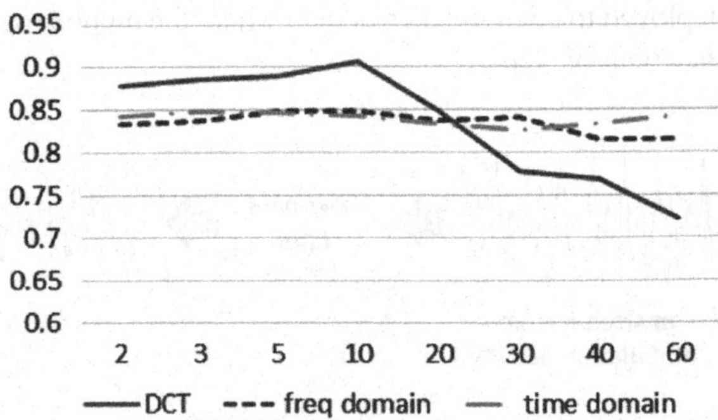

Fig. 8 Activity recognition performance at different window sizes

Table 2 Confusion matrix for 9-class activity classification

	Lying	Sitting	Standing	Jogging	Upstairs	Down stairs	Walk fast	Walk normal	Walk slow
Lying	115	127	8	0	0	0	0	0	1
Sitting	84	161	4	2	1	0	0	0	0
Standing	6	6	212	2	2	0	0	0	11
Jogging	0	0	0	284	1	0	0	1	0
Up stairs	0	0	3	8	92	7	7	11	30
Down stairs	1	0	1	5	31	89	5	4	8
Walk fast	0	0	1	21	3	1	157	53	6
Walk normal	0	0	1	4	11	3	48	141	41
Walk slow	0	1	7	3	23	4	0	32	181

5.2 *Classification Granularity*

Recall from Sect. 4.2 that data was collected relative to 9 different activities. Here we analyse classification accuracy with focus on inter-class relationships. In particular we study the separability of classes to establish which specific classes are best considered under a more general class of activity.

Overall F1 score for activity classification using 9 classes remains low at 0.688. Its confusion matrix is provided in Table 2, where the columns represent the predicted classes and the rows represent the actual classes. Close examination of the matrix shows that the main contributors to this low F1 score are due to classification errors involving activities lying, walking normal and upstairs. For instance we can see that for the activity class lying, only 115 instances are correctly classified and 125 instances are incorrectly classified as sitting. Similarly, 84 instances of sitting are incorrectly classified as lying. This indicates a greater discrimination confusion between lying and sitting which can be explained by wrist movement alone being insufficient to differentiate between these activities with a wrist worn accelerometer. However, both sitting and lying do represent sedentary behaviour and as such could naturally be categorised under the more general Sedentary class. A similar explanation follows for walking normal, where 48 instances are incorrectly classified as walking fast and 41 as walking slow. Accelerometer data for walking at different speeds will naturally be very similar. Also, the same walking speed is likely to be different between participants due to the subjectivity inherent in users judgment about their walking speeds. In addition, a user may unnaturally vary their pace while trying to adhere to a specific walking speed under data collection conditions. Again these reasons make it more useful to have the three walking speeds combined into one general class called Walking and have walking speed computed as a separate function of step rate. Regarding walking upstairs, we can see that it is most confused with walking slow but also suggests difficulties with differentiating between walking normal, walking fast and jogging. Many of these errors are likely to be addressed by

Table 3 Confusion matrix for 5-class activity classification

	Sedentary	Standing	Jogging	Stairs	Walking
Sedentary	490	7	2	1	3
Standing	17	205	2	1	14
Jogging	0	0	283	0	3
Stairs	3	0	9	0 223	67
Walking	3	5	24	31	679

taking into account inter-class relationships to form more general classes instead of having too many specialised classes.

Accordingly with the 5 class problem we have attempted to organise class membership under more general classes to avoid the inherent challenge of discriminating between specialised classes (e.g. between normal and fast walking). Therefore, there is a sedentary class combining sitting and lying classes; a stairs class to cover both upstairs and downstairs and a single walking class bringing together all different paces of walking speeds (See Fig. 5). Jogging and Standing remain as distinct classes as before.

As expected results in Table 3 shows that, 4 of the 5 classes have F1 scores greater than 0.9 with only Stairs achieving a score of 0.8. This result is far more acceptable than that achieved with the 9 class problem. The relatively lower F1 score with Stairs is due to 67 instances being incorrectly classified as Walking. This highlights the difficulty with differentiating between walking on a flat surface versus walking up or down stairs. However apart from the inclination of the surface there is no other characteristic that can help to differentiate these seemingly similar movements.

5.3 *Step Counting*

This final sub-section presents an evaluation of our step counting algorithms. For this, we collected a separate set of walking and running data with known actual step counts. This was necessary because actual counts of steps were not recorded for the initial dataset collected. In total, 19 data instances were collected for walking and 11 for running. For walking, participants were asked to walk up and down a corridor while counting the number of steps they took from start to finish. Reported step counts for walking range from 244 to 293. Participants performed a number of different hand positions which included walking with normal hand movement, with hands in trouser pocket and carrying a book or coffee mug. Walking data also included one instance of walking down a set of stairs (82 steps) and one instance of walking up a set of stairs (78 steps).

Table 4 Performance of step counting approaches measured using Root Mean Squared Error

Step counting approach	RMSE walking	RMSE running
Frequency analysis	11.245	6.250
Peak counting	**6.374**	**5.576**

Running data was collected on a treadmill. Participants were requested to run on a treadmill at a self-selected speed for a self-selected duration of time. Here also, three claps were used to mark the start and end of the running session. Two participants standing on the side were asked to count the steps in addition to the runner, due to the difficulty that may be involved in running and counting steps at the same time. Reported step counts for running range from 150 to 210.

The objective of this evaluation is to match, for each data instance, the count of steps predicted by each algorithm, to the actual step counts recorded. Root means squared error (RMSE) is used to measure performance. Because both step counting algorithms do not require any training, all 30 data instances are used for testing. Evaluation results are presented in Table 4. Generally it is useful to have mean squared error values that are below 10 for step counts. Overall we can see that better performance is observed from the Peak Counting method, thus this has been set as the default step counting approach for the SELFBACK system.

6 Conclusion

This paper focuses on the activity recognition part of the SELFBACK system which helps to monitor how well users are adhering to recommended daily physical activity for self-management of low back pain. The input into the activity recognition system is tri-axial accelerometer data from a wrist-worn sensor.

Activity recognition from the input is achieved using a supervised machine learning approach. This is composed of 4 stages: windowing, feature extraction, labelling and classifier training. Our results show that a window size of 10 s is best for identifying SELFBACK activity classes and highlighted the inherent challenge in differentiating between similar movement classes (such as lying with sitting and different paces of walking) using a wrist-worn sensor. Our approach to using Discrete Cosine Transform to represent instances achieved a 5 % classification performance gain over time and frequency domain feature representations. Algorithms to infer step counts from ambulation data suggests a simple peak counting approach following a low pass filter applied to the magnitude of the tri-axial data to be best. Future work will explore techniques for recognising a larger set of dynamically changing activities using incremental learning and semi-supervised approaches.

References

1. Abel, M., Hannon, J., Mullineaux, D., Beighle, A., et al.: Determination of step rate thresholds corresponding to physical activity intensity classifications in adults. J. Phys. Activity Health **8**(1), 45–51 (2011)
2. Ahanathapillai, V., Amor, J.D., Goodwin, Z., James, C.J.: Preliminary study on activity monitoring using an android smart-watch. Healthc. Technol. Lett. **2**(1), 34–39 (2015)
3. Airaksinen, O., Brox, J., Cedraschi, C.O., Hildebrandt, J., Klaber-Moffett, J., Kovacs, F., Mannion, A., Reis, S., Staal, J., Ursin, H., et al.: Chapter 4 european guidelines for the management of chronic nonspecific low back pain. Eur. Spine J. **15**, s192–s300 (2006)
4. Bao, L., Intille, S.S.: Activity recognition from user-annotated acceleration data. In: Pervasive Computing, pp. 1–17. Springer (2004)
5. Buchbinder, R., Blyth, F., March, L., Brooks, P., Woolf, A., Hoy, D.: Placing the global burden of low back pain in context. Best Pract. Res. Clin. Rheumatol. **27**, 575–589 (2013)
6. Choi, B.C., Pak, A.W., Choi, J.C.: Daily step goal of 10,000 steps: a literature review. Clin. Investig. Med. **30**(3), 146–151 (2007)
7. Dagenais, S., Caro, J., Haldeman, S.: A systematic review of low back pain cost of illness studies in the united states and internationally. Spine J. **8**(1), 8–20 (2008)
8. Figo, D., Diniz, P.C., Ferreira, D.R., Cardoso, J.M.: Preprocessing techniques for context recognition from accelerometer data. Pers. Ubiquit. Comput. **14**(7), 645–662 (2010)
9. Gao, L., Bourke, A., Nelson, J.: Evaluation of accelerometer based multi-sensor versus single-sensor activity recognition systems. Med. Eng. Phys. **36**(6), 779–785 (2014)
10. He, Z., Jin, L.: Activity recognition from acceleration data based on discrete consine transform and svm. In: Systems, Man and Cybernetics, 2009. SMC 2009. IEEE International Conference on, pp. 5041–5044. IEEE (2009)
11. Jordan, K.P., Kadam, U.T., Hayward, R., Porcheret, M., Young, C., Croft, P.: Annual consultation prevalence of regional musculoskeletal problems in primary care: an observational study. BMC Musculoskelet. Disord. **11**(1), 1 (2010)
12. Kent, P.M., Keating, J.L.: The epidemiology of low back pain in primary care. Chiropractic Osteopathy **13**(1), 1 (2005)
13. Lara, O.D., Labrador, M.A.: A survey on human activity recognition using wearable sensors. IEEE Commun. Surv. Tutorials **15**(3), 1192–1209 (2013)
14. Mannini, A., Intille, S.S., Rosenberger, M., Sabatini, A.M., Haskell, W.: Activity recognition using a single accelerometer placed at the wrist or ankle. Med. Sci. Sports Exerc. **45**(11), 2193 (2013)
15. Mäntyjärvi, J., Himberg, J., Seppänen, T.: Recognizing human motion with multiple acceleration sensors. In: 2001 IEEE International Conference on Systems, Man, and Cybernetics, vol. 2, pp. 747–752. IEEE (2001)
16. Maurer, U., Smailagic, A., Siewiorek, D.P., Deisher, M.: Activity recognition and monitoring using multiple sensors on different body positions. In: International Workshop on Wearable and Implantable Body Sensor Networks. BSN 2006, pp. 4–pp. IEEE (2006)
17. Shoaib, M., Bosch, S., Incel, O.D., Scholten, H., Havinga, P.J.: Fusion of smartphone motion sensors for physical activity recognition. Sensors **14**(6), 10146–10176 (2014)
18. Tapia, E.M., Intille, S.S., Haskell, W., Larson, K., Wright, J., King, A., Friedman, R.: Real-time recognition of physical activities and their intensities using wireless accelerometers and a heart rate monitor. In: 2007 11th IEEE International Symposium on Wearable Computers, pp. 37–40. IEEE (2007)
19. Zheng, Y., Wong, W.K., Guan, X., Trost, S.: Physical activity recognition from accelerometer data using a multi-scale ensemble method. In: IAAI (2013)

Automated Sequence Tagging: Applications in Financial Hybrid Systems

Peter Hampton, Hui Wang, William Blackburn and Zhiwei Lin

Abstract Internal data published by a firm regarding their financial position, governance, people and reaction to market conditions are all believed to impact the underlying company's valuation. An abundance of heterogeneous information coupled with the ever increasing processing power of machines, narrow AI applications are now managing investment positions and making decisions on behalf of humans. As unstructured data becomes more common, disambiguating structure from text-based documents remains an attractive research goal in the Finance and Investment industry. It has been found that statistical approaches are considered high risk in industrial applications and deterministic methods are typically preferred. In this paper we experiment with hybrid (ensemble) approaches for Named Entity Recognition to reduce implementation and run time risk involved with modern stochastic methods.

Keywords Finance · XBRL · Information Extraction · Representation

1 Introduction

As human beings, we regard the ability to produce and understand language as a central product of our intelligence. Due to various linguistic phenomena and the nature of mankind, creating machines to emulate the *understanding* of language is more complicated than many might believe. As economic actors generate new information, increased data is made available for investment decision support. Many

P. Hampton (✉) · H. Wang · W. Blackburn · Z. Lin
Artificial Intelligence and Applications Research Group, Ulster University,
Newtownabbey BT37 0QB, UK
e-mail: hampton-p1@email.ulster.ac.uk

H. Wang
e-mail: h.wang@ulster.ac.uk

W. Blackburn
e-mail: wt.blackburn@ulster.ac.uk

Z. Lin
e-mail: z.lin@ulster.ac.uk

M. Bramer and M. Petridis (eds.), *Research and Development in Intelligent Systems XXXIII*, DOI 10.1007/978-3-319-47175-4_22

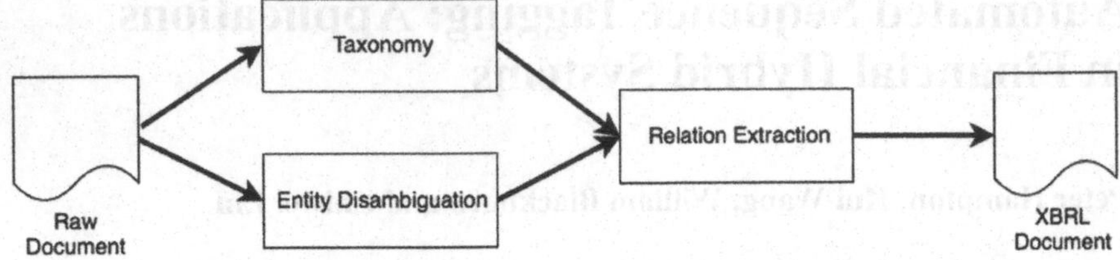

Fig. 1 The overall extraction process. The conversion of a raw document into an XBRL document. The first step takes a raw document and then in parallel identifies the appropriate taxonomy while recognising the entities. Once complete, the entities and their relations are extracted to compose the final XBRL document. This study focuses on the entity disambiguation stage

of this information is delivered in a text-based unstructured format. This information gain has increased so much over the past decade that decision makers are relying on emerging technologies from the field of Artificial Intelligence (AI) to model and analyze natural language to support investment positions [1–3]. Natural Language Processing (NLP) has become an integral part of next generation technologies for working with unstructured data. In turn, research in this area is telling us more about how humans understand and interact with language [4, 5] (Fig.1).

Advances in AI have enabled the automatic or semi-automatic collection, extraction, aggregation and classification of various text based documents [6]. This has undoubtedly had a large impact on how companies retrieve and store information, but further processing and analysis of such information is still an on-going research challenge for academic researchers and industry practitioners alike. The current inability for machines to understand language has led to increased interest in creating structure out of unstructured text. Financial systems that mimic an understanding of language are becoming increasingly important to decision-making processes when buying and selling financial instruments [7, 8]. Yet little is published on how to analyse documents containing various information formatted in various ways.

A first step to this *artificial* understanding is identifying and disambiguating Named Entities within a text document. Recent advances in this field have focused on statistical methods but remain problematic for implementing. Our paper surveys related works and methods that can be used to extract critical financial information from annual (10-K) accounts. We then build on previous work using a Hybrid Conditional Random Fields contrary to [9]. We conclude this paper by discussing the advantages and limitations of such Hybrid-CRFs and propose a future research direction for recognising relevant entities in financial literature.

2 Background

New technologies and entire sectors are evolving around the idea that is *big data* management and analysis. One area in the research domain that remains relatively untouched is the analysis of financial reports [10]. We analysed a small sample of 6 companies and found that on average between the years of 2011–2015 that the average length of an annual report was 97 pages. These documents are information rich, have inconsistent formatting and contain a mix of factual and subjective information from internal company. With so much information, it is important to condense this information as much as possible for the ease of human consumption. It has been found that firms can prefer rule based over statistical methods due to risk. Rule based approaches tend to be rigid, over-engineered and hard to scale. Statistical methods also have implementation downsides as they are difficult to adapt to new domains, need manual feature engineering and very large amounts of data.

Early NER practitioners used rule based methods to interact with and understand language. That is hand crafting algorithms to fit the feature space. This restrictive approach led researchers to focus on shallow methods in statistical machine learning which make heavy use of feature engineering (now considered the state of the art). With increased computing power and various parametric model limitations, unsupervised methods have become an interesting area of research after experiencing their own AI winter [11]. Named Entity Recognition is itself a classification task, which is the process of categorizing a set of n-grams while only using a non-exhaustive list of data features that describe them. Early work in this area found that it is imperative to identify meaningful information units such as people, organizations, locations, and numeric expressions including time, date, money and percent expressions [12–15]. To date, practitioners in this area have been using word level features to identify Named Entities which can be seen in Table 1. Our initial hypothesis was that we can offset some of the risk from the statistical methods with rule-based methods [9].

Table 1 Natural and engineered word-level features

Feature	Type[a]	Description
Case	Natural	The casing reflects the grammatical function performed by a noun or pronoun in a phrase, clause, or sentence in English and some other languages
Punctuation	Natural	Used in text to separate sentences and their elements. Other punctuation marks are used to clarify meaning
Digit	Natural	These are numbers in text which can include digit patterns, cardinal and ordinal numbers, roman numbers and words with digits
Character	Natural	These features tend to be possessive marks or greek letters
Morphological Feature	Engineered	Morphological features include prefixes, suffixes and stems of words
Part-of-Speech Tag	Engineered	The category assigned in accordance with a words syntactic function such as a noun or verb
Function	Engineered	Typically functions defined by the programmer, such as token/phrase length, case-insensitivity, hand crafted patterns and so on

[a] *Natural* types are words that can be identified using morphological functions whereas *Engineered* requires a system to generate the feature in a preprocessing stage

2.1 Financial Reporting

The aim of this project is to computationally *understand* the language contained within financial reports with Information Extraction as the means. The typical objective of a financial statement is to give an overview of the financial position, financial performance, and cash flows of a business that is useful to a wide range of users, human or machine, in making economic decisions.

Strategies The strategic part of accounts contains information such as the Chairmans and Chief Executives professional analysis of the business. This information is typically presented in an unstructured nature.

Governance The governance section of a financial report looks at information such as information about the board of directors, discussions around corporate governance, etc. This information is usually presented in an unstructured nature.

Financials Financial information includes sections as previously discussed such as the cash flow statement, balance sheet, specific types of incomes and outgoings. The information is typically presented in a structured nature, however, the format is unpredictable.

There are various types of information that would be of interest to a decision maker. Some that would be of interest include:

From the above short descriptions it is reasonable to infer that a financial document such as a 10-K (Annual Statement) or 10-Q (Quarterly Statement) [16] is a long and complex document with diverse contents. Take the following excerpt from the First Derivatives PLC (AIM: FDP.L) annual released in 2016:

> Revenue for the year increased by 40.6 % to 117.0m (2015: 83.2m), while adjusted EBITDA rose by 50.5 % to 23.3m (2015: 15.5m) and adjusted earnings per share increased by 33.2 % to 51.7p (2015: 38.8p).

Annotated, this **sentence** would look very different with 3 % entities, 7 money entities, 2 year (time based) entities and 3 accounting entities. From this snippet, the immediate knowledge appears to be dense with rich information tightly coupled together in financial reports. Extracting information into predefined ontological templates as previously practiced would be incredibly difficult, if not impossible.

Even with an expert or team of experts, designing a template for the extraction would be a very long and very complicated task. Using the shallow parsing described in most Named Entity Classification experiments would also prove incredibly challenging due to the heterogeneous markup of the document ranging from free-flowing sentences to tabular formatted data, with different table formats [6].

Initial work in this area comes from the work of the EU Musings project which experimented with Ontology Based Information Extraction (OBIE) as a means to creating business intelligence from static documents. The areas their research focused on where financial risk management, internationalization and IT operational risk management. The end result is a populated ontology (Knowledge Base) that can be queried. Although this is semantically guided, our approach does not take a ontological approach. Through our previous experiments with OBIE, we found the basis of the ontology to be far too restrictive although it proves to be very successful in narrow applications. Our approach deviates from their approach by leveraging the XBRL taxonomy as an ontology, discarding relations for *atomic facts*, or terms.

2.2 Semantic Representation

From a computational standpoint (and the end goal of this project), this is a very challenging task regardless of the documents integrity. Such reasons include:

- Entites can have a non-numerical relation to another entity
- Some facts might relate to non-financial facts.
- Entities have fidelity
- There are many namespaces to choose from.
- Not all entities can be represented atomically

XBRL (eXtensible Business Reporting Language) is a relatively new standard for exchanging business information in an XML like format. XBRL allows the expression of semantic meaning commonly required in financial reporting. It is believed XBRL is a suitable way to communicate and exchange business information between different business systems [17]. These communications are defined by metadata set out in taxonomies, which capture the definition of individual reporting concepts as well as the relationships between concepts and other semantic meaning. Information being communicated or exchanged is provided within an XBRL instance. One use of XBRL is to define and exchange financial information, such as a financial statement. An short snippet of XBRL is displayed below for easy conceptualization using the **us-gaap** namespace.

```
<!-- revenue for the years 2007 / 08 --> <us-gaap:OperatingRevenue
contextRef="FY08Q1"
unitRef="USD">1376200000</us-gaap:OperatingRevenue>
<us-gaap:OperatingRevenue contextRef="FY07Q1"
unitRef="USD">1081100000</us-gaap:OperatingRevenue>

<!-- Cost of revenue for the years 2007 / 08 -->
<us-gaap:CostOfRevenue contextRef="FY08Q1"
unitRef="USD">2675000000</us-gaap:OperatingRevenue>
<us-gaap:CostOfRevenue contextRef="FY07Q1"
unitRef="USD">1696000000</us-gaap:OperatingRevenue>
```

The goal is eventually to be able to detect the appropriate XBRL taxonomy for a given document, and use the XBRL namespace entities as the filling template for the extracted information of the document. This paper focuses on identifying entities, rather than extracting them. The complete process however is discussed in detail in Sect. 7.

3 Methodologies

Information Extraction typically involves several stages: Preprocessing, Named Entity Recognition, Relation Extraction and Template Filling [18]. Named Entity Recognition is the process of identifying n-grams of interest that belong to some predefined category of interest. Information Extraction is an eclectic field that focuses on extracting structure from heterogeneous documents. Typical approaches are to directly lift information from a document and fill a template. With the continued evolution of the World Wide Web into areas such as offline-first and the Internet of Things (IoT), more potential application areas for Information Extraction now exist. Although the AI sub-field of Natural Language Processing typically focuses on unstructured data, extracting information from structured and semi-structured sources can still prove a very challenging research goal.

Common approaches include human designed algorithmic and statistically driven methodology in Information Extraction. Both approaches exhibit promising results and strengths that have inspired recent uses described in Sect. 3.3. [19] identify information extraction on several different dimensions: deterministic to stochastic, domain-specific to general, rule based to learned, and narrow to broad test situations.

3.1 *Rule Based*

Rule based approaches have been the most popular approach in industry and are largely considered obsolete in academic research. Authors [20] found that only 6 publications relied solely on rules across a 10-year period reflecting the popularity for machine learning. This reflects the growing popularity of statistical techniques. Rule based method typically rely on pattern matching or list based methods as listed in Table 2.

It is believed that manually constructing such IE systems is a laborious and error prone task. This has led researchers to believe that a more appropriate solution would be to implement machine learning instead of expert defined rules and dictionary. One domain that heavily uses gazetteers in Named Entity Recognition tasks includes the BioMathematics and BioInformaitics (BMI). The approach outlined by [21] uses a similar approach to this paper implementing a Conditional Random Fields with a dictionary to extract chemical names, a research topic that largely remains virgin territory.

Table 2 List Based Features

List type	Description
Gazetteer	A **generalized** collection of words such as stopwords
Entity list	Lists of organization names, people names and countries
Entity cues	Cues that include things such as name prefixes, post-nominal letters, the location of a typical word

3.2 *Statistical*

Machine Learning (Shallow Parsing) approaches to Named Entity Recognition are currently the most prominent form of research in academia. [22] believe that the domain specific knowledge and system expertise needed for this task makes the area a prime target for machine learning. The idea is to let machines self discover and learn from large corpora; annotated and real time text data, say from corporate websites. However, IBMs industrial survey found that only one third of users relied exclusively on Machine Learning techniques for extraction products [23]. Here we briefly review the two of the most successful machine learning methods to date in the area of Named Entity Recognition: Hidden Markov Models (HMMs) and Conditional Random Fields (CRFs) [24, 25].

3.2.1 Hidden Markov Models

A Hidden Markov Model (HMM) is a statistical model in which the system being modeled is assumed to be a Markov process with unobserved (hidden) states and have been used liberally in research with sequential and temporal data which tends to serve as a strong baseline [14, 20]. Hidden Markov Models are famous for playing a big role in the Human Genome Project and various speech recognition studies before Deep Neural Networks became the state of the art (Fig. 2).

$$z_1 \ldots z_n \varepsilon \{1, \ldots, n\} \tag{1}$$

$$x_1, \ldots, x_n \varepsilon \chi \tag{2}$$

$$p(x_1 \ldots x_n, z_1 \ldots z_n) = p(z_1)p(x_1|z_1) \prod_{k=2}^{n} p(z_k|z_{k-1})p(x_k|z_k) \tag{3}$$

(1) Shows the hidden states whereas (2) shows the depicts the observed data. The joint distribution (3) of all the following factors in the following way can be easily conceptualized in Fig. 3.

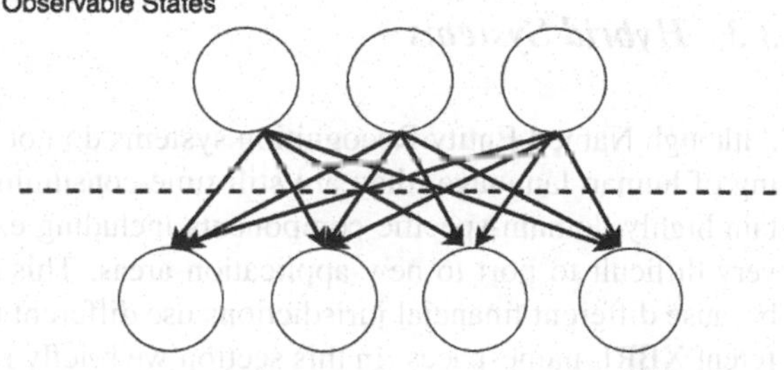

Fig. 2 A simple depiction of a Hidden Markov Model (HMM)

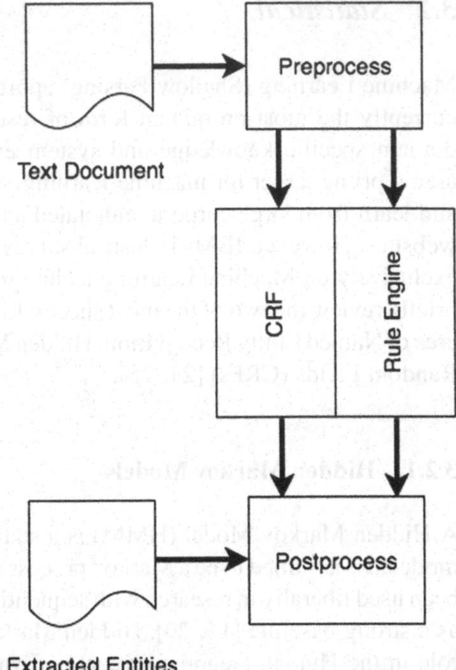

Fig. 3 The text document (PDF) is sent for preprocessing (OCR [Text Extraction], tokenization, Part-of-Speech tagged before being classified by the CRF and Rule base in parallel. The rule engine is a recursive tree-like structure that parse each token sequentially, firing a callback function if the pattern is found in the pattern database. Once both models are finished, the document entities are merged and the annotated document is outputted

3.2.2 Conditional Random Fields

When examining the results of various Named Entity Recognition competitions such as CoNLL and Biocreative it's clear that are Conditional Random Fields or similar conditional taggers are the state of the art. [25] introduced this sequence modelling framework that has all the advantages of a Hidden Markov Model and manages to overcome the labelling bias problem. The authors describe their creation as a *finite state model with normalized transition probabilities* that is trained by maximum likelihood estimation.

3.3 Hybrid Systems

Although Named Entity Recognition systems do not attempt a holistic understanding of human language, they are still time-consuming to build and generally contain highly domain-specific components including expert bias, making the models very difficult to port to new application areas. This is troublesome in this project because different financial jurisdictions use different terminology and therefore different XBRL namespaces. In this section we briefly review methodologies believed to rapidly evolve over the coming years and overcome various limitations in rigid rules and statistical estimators.

Hybrid Systems are dynamic systems that can exhibit both continuous and discrete dynamic behavior a system that can both flow (described by a differential equation) and jump (described by a difference equation or control graph). Hybrid systems are an attractive alternative for Named Entity Recognition tasks in a Financial setting, as it is easily adoptable, scalable, and has transparent debugging features compared to machine learning. A common approach is to complement a black-box, machine learning based approach with rules to handle exceptions, employ a hybrid approach where rules are used to correct classifier mistakes in order to increase the precision of the overall system.

4 Architecture

We implement a rule-based, recursive pattern matching module alongside a machine-learning algorithm to improve the overall results for NER. We expect that this will yield significantly better results for entities that are typically expressed numerically in text such as monetary values, percentages, dates and times based on the [9].

We run the deterministic and stochastic processes concurrently. The rule based pattern matching acts much faster than the CRF model, which usually finishes first, but blocks the pipeline until finished. From here, the documents are merged. If there is an annotation collision, the entity is annotated with the rules as priority, that is because the pattern matching acts a binary switch, 0 if not and 1 is positive.

5 Experiments

When evaluating Information Extraction tasks, researchers use the Precision, Recall and F-Score metrics and a train/test split. There are 4 states which we can separate the extracted output into; *True Positive, False Positive, False Negative, False Positive*. Precision and Recall is a measure to understand the relevance of extracted text.

- **True Positive**: Selected and Correctly Classified.
- **True Negative**: Not Selected and Not Correct
- **False Positive**: Selected, but incorrect.
- **False Negative**: Correct, but not selected.

The precision is the instances classified that are correct. That is the true positives divided by the sum of the true positives plus the false positives.

The recall, or sensitivity in this case, is the fraction of the entities that were classified as relevant. The F Score is weighted average of the precision and recall can tell us more about the performance of the classifier performance (Table 3).

Table 3 The result of the hybrid CRF

	Hyrbid MEMM[a] (%)	Hybrid CRF[b] (%)
F-Score	77.86	84.81

[a]The baseline used in the initial experiments [9]
[b]The results of the implementation discussed in Sect. 4

6 Discussion

Although conditional models such as CRFs are unbiased (typically), and high variance. Quite the opposite is true for the rule base which exhibit high amounts of bias and low variance. It can be concluded that the method presented in this paper is good at trading of bias/variance where appropriate by combining the two methods. Further, [26] describe in their paper that CRFs generalize easily to analogues of stochastic free-grammars which could prove an interesting research direction. When developing and evaluating the Named Entity Recognition part of our work, we found the following advantages and disadvantages in our hybrid implementation:

Advantages:

1. The rules enhance the performance of the CRF.
2. It is simple to maintain, and retrain if needed.
3. The rules are portable.
4. the rules are cross domain.
5. The rules are easy to maintain
6. The rules can be expanded by any novice programmer with limited proficiency in data structures and regular expressions.

Disadvantages:

1. CRF tuning require tedious and laborious feature engineering to outperform a Hidden Markov Model.
2. It's typical training method makes training and model inference extremely computationally intensive.
3. And finally, very high model complexity and variance.
4. CRF domain portability is nearly impossible because of the training set and hand-tuned features.

To conclude, researchers focusing on the automatic generation of XBRL documents have struggled to deal with the imprecise and ambiguous nature of natural languages. Many have relied on semi-structured documents which are fairly easy to build a semantic web service to interpret and act on these documents. Others have used an ontological approach, hard wiring expert-guided knowledge to guide the extraction process. Recent works have focused on merging methods together such as the work of [6] where deterministic rules are used to gauge the predictable nature of certain entities such as dates and money, and introduce stochastic methods to handle

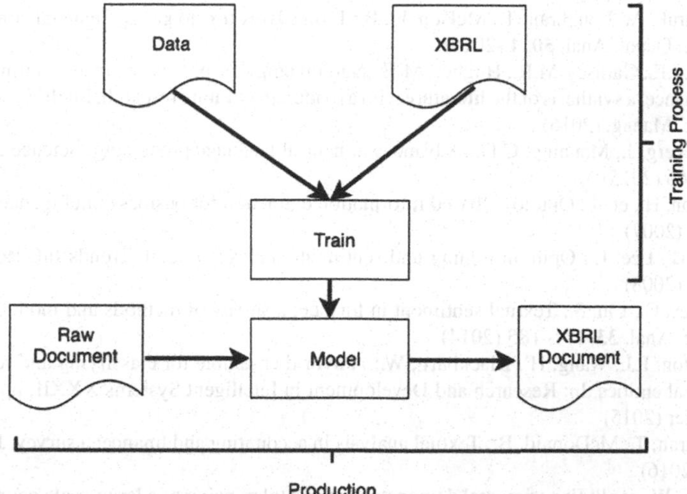

Fig. 4 The overall (simplified) desired pipeline of extracting entities into an XBRL namespace setup. The training process takes a pair of raw accounts (data points) and XBRL documents (target) and fits their together to create a model. In the production process, a raw document can be delivered to the model and outputs an XBRL document

imperfections in written text. From this and previous works, we believe a hybrid approach to NER to be the best approach to preprocessing for Relation Extraction within the financial domain.

7 Further Research

This may be wise and a strong ground for inspiration for all future work in this area. However, the case for hybrid systems shouldn't be discarded or ignored. Because the XBRL taxonomy is a human designed discrete space, a predefined template influenced by various economic and sociological factors such as tax jurisdiction, turnover, formation type, and the use of hybrid characteristics or reinforcement learning. One possible further direction would be to train a model on large amounts of unlabeled but mapped data from financial accounts as depicted in Fig. 4.

References

1. Li, N., Desheng Dash, W.: Using text mining and sentiment analysis for online forums hotspot detection and forecast. Decis. Support Syst. **48**(2), 354–368 (2010)
2. Loughran, T., McDonald, B.: When is a liability not a liability? Textual analysis, dictionaries, and 10Ks. J. Financ. **66**(1), 35–65 (2011)

3. Bodnaruk, A., Loughran, T., McDonald, B.: Using 10-K text to gauge financial constraints. J. Financ. Quant. Anal. **50**, 4 (2015)
4. Fisher, I.E., Garnsey, M.R., Hughes, M.E.: Natural language processing in accounting, auditing and finance: a synthesis of the literature with a roadmap for future research. Intell. Syst. Account. Financ. Manag. (2016)
5. Hirschberg, J., Manning, C.D.: Advances in natural language processing. Science **349**(6245), 261–266 (2015)
6. Saggion, H., et al.: Ontology-based information extraction for business intelligence. Springer, Berlin (2007)
7. Pang, B., Lee, L.: Opinion mining and sentiment analysis. Found. Trends Inf. Retr. **2**(1–2), 1–135 (2008)
8. Kearney, C., Liu, S.: Textual sentiment in finance: a survey of methods and models. Int. Rev. Financ. Anal. **33**, 171–185 (2014)
9. Hampton, P.J., Wang, H., Blackburn, W.: A hybrid ensemble for classifying and repurposing financial entities. In: Research and Development in Intelligent Systems XXXII, pp. 197–202. Springer (2015)
10. Loughran, T., McDonald, B.: Textual analysis in accounting and finance: a survey. J. Account. Res. (2016)
11. Burr, G.W., et al.: Experimental demonstration and tolerancing of a large-scale neural network (165 000 Synapses) using phase-change memory as the synaptic weight element. IEEE Trans. Electron Dev. **62**(11), 3498–3507 (2015)
12. Shaalan, K.: A survey of arabic named entity recognition and classification. Comput. Linguist. **40**(2), 469–510 (2014)
13. Marrero, M., et al.: Named entity recognition: fallacies, challenges and opportunities. Comput. Stand. Interfaces **35**(5), 482–489 (2013)
14. Nadeau, D., Sekine, S.: A survey of named entity recognition and classification. Lingvist. Investig. **30**(1), 3–26 (2007)
15. Reeve, L., Han, H.: Survey of semantic annotation platforms. In: Proceedings of the 2005 ACM Symposium on Applied computing. ACM (2005)
16. Loughran, T., McDonald, B.: When is a liability not a liability? Textual analysis, dictionaries, and 10Ks. The Journal of Finance **66**(1), 35–65 (2011)
17. Wisniewski, T.P., Yekini, L.S.: Stock market returns and the content of annual report narratives. In: Accounting Forum, vol. 39, no. 4. Elsevier (2015)
18. Troshani, I., Parker, L.D., Lymer, A.: Institutionalising XBRL for financial reporting: resorting to regulation. Account. Bus. Res. **45**(2), 196–228 (2015)
19. Burdick, D., et al.: Extracting, linking and integrating data from public sources: a financial case study. SSRN (2015)
20. Chiticariu, L., Li, Y., Reiss, F.R.: Rule-based information extraction is dead! Long live rule-based information extraction systems!. In: EMNLP, no. 10 (2013)
21. Rao, D., McNamee, P., Dredze, M.: Entity linking: finding extracted entities in a knowledge base. In: Multi-source, Multilingual Information Extraction and Summarization, pp. 93–115. Springer, Berlin (2013)
22. Russell, S., Norvig, P.: Artificial Intelligence—A Modern Approach. Prentice-Hall, Egnlewood Cliffs, vol. 25, p. 27 (2013)
23. Rocktschel, T., Weidlich, M., Leser, U.: ChemSpot: a hybrid system for chemical named entity recognition. Bioinformatics **28**(12), 1633–1640 (2012)
24. Califf, M.E., Mooney, R.J.: Bottom-up relational learning of pattern matching rules for information extraction. J. Mach. Learn. Res. **4**, 177–210 (2003)
25. Morwal, S., Jahan, N., Chopra, D.: Named entity recognition using hidden Markov model (HMM). Int. J. Nat. Lang. Comput. (IJNLC) **1**(4), 15–23 (2012)
26. Lafferty, J., McCallum, A., Pereira, F.C.N.: Conditional random fields: probabilistic models for segmenting and labeling sequence data (2001)

Telecoms and E-Learning

A Method of Rule Induction for Predicting and Describing Future Alarms in a Telecommunication Network

Chris Wrench, Frederic Stahl, Thien Le, Giuseppe Di Fatta, Vidhyalakshmi Karthikeyan and Detlef Nauck

Abstract In order to gain insights into events and issues that may cause alarms in parts of IP networks, intelligent methods that capture and express causal relationships are needed. Methods that are predictive and descriptive are rare and those that do predict are often limited to using a single feature from a vast data set. This paper follows the progression of a Rule Induction Algorithm that produces rules with strong causal links that are both descriptive and predict events ahead of time. The algorithm is based on an information theoretic approach to extract rules comprising of a conjunction of network events that are significant prior to network alarms. An empirical evaluation of the algorithm is provided.

Keywords Rule Induction · Event Mining · Network Alarm-prediction

1 Introduction

The reliance on Telecommunications in our personal and business lives makes alarm prediction in this domain an extremely important field of research with the potential of saving businesses a great deal of money; personal users a large amount of inconvenience; and, in crisis situations, may save lives [1]. The goal of this work is to produce a Rule Induction Algorithm to produce human readable rules that predict the occurrence of alarms in a telecommunication network ahead of time. These are

C. Wrench (✉) · F. Stahl · T. Le · G. Di Fatta
Department of Computer Science, University of Reading,
PO Box 225, Whiteknights, Reading RG6 6AY, UK
e-mail: C.Wrench@pgr.reading.ac.uk

F. Stahl
e-mail: F.T.Stahl@reading.ac.uk

G. Di Fatta
e-mail: G.DiFatta@reading.ac.uk

V. Karthikeyan · D. Nauck
BT Research and Innovation, Adastral Park, Ipswich IP5 3RE, UK
e-mail: Detlef.Nauck@bt.com

M. Bramer and M. Petridis (eds.), *Research and Development in Intelligent Systems XXXIII*, DOI 10.1007/978-3-319-47175-4_23

advantageous as the prediction is accompanied with [9] rationale of why the prediction was made which may give valuable insight to domain experts. The domain experts are then better equipped to correct the issue as well as mitigate it in the future. Furthermore, it is desirable to produce these in such a way as to utilise as much context information from the data rich events that make up the data set. The paper focusses on a data set provided by BT of events gathered from a national Telecommunication network over a period of two months. Events are often generated in a non-uniform, bursty manner but contain a large number of attributes describing the cause of AND conditions that led to the target events being generated [2]. There are algorithms that are able to predict the occurrence of an event with some accuracy [3] but they are limited to using a market basket approach which discards the majority of data. The terms *alarm* and *events* are used throughout this paper, the data set that forms the focus of this research consists of alarm data, these are a special form of events as studied in event mining. Whilst an event can be anything that has happened, an alarm further indicates that something has happened that was *unexpected* [4].

The paper is laid out with a summary of existing Rule Induction techniques and other works focussed on Telecommunication data in Sects. 2 and 3 outlines the preprocessing steps taken to refine the data set with a description of how the temporal (timestamps) and geolocation attributes were exploited in Sect. 3.2. Section 4 details the progression towards the current state of the Rule Induction Algorithm and Sect. 5 contains the results. Finally Sect. 6 describes some ongoing work and Sect. 7 contains a conclusion.

2 Related Work

2.1 Event Mining in Telecommunication Networks

Telecommunications were amongst the first to use Data Mining and Data Stream Mining for applications that, as well as Fault Prediction, include Marketing (detecting likely adopters of new services or value customers) and Fraud Detection (identifying unscrupulous accounts to prevent losses to both the company and victimised customers), as categorised by [5]. The following is a collection of algorithms involved in fault detection in both telecommunication networks and networks in general.

In [6] Ant Colony Optimisation is used to produce time based rules. The approach yields a high accuracy, however, it is limited to predicting the class value of an instance rather than predicting a future alarm. Decision Trees (a variant of C4.5) are used in [7] for alarm detection using log data by presenting event types (i.e. one key attribute of the event) as an item set. The Decision Tree is passed an item set that consists of the previous observations and the current observation. The rules produced are then filtered to retain only those whose terms obey temporal ordering (the conditional is made of those items that appeared before the consequent).

TP Mining from [8] searches for repeated event patterns within a time window and promotes those with a high Topographical Proximity (TP), a metric derived from the

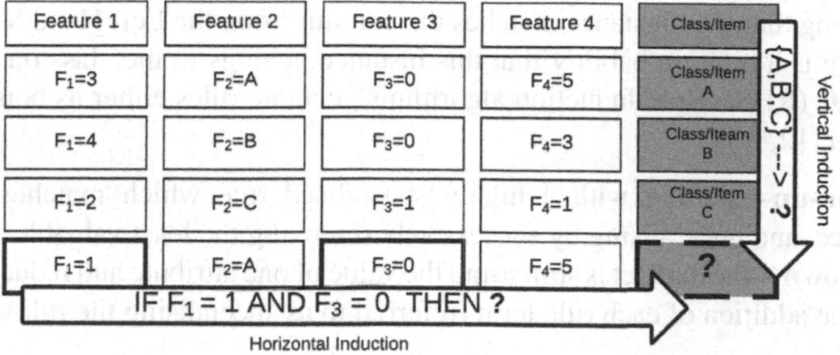

Fig. 1 Data utilisation in prediction and description, existing algorithms predict a value (or event) along just one dimension

relative position of a source device to other devices. The authors of [9] produced a genetic algorithm named *Timeweaver* that specialises in predicting rare events from a telecommunications alarm data set.

Several algorithms are based on, or incorporate, Association Rule Mining and the Apriori [10] approach to make their predictions. The authors of [11] focus their work around alarm prediction in the Pakistan Telecom (PTLC) network. They approach the problem with an adaptation of Association Rules (along with Decision Trees and Neural Networks). It allows the prediction of alarms based on a sequence, much like the target of this work. The rules are non-descriptive, however, and the restriction to producing rules by device type is a narrower problem than the one we are presented with. The authors of [4] investigate an enhancement to alarm correlation. The algorithm TASA (Telecommunication Alarm Sequence Analyser) uses sliding windows to find both Association Rules and Episodic rules (frequently occurring sequences of event types that occur in a time interval). The rules produced are validated by domain experts.

The above algorithms can collectively predict an instance's target class using its own attributes or predict the value of a future instance by focussing on one target feature from the preceding instances and grouping these into an item set, see Fig. 1. The approaches described above are effective for prediction along one axis, either vertically or horizontally, but not both. Utilising both methods could discover alternative rule sets and make more interesting causal relations, which is the focus of the work presented in this paper.

2.2 *Rule Induction*

The purpose of Rule Induction is to generate a series of human readable rules that classify an instance and provide an explanation behind the classification:

$$\text{IF } A \text{ AND } B \text{ THEN } C_i$$

Meaning that if an instance matches the conditions on the Left Hand Side (LHS) then there is a high probability that this instance belongs to the class on the Right Hand Side (RHS). Rule Induction algorithms generate rules either as bottom-up or top-down [12, 13].

- **Bottom-up**—starting with a highly specialised rule which matches a single instance, and generalising by successively removing the least valuable attributes
- **Top-down**—the data set is split using the value of one attribute and reduced further with the addition of each rule term (referred to as specialising the rule).

Rules can be generated through the application of pre-existing forms of machine learning: Decision Trees [14, 15], Covering algorithms [16–18], and the extraction of rules from black box algorithms such as Neural [19] or Bayesian networks [20]. The goal is always to produce a rationale along with a classification that is easily understood by a human reader. The more general a rule is (the less specific) the greater coverage it will have and the rules produced will have a greater resilience to noise. The goal of this work is to produce expressive rules, such as those described above, but with the additional ability to predict, ahead of time, events occurring in the near future.

3 Data Preparation

3.1 Pre-processing

Feature Selection was conducted manually using input from domain experts. Strings, particularly those relating to location, were cleaned of white spaces and trailing characters. These could then be matched using Levenshtein distance [21] to reduce the number of unique string values in the dataset. Numerical attributes (excepting timestamps) were binned and instances with a high number of missing values were removed. If the majority of an attribute's values were missing it was replaced with a boolean attribute to indicate if the field is populated or not.

One of the key features in the data set is Event Name which contains over 180 distinct string values. To simplify the data set it was decided to reduce these. This was done by dividing the data set into 2000 time windows, making each window 2547 s long, and counting the occurrences of each value in every window. This effectively creates 180 time series, one for the occurrence of every event type. These time series are then comparable through Dynamic Time Warping (DTW) [22] producing a distance matrix. The final step is to perform hierarchical clustering using these distances producing a dendrogram that clearly indicates the groups that occur alongside each other. Many of the events in these groups have a high string similarity suggesting they are semantically similar. The process of aggregating event counts across time windows has also demonstrated that there are events that occur regularly across all time windows, indicating that these are largely noise and will need to be addressed.

3.2 Co-occurrence Detection

Included in the data set are several features that relate to the temporal or physical location of the event origin that can be used for Co-Occurrence detection. Co-Occurrence is used to increase the likelihood that two events are causally linked by identifying events that occur within physical and temporal proximity to each other. The utilisation of both kinds of proximity are described in Sects. 3.2.1 and 3.2.2.

3.2.1 Geo-location Clustering

The national data set was divided into subsets based on the event's accompanying northing and easting values. This was done by first normalising these values, to account for the disproportion in the UK's easting and northing, then breaking the data set into time windows and running each window through a clustering algorithm.

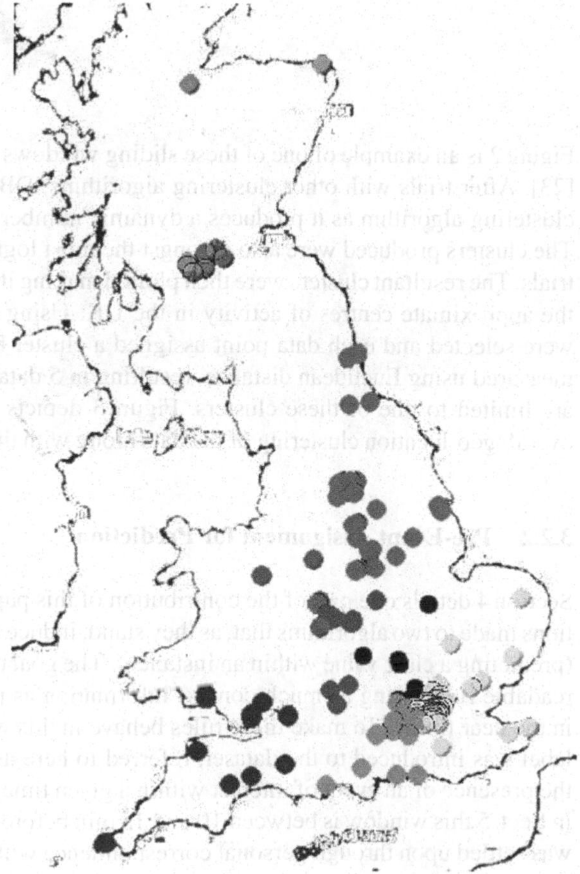

Fig. 2 Clustering of geo-location events map

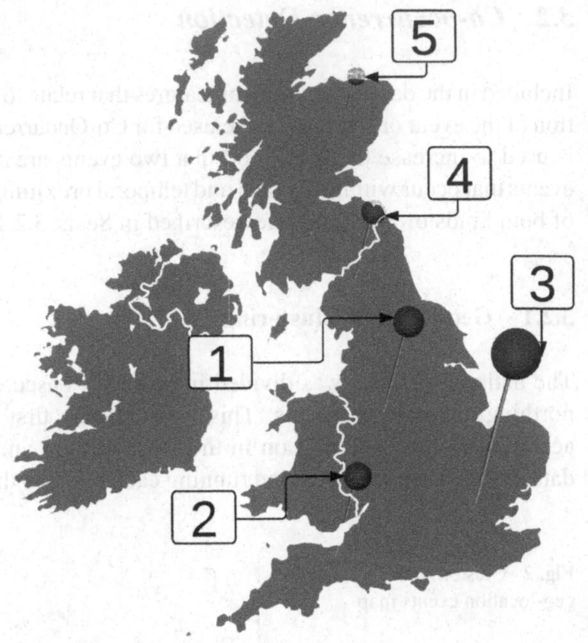

Fig. 3 Cluster centroids and relative sizes

Figure 2 is an example of one of these sliding windows plotted over a map of the UK [23]. After trials with other clustering algorithms, DBScan [24] was chosen as the clustering algorithm as it produces a dynamic number of clusters based on density. The clusters produced were also amongst the most logically defined out of the other trials. The resultant clusters were then plotted making it possible to manually identify the approximate centres of activity in the UK. Using this information, 5 centroids were selected and each data point assigned a cluster based on its nearest centroid, measured using Euclidean distance, resulting in 5 data sets. Experiments from here are limited to one of these clusters. Figure 3 depicts the chosen centroids for the overall geo-location clustering of the data along with the relative size of each cluster.

3.2.2 Pre-Event Assignment for Prediction

Section 4 details one part of the contribution of this paper, specifically the modifications made to two algorithms that, as they stand, induce rules along the horizontal axis (predicting a class value within an instance). The goal of this work is to make human readable rules using as much context information as possible that predict an event in the near future. To make these rules behave in this way an additional binary class label was introduced to the dataset, referred to here as a *Pre-Event*. These indicate the presence of an event of interest within a given time window. For the experiments in Sect. 5 this window is between 10 and 15 min before the event occurs, this interval was settled upon through personal correspondence with BT. The data set is parsed in

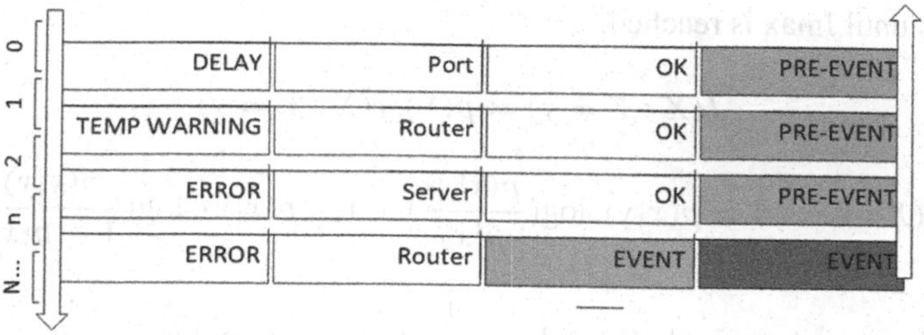

Fig. 4 Creating a *Pre-Event* class used for prediction

reverse order (starting with the most recent) and events that occur within the window are marked as positive, see Fig. 4. A problem with this method is that it skews class bias, this will be addressed in future work.

4 Alarm Prediction and Description

This Section outlines the modifications and optimisation of a rule induction algorithm used on the dataset from two ITRULE based approaches to a covering approach. The first algorithm used is a basic implementation of ITRULE with minor alterations to combat over specialisation, used here as a baseline for comparison. The second algorithm is an adaptation of ITRULE first introduced in a previous work [25] featuring a pruning method to correct partial rule dominance, developed originally for streaming data. Following this, a covering method was developed based on the PRISM method but using J-measure to induce rules and later incorporating pruning based on confidence. Performance increases in some areas are accompanied by each new algorithm.

ITRULE was initially chosen for adaptation because of it's strong statistical foundation, faculty to produce expressive rules, and it's resistance to over-specialisation through the property Jmax, explained later in this section. ITRULE was developed by Goodman and Smyth [26] and produces generalised rules from batch data consisting of many nominal attributes. It evaluates every combination of possible rule terms using the resultant rule's theoretical information content, known as it's J-measure, see Eq. 1. This is calculated from the product of the LHS's probability (*p(Y)*) and the cross entropy of the rule (*j(X:Y=y)*) see Eq. 2. A very useful property of ITRULE is that the maximum information content of a given rule (it's maximum J-measure) is bounded by the property Jmax, see Eq. 3. This enables ITRULE to stop specialising a rule as it nears Jmax as no further gain in J-measure can be obtained.

It uses Beam Search to keep the search space to a manageable size, only selecting the top *N* rules from each iteration to expand upon. The rule is then specialised by appending further rule terms to the LHS of the rule. When every combination of the next phase of rules have been produced, the top *N* are again selected and the process continues until Jmax is reached.

$$J(X : Y = y) = p(Y).j(X : Y = y) \tag{1}$$

$$j(X : Y = y) = p(x|y).\log(\frac{p(x|y)}{p(x)}) + 1 - p(x|y).\log(\frac{p(x|y)}{1 - p(x)}) \tag{2}$$

$$J_{max} = p(y).\max\{p(x|y).\log(\frac{1}{p(x)}), 1 - p(x|y).\log(\frac{1}{1 - p(x)})\} \tag{3}$$

Beam Search is a greedy algorithm and frequently produces rule sets with little variation in the rule terms used, a problem known as partial rule dominance. This is particularly evident in cases where there is a large disparity between the assigned worth, or *goodness*, of attributes. In this case there is a great disparity between the number of distinct values belonging to an attribute, even after pre-processing. As the J-measure is a product of the probability of an instance matching the LHS of a rule, those features with fewer distinct values tend to have a much higher J-measure than those with many values. If not accounted for, these feature values will repeatedly be selected for further specialisation to the extent that they dominate the beam width. This in turn leads to rules which share very similar conditionals and cover a very narrow range of features, often repeating similar tests [13]. For example, in the below, feature values A_1 and B_1 have a high probability of both occurring and being selected in the top N rules. This leads to very similar rules with low coverage:

IF A_1 AND B_1 AND C_1 THEN X
IF A_1 AND B_1 AND C_2 THEN X
IF A_1 AND B_1 AND C_3 THEN X

This rule set has a large amount of redundant information and could instead be presented as a more general rule leaving space in the beam for more interesting rules:

IF A_1 AND B_1 THEN X

This can be addressed either as rules are induced or afterwards (pre-pruning or post-pruning).

To mitigate against partial rule dominance a form of pre-pruning was incorporated. When selecting the candidates for the next iteration of rule specialisation, the rules are ranked according to their respective J-measures and the top N are added into the beam for specialisation. When a rule is added a distribution of the features within the beam is updated. Here a rule is only added if this rule's individual member features do not surpass a threshold proportion of the beam (set to 40 % for these experiments). Otherwise it is not included and the next best rule term is tested in the same way. It is possible to exhaust the list of candidate rule terms with this method if the beam size is sufficiently large or the candidate list is sufficiently small. This being the case, the list (minus the rules already selected) is iterated again until a sufficient number of rules have been selected. This approach leads to a more varied rule set though it has the disadvantage of potentially losing a rule with high a J-measure in preference for a new minority rule term with a lower J-measure. The rule set stands a greater chance of producing rules that are of interest to the domain experts.

An alternative to using the algorithms original Beam Search is to adopt the Separate and Conquer method used in covering algorithms such as AQ [18] and PRISM [16]. PRISM was developed as a means to overcome the repeated sub-tree problem inherent with decision trees whereby parts of the tree are repeated leading to the same tests being carried out multiple times. It does this by selecting the rule term so that the conditional probability of covering the target class is maximised before separating out the instances it covers from the rest of the data set. From this new, smaller data set another rule term is selected (the rule is specialised) and the data set is further reduced, those not falling under the longer rule are returned to the original dataset. This continues until all instances in the set have the same class value and the rule is finalised. The process repeats until all instances are covered by a rule.

A new algorithm was developed as a hybrid between PRISM and ITRULE by replacing the beam search approach with a covering technique which removes any partial rule dominance issues.

Algorithm 1 Prism-ITRULE Hybrid—where J is calculable using Eqs. 1 and 2 and J_{max} is calculable using Eq. 3

```
Dataset D with Target Classes C_n and Attributes A_n
for Every Class C_i in D do
  while D contains classes other than C_i do
    for Every Attribute A_i in D do
      for Every Attribute Value A_iv in A_i do
        Generate Rule R_n with Rule Term A_iv
        Calculate j
        J_max ← J_max * ThresholdT
        add R_n to Candidate Rule Set
      end for
    end for
    Select R_n where j(R_n) is maximised
    Remove Instances not covered by R_n
    if j(R_n) > J_max(R_n) then
      Rule Complete
      Break
    end if
  end while
  for all Rule Terms RT_i : in Rule R do
    if Confidence( RT_i) <Confidence(RT_{i-1}) then
      Remove RT_i from R
    end if
  end for
  C_i ← Remove Instances Covered by R_n from D
end for
```

PRISM uses conditional probability, designed to maximise coverage, to induce rules. Here, in place of this, the J-measure is used, inducing rules based on their information content, and Jmax is retained as a stopping criteria to avoid over-fitting. The resultant algorithm is referred to here as a *Prism-ITRULE Hybrid*. A pruning step based on confidence was later incorporated after rule induction, the algorithm is described in Algorithm 1.

5 Results

This Section details results from the experiments performed with the algorithms outlined in Sect. 4. The following tests were conducted using three different alarm types as their Pre-Event target class (i.e. for each data set one of three different alarm types was used for Pre-Event generation). These data sets were split into 3 smaller data sets of 30,000 instances based on the time of occurrence and divided randomly into test and training sets. The training and test sets consist of 10,000 and 20,000 instances respectively.

Table 1 shows the percentage of instances marked as positive (the target attribute value) in each data set. 3 values of the feature Event Name were selected as target classes (from which Pre-Events are made), these will be referred to as events *A*, *B* and *C*. Target class *A* is of particular importance to BT (determined from personal communication with BT) whilst B and C were selected for their contrasting distributions.

Several tests were carried out using the algorithms described in Sect. 4 on a sample of the data. Tables 2 and 3 detail some of the recorded metrics from these experiments, including Abstain Rate and Tentative Accuracy. A brief description of these is provided below:

- Accuracy % (Acc)—the percentage of instances that were correctly classified from the whole data set.

Table 1 Class proportion of target events

Data set	Class breakdown	Percentage target class
A	A_i	7.41
	A_{ii}	6.92
	A_{iii}	4.24
B	B_i	69.55
	B_{ii}	35.50
	B_{iii}	66.16
C	C_i	0.68
	C_{ii}	0.73
	C_{iii}	0.51

Table 2 Results for ITRULE and ITRULE with partial rule dominance pruning

	Basic ITRULE				Partial rule dominance pruning			
Data set	Acc %	Tent-Acc %	Abs rate %	Time (s)	Acc %	Tent-Acc %	Abs rate %	Time (s)
A_i	23.03	93.75	75.44	68.41	36.63	92.87	60.57	41.83
A_{ii}	35.92	92.94	61.35	31.35	49.07	95.93	48.85	20.18
A_{iii}	33.73	93.47	63.92	41.89	30.51	93.49	67.37	40.54
B_i	99.16	99.31	0.15	7.91	76.83	99.14	22.51	7.75
B_{ii}	99.49	99.49	0.00	8.38	92.56	99.47	6.96	6.69
B_{iii}	95.13	99.26	4.17	6.71	43.54	99.16	56.09	12.20
C_i	0.92	45.75	98.00	11.72	5.43	68.52	92.08	65.48
C_{ii}	0.37	18.25	98.00	30.82	0.42	18.66	97.75	19.42
C_{iii}	1.20	59.01	97.98	0.99	0.78	37.05	97.90	26.13

Table 3 Results for the Prism-ITRULE hybrid with an without pruning using confidence

	Prism-ITRULE hybrid				Prism-ITRULE hybrid with pruning			
Data set	Acc %	Tent-Acc %	Abs rate %	Time(s)	Acc %	Tent-Acc %	Abs rate %	Time(s)
A_i	94.29	94.29	0.00	79.23	94.20	94.20	0.00	78.75
A_{ii}	95.91	95.91	0.00	77.95	95.90	95.90	0.00	80.56
A_{iii}	94.23	94.23	0.00	76.07	94.23	94.23	0.00	78.51
B_i	99.31	99.31	0.00	82.13	99.31	99.31	0.00	84.68
B_{ii}	98.50	99.43	0.94	78.03	99.46	99.46	0.00	79.42
B_{iii}	99.22	99.24	0.02	74.32	99.29	99.29	0.00	76.99
C_i	17.68	78.85	89.87	75.12	8.70	70.36	98.55	78.91
C_{ii}	40.40	87.46	53.81	74.50	42.14	88.98	52.65	77.58
C_{iii}	17.68	78.85	77.59	74.89	8.88	86.17	89.70	75.70

- Tentative Accuracy % (Tent-Acc)—the percentage of instances that were correctly classified from those covered by rule set.
- Abstain Rate % (Abs Rate)—the percentage of instances in the data set that were not covered by a rule and no attempt at classifying was made.
- Time (s)—the time taken for training on the data set taken from the mean of three runs.

Tentative Accuracy is included here as this problem does not require total classification of all instances, more important is the correct classification of the rule firing. The data set now uses a binary class (an event is marked as a Pre-Event or not) and rules are produced for both class types, as such the accuracies apply, as well, to both classes.

5.1 ITRULE

The first tests were run on the original version of ITRULE and serve as a baseline for comparison, see Table 2. For these experiments a beam width of 45 was used and the value of Jmax was set to 80 % of its full value to combat over fitting. These values were found to yield the highest accuracies in previous experiments. Table 2 also contains the results of experiments using Partial Rule Dominance Pruning. In terms of accuracy and tentative accuracy the results are very similar, the accuracy for predicting A events are higher due to a much reduced abstain rate. The trade-off in the variance in execution time is higher in all cases, this is due to a large amount of additional tests needed to populate the beam size in such a heavily skewed data set. Even with the inclusion of Partial Rule Dominance Pruning the overall accuracy is low and the abstain rate is high.

5.2 Prism-ITRULE Hybrid

In Table 3 are the results of experiments done with two implementations of a Prism-ITRULE Hybrid, one with and one without confidence pruning. Again with a Jmax threshold set to 80 % of its full value. There are many evident benefits over beam search as the abstain rate is far lower, boosting the accuracy in turn. The tentative accuracy has increased too even though it is independent of the latter two metrics. It does, however, suffer on the smaller target class set, B. It can be seen that incorporating the pruning step yields an increase in overall accuracy. The execution time has increased as expected but, as pruning only takes place on the selected candidate rule, the percentage increase is minimal.

5.3 Comparison of Results

Table 4 contains the confidence, support, J-measure, Jmax and J-distance of 3 runs from each iteration of the algorithm for comparison. J-distance is the distance between the final J-measure of a rule and the Jmax, a low distance indicates that the rules were nearly optimised when they reach their finished form. Confidence is only used by the algorithm for pruning in the final stage of the Prism-ITRULE hybrid and support is recorded but not used at all in producing rules. Each result is the average of the top 10 rules produced by the algorithm when ranked by their J-measures.

It can be seen that there has been a steady rise in the support of the rules over each iteration. There is a drop in the confidence between the switch from Beam Search to Separate and Conquer, this is misleading however as the high confidence of the original Beam Search approach is due to the over general form of the rules, leading to

Table 4 Support, confidence and J-measure values for each algorithm run on the A data sets

	Set	Supp.	Conf.	J-measure	Jmax	Jdist
Prism-ITRULE hybrid with confidence pruning	A_i	0.38	0.68	0.38	0.51	0.13
	A_{ii}	0.29	0.59	0.34	0.47	0.13
	A_{iii}	0.32	0.59	0.33	0.47	0.14
	Avg	0.33	0.62	0.35	0.48	0.13
Prism-ITRULE hybrid	A_i	0.29	0.68	0.30	0.35	0.06
	A_{ii}	0.22	0.54	0.27	0.30	0.03
	A_{iii}	0.25	0.59	0.28	0.39	0.11
	Avg	0.25	0.60	0.28	0.35	0.07
ITRULE with partial rule dominance pruning	A_i	0.18	0.49	0.33	0.51	0.18
	A_{ii}	0.35	0.98	0.74	1.08	0.34
	A_{iii}	0.36	0.96	0.66	1.02	0.36
	Avg	0.30	0.81	0.58	0.87	0.29
Basic ITRULE	A_i	0.15	0.94	0.46	0.56	0.10
	A_{ii}	0.15	0.95	0.51	0.61	0.10
	A_{iii}	0.17	0.94	0.50	0.62	0.12
	Avg	0.16	0.94	0.49	0.60	0.11

the very low support values. The beam search approach produces rules with a higher average J-measures and Jmax which is likely due to the less restricted data set used at each phase of rule induction.

Introducing confidence pruning into the algorithm increases the distance between the rule's J-measures and Jmaxs as would be expected. The J-measure of a rule can increase or decrease with the addition of any term (with the exception that it is bounded by 0 and Jmax). In one instance the J-measure has increased after applying pruning from 0.27 to 0.34, demonstrating that pruning to increase confidence is not necessarily at the cost of the other metrics used.

Particularly in terms of accuracy, the algorithm developed in this work has seen dramatic improvement whilst its other recorded metrics have remained steady. It follows that the resultant algorithm is effective at predicting and describing alarms ahead of time in line with this paper's goals.

6 Ongoing Work

In Sect. 3.2.2 it was mentioned that the class bias of the data set has been largely skewed through the introduction of Pre-Events. A further area of research is to be conducted on producing a Rule Induction algorithm that works with multi-label data (i.e. where an instance can belong to more than one target classes at the same time) using the generated Pre-Events [27, 28]. This would directly address the class bias problem and lead to more expressive rules being produced. As this work represents an extension to algorithms that, in their base form, are limited to one class attribute, this further extension has not been included here. There is also an additional data set containing the underlying IP network performance data from which the events are generated, it is hoped that by incorporating this additional data more interesting rule sets can be produced.

7 Conclusion

This paper outlines the progression of a Rule Induction Algorithm to produce rules that predict target events in the IP network. It does this by utilising the information available in the events rather than discarding this in favour of an Association Rule Mining Approach. A great deal of pre-processing has gone into the data set to clean the data and exploit the geolocation and temporal properties of the data set to increase the probability of producing rules that have strong causal links.

The ITRULE algorithm initially chosen for the project has been altered significantly due to the issues with using Beam Search on such a highly disproportionate distribution of feature values. Partial rule dominance led to producing a rule set with little variation in the rules and a tendency to favour numerical attributes over nominal ones. This was addressed with Partial Rule Dominance Pruning, however, the abstain rates are still high and accuracies were still low. Replacing Beam Search with a Separate and Conquer technique produced a large increase in accuracy which was higher still with the inclusion of confidence pruning though there are still issues with producing accurate rules from the dataset with the most sparse target class.

References

1. CPNI/Centre for the Protection of National Infrastructure. Telecommunications Resilience Good Practice Guide Version 4. Technical Report March, Centre for the Protection of National Infrastructure (2006)
2. Perrochon, L., Mann, W., Kasriel, S., Luckham, D.C.: Event mining with event processing networks. In: Methodologies for Knowledge Discovery and Data Mining. Third Pacific-Asia Conference, PAKDD-99 Beijing, China, April 2628, 1999 Proceedings, pp. 474–478 (1999)
3. Fülöp, L.J., Tóth, G., Rácz, R., Pánczél, J., Gergely, T., Beszédes, Á., Farkas, L.: Survey on complex event processing and predictive analytics. In: Proceedings of the Fifth Balkan Conference in Informatics, pp. 26–31 (2010)
4. Klemettinen, M., Heikki, M., Toivonen, H.: Rule discovery in telecommunication alarm data. J. Netw. Syst. Manage. **7**(4) (1999)
5. Weiss, G.M.: Data mining in the telecommunications industry. In: Data Mining and Knowledge Discovery Handbook, pp. 1189–1201 (2005)
6. Khan, I., Huang, J.Z., Tung, N.T.: Learning time-based rules for prediction of alarms from telecom alarm data using ant colony optimization. Int. J. Comput. Inf. Technol. ISSN: 2279 0764

7. Karimi, K., Hamilton, H.J.: TimeSleuth: a tool for discovering causal and temporal rules. In: 14th IEEE International Conference on Tools with Artificial Intelligence, 2002. (ICTAI 2002). Proceedings, pp. 375–380 (2002)
8. Devitt, A., Duffin, J., Moloney, R.: Topographical proximity for mining network alarm data. In: Proceeding of the 2005 ACM SIGCOMM Workshop on Mining Network Data—MineNet'05, p. 179 (2005)
9. Weiss, G., Hirsh, H.: Learning to predict rare events in event sequences. In: Kdd-98, pp. 359–363 (1998)
10. Agrawal, R., Srikant, R.: Fast algorithms for mining association rules. In: Proceedings of 20th International Conference on VLDB, pp. 487–499 (1994)
11. Jaudet, M., Iqbal, N.: Neural networks for fault-prediction in a telecommunications network. In: 8th International Multitopic Conference, 2004. Proceedings of INMIC 2004, pp. 315–320 (2004)
12. Fürnkranz, J., Gamberger, D., Lavrač, N.: Foundations of Rule Learning. Cognitive Technologies. Springer, Berlin (2012)
13. Fürnkranz, J.: A pathology of bottom-up hill-climbing in inductive rule learning. In: Algorithmic Learning Theory, vol. 2533(Section 2), pp. 263–277 (2002)
14. Leech, W.J.: A rule-based process control method with feedback. ISA Trans. **26**, 73–78 (1986)
15. Quinlan, J.R.: C4.5: Programs for Machine Learning. Morgan Kaufmann, San Francisco (1993)
16. Cendrowska, J.: PRISM: an algorithm for inducing modular rules. Int. J. Man-Mach. Stud. **27**, 349–370 (1987)
17. Chaudhry, N.: Introduction to stream data management. In: Chaudhry, N., Shaw, K., Abdelguerfi, M. (eds.) Stream Data Management. Database, pp. 1–11. Springer, US (2006)
18. Michalski, R.S.: On the quasi-minimal solution of the general covering problem (1969)
19. Taylor, B.J., Darrah, M.A.: Rule extraction as a formal method for the verification and validation of neural networks. In: Proceedings of the International Joint Conference on Neural Networks, vol. 5, pp. 2915–2920 (2005)
20. Gopalakrishnan, V., Lustgarten, J.L., Visweswaran, S., Cooper, G.F.: Bayesian rule learning for biomedical data mining. Bioinformatics **26**(5), 668–675 (2010)
21. Gusfield, D.: Algorithms on Strings, Trees and Sequences: Computer Science and Computational Biology (1997)
22. Yi, B.-K., Jagadish, H.V., Faloutsos, C.: Efficient retrieval of similar time sequences under time warping. In: 14th International Conference on Data Engineering, 1998. Proceedings, pp. 201–208. IEE (1998)
23. Map Data @ 2016 Geo-Basis-DE/BKG and Google. Google Maps (2016)
24. Cao, F., Ester, M., Qian, W., Zhou, A.: Density-based clustering over an evolving data stream with noise. In: SDM, vol. 6, pp. 328–339 (2006)
25. Wrench, C., Stahl, F., Di Fatta, G., Karthikeyan, V., Nauck, D.: Towards expressive rule induction on IP network event streams. In: Research and Development in Intelligent Systems XXXII: Incorporating Applications and Innovations in Intelligent Systems XXIII, pp. 191–196. Springer, Cham (2015)
26. Smyth, P., Goodman, R.M.: An Information Theoretic Approach to Rule Induction from Databases (1992)
27. Menc, E.L., Janssen, F.: Towards Multilabel Rule Learning (2008)
28. Zhang, M., Zhou, Z.: A review on multi-label learning algorithms **26**(8), 1819–1837 (2014)

Towards Keystroke Continuous Authentication Using Time Series Analytics

Abdullah Alshehri, Frans Coenen and Danushka Bollegala

Abstract An approach to Keystroke Continuous Authentication (KCA) is described founded on a time series analysis based approach that, unlike previous work on KCA (using feature vector representations), takes the sequencing of keystrokes into consideration. The significance of KCA is in the context of online assessments and examinations used in eLearning environments and MOOCs, which are becoming increasingly popular. The process is fully described and analysed, including comparison with established feature vector approaches. Our proposed method outperforms these other approaches to KCA (with a detection accuracy of 94 %, compared to 79.53 %), a clear indicator that the proposed time series analysis based KCA has significant potential.

Keywords Keystroke Pattern Recognition · Keystroke Time Series · Behavioural Biometrics · Continuous Authentication

1 Introduction

Keystroke patterns (typing patterns) are a recognised behavioural biometric for establishing the security credentials of users in the context of static user authentication. The fundamental idea is that the rhythm of typing a predefined text by a legitimate user can be learned, and consequently used for authentication purposes [1]. Keystroke Static Authentication (KSA) has been applied with respect to applications such as password, username and pin number authentication [2–5]. However, KSA is unsuited to applications that require continuous authentication such as in the context of the online assessments and examinations frequently used in eLearning environments

A. Alshehri (✉) · F. Coenen · D. Bollegala
Department of Computer Science, University of Liverpool, Liverpool L69 3BX, UK
e-mail: a.a.alshehri@liverpool.ac.uk

F. Coenen
e-mail: coenen@liverpool.ac.uk

D. Bollegala
e-mail: danushka.bollegala@liverpool.ac.uk

M. Bramer and M. Petridis (eds.), *Research and Development in Intelligent Systems XXXIII*, DOI 10.1007/978-3-319-47175-4_24

and MOOCs, where Keystroke Continuous Authentication (KCA) is required. KCA is significantly more challenging than KSA because the process relies on detecting patterns from free text (unlike in the case of KSA where we are looking for a single fixed pattern).

Work on KCA to date has been predominantly focussed on feature vector based binary classification where the features are statistics, such as the average hold time (duration of a key press) and digraph latency (duration between the start or end of pairs of common consecutive key presses) [6–11]. These systems operate by continuously measuring the similarity between a learnt user statistical profile and previously unseen profiles presented in the form of a data stream. However, there is a great deal of variability in the statistical features used to make up the feature vectors and consequently the reported results to date have tended to not be as good as anticipated. The overriding disadvantage of the feature vector based approaches is that the sequencing of key presses is largely lost. In addition, classifiers (predictors) need to be built for each user and this in turn adversely affects the efficiency of the application of KCA in real environments. The idea presented in this paper is to conceptualise the keystroke process in terms of a ongoing time series from which KCA can be realised through a time series analysis process rather than using a feature vector based classification approach. More specifically the idea is to view keystrokes in terms of press-and-release temporal events such that a series of successive events can be recorded. Each keystroke is defined in terms of a pair $P = (t, k)$, where t is a time stamp or temporal identifier of some form; and k is some keystroke attribute (such as flight time between keys or key-hold length). The intuition is that the time series paradigm can be more readily used to dynamically identify "suspect behaviour", in real time, because it serves to capture keystroke sequences (unlike in the case of statistical techniques).

The rest of this paper is organised as follows. In Sect. 2 a brief state-of-the-art review of KCA is presented. This is followed in Sect. 3 with a description of the proposed keystroke time series representation, while in Sect. 4 we introduce the proposed KCA approach. Keystroke time series similarity is then discussed in Sect. 5. The evaluation and comparison of the proposed approach is reported on in Sect. 6. Finally, the paper is concluded with a summary and some recommendations for future work in Sect. 7.

2 Previous Work

Dealing with keystroke patterns, especially KCA, in terms of time series has received little attention in the literature. Reference is made in [12] where a sequence pattern mining algorithm is presented for which a potential suggested application is KCA; an idea that has some similarity with the time series approach proposed in this paper. The majority of work on keyboard usage authentication has been directed at the idea of using statistical feature vectors to recognise keystroke patterns. As mentioned in the introduction to this paper, keystrokes have a range of timing features associated with them including key-down, key-up and hold time. Also, given sequences of pairs of keystrokes, the flight time between n successive keystrokes can also be considered. These features have been represented in terms of feature vectors by computing statistical quantitative equivalents. Such feature vectors have then been used to recognise typing patterns with a view to keyboard usage authentication. The similarity, between (say) two typing profiles may be measured, for example, in terms of the extent of the distance between two vectors.

One of the earliest studies that have considered the idea of keyboard authentication is [13] where the authors utilized the concept of digraph latency for the feature vector representation from which a binary classifier was generated. The classifier operated using the mean and standard deviation of digraph occurrences in a training profile. Only digraphs that satisfied a predefined threshold of occurrences were used; not all digraphs were included. The principal disadvantages of the approach were that: (i) to achieve a reasonable classification performance a substantial amount of data was required with which to train the classifier, the classifier requires an average of 6,390 digraphs to recognise a useful pattern, and (ii) a dedicated classifier was required for each individual. The approach would thus be difficult to apply in the context of eLearning platforms and MOOCs. An alternative approach was presented in Gunetti and Picardi [9] where the average time for pressing frequent key sequences (n-graphs) was recorded and stored in arrays, one per n-graph. Common n-graphs were extracted for corresponding samples (reference and test). The elements of the arrays were then ordered and the distance between sample pairs computed by comparing the ordering in the reference array with the ordering in the test array. This measure was referred to as "the degree of disorder". However, learning a reference sample depends on all other samples in the reference profile. This can cause an efficiency issue when dealing with large numbers of samples. The training of legitimate users profiles is thus as time consuming as in [13]. Ahmed et al. [6] have used key-down time information and the average of digraph flight time and monograph to represent features. An Artificial Neural Network classifier was used. This mechanism worked reasonably well in a controlled experimental setting, typing of the same text using the same keyboard layout in an allocated environment. This is not the situation we would find in uncontrolled environments, such as those used for conducting eLearning assessments and examinations.

From the above, most KCA studies have been directed at the usage of quantitative statistical measures to represent keystroke features founded on a feature vector representation. However, it is argued here that the feature vector representation may not necessarily be representative of useful typing patterns. Therefore, it is conjectured that representing keystroke features as a time series can lead to a better interpretation of typing patterns with respect to KCA.

3 Keystroke Time Series Representation

Keyboard usage is typically undertaken in a sequential manner key-press by key-press, on occasion two keys may be pressed together (for example using the shift or control keys). Thus typing action is well suited to representation in terms of time series where each key press describes a point (event) within the time series. More formally a Keystroke time series K_{ts} is a sequential ordering of a set of data points that occur within a specified interval of time, $K_{ts} = \{p_1, p_2, \ldots, p_n\}$ where each point p_i corresponds to a tuple of the form $\langle t_i, k_i \rangle$ where t_i is a temporal identifier of some description and k_i is some associated attribute value. Thus $K_{ts} = \{\langle t_1, k_1 \rangle, \langle t_2, k_2 \rangle, \ldots, \langle t_n, k_n \rangle\}$. As such, a time series can be viewed as a 2D plot with time along the x-axis and attribute value along the y-axis. The value for t_i can be either: (i) key-down time KD^t, (ii) key-up time KU^t or (iii) a sequential ID number KN per keystroke. However, when using KD^t or KU^t the "ticks" along the time series x-axis are typically irregularly spaced which in turn tends to hinder the time series analysis. Therefore, in the context of the work presented in this paper, KN was used for t. For k_i we have used flight time F^t. Flight time serves to capture the duration between keystrokes which is lost if we use KD^t or KU^t for t. Thus, F^t was adopted, $K_{ts} = \{\langle KN_1, F_1^t \rangle, \langle KN_2, F_2^t \rangle, \ldots\}$. With respect to the forgoing the following definitions should be noted:

Definition 1 A Keystroke Time Series K_{ts} is an ordered discrete sequence of points; $K_{ts} = [p_1, p_2, \ldots, p_i, \ldots, p_n]$ where $n \in \mathbb{N}$ is the length of the series, and p_i is a tuple corresponding to a feature pairing.

Definition 2 A keystroke time series sub-sequence s_k, of length l, is generated from K_{ts} and starts at the position i, $s_k = [p_{ki}, p_{k(i+1)}, \ldots, p_{k(i+j)}]$, where $k \in \mathbb{N}$ is the identifier of current sub-sequence, $j = l - 1$ and $1 < l < n - l$. Thus an ordered subset of K_{ts} can be indicated using the notation $s_k \preceq K_{ts}$ $(\forall p_{ki} \in s_k, \exists p_i \in K_{ts})$.

Definition 3 A profile P is a set of k keystroke time series sub-sequences $P = \{s_1, s_2, \ldots, s_k\}$.

In practice each K_{ts} describes a task dependent keyboard session. For example, in the context of an *online* assessment where a student is answering an assessment question, the generated K_{ts} should represent that keyboard session associated with the student conducting this task. For KCA to operate each time series K_{ts} needs to be evaluated at the start (by comparison of an initial sub-sequence s_1 of the time series with P, a previously stored "bank" of sub-sequences for the subject). As the session proceeds continuous comparison needs to be undertaken by comparing the most recent sub-sequence s_i with earlier collected sub-sequences $S = \{s_1, s_2, \ldots, s_{i-1}\}$ (of the time series K_{ts}).

The fundamental idea proposed in this paper is illustrated in Fig. 1 generated using several randomly selected time series from the sample data used for evaluation purposes as presented later in this paper (see Sect. 6). The figure shows four keystroke time series sub-sequences representing two subjects, two sub-sequences from each subject (Subjects 2 and 9). From the figure it can be seen that there are clear similarities in the keystroke sub-sequences associated with the same subjects (despite the sub-sequences being related to different texts), and clear dissimilarities in the keystroke sub-sequences associated with different subjects. It is thus argued here that such time series can be fruitfully used for KCA.

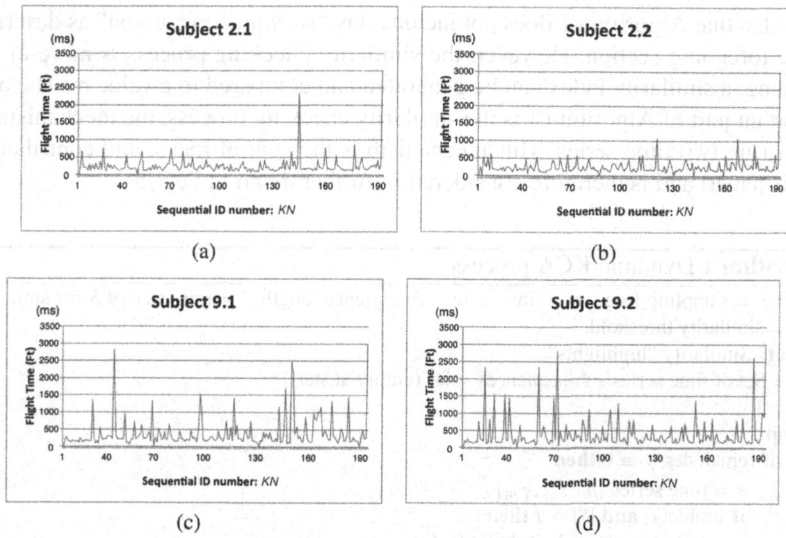

Fig. 1 Examples of keystroke time series: **a** and **b**, time series for Subject 2 writing two different texts; **c** and **d**, time series for Subject 9 writing two different texts

4 Proposed KCA Approach

The proposed KCA process is presented in Algorithm 1. The inputs are: (i) a desired sampling frequency f, (ii) a desired keystroke time series sub-sequence length m and (iii) a minimum size l for S (a set in which to hold collected sub-sequences) before similarity measurements can be made and (iv) a similarity threshold σ value. The process operates on a continuous loop; after every f ticks (line 4) a keyboard time series sub-sequence s of length m is constructed (line 5). Not every time series is usable, for example, there may be sizeable durations between key presses indicating "away from keyboard" events, thus the generated subsequence s needs to be verified for its usability (line 6). Recall that for the time stamps keypress indexes, not the actual time of the key press (for reasons presented earlier) were used. The usability of a time series can be simply identified from the presence of an excessive flight time value. If s is usable and we have collected a sufficient number of sub-sequences ($|S| > l$) authentication can be undertaken (line 7). This is done by calculating a similarity index (simIndex); the simplest way of doing this is to obtain an average of the similarity values between s and the sub-sequences in S. If $simIndex \geq \sigma$, then an authentication error has occurred (lines 8 to 10). Note that in a similar manner to plagiarism checkers (such as Turnitin[1]) the proposed KCA is essentially a similarity checker, thus when dissimilarity is found this is an indicator of further investigation being required. The most recent time series s is then added to S (line 11) and the process continuous until a data stream end signal is received (line 19).

[1] http://www.turnitinuk.com/.

Note also that Algorithm 1 does not include any "start up comparison" as described in the foregoing section. However, the similarity checking process is more-or-less the same; a similarity index can be generated and compared to a value σ. The most important part of Algorithm 1 is the similarity checking process, the mechanism for comparing two time series. This mechanism is the central focus, and contribution, of this paper; and is therefore considered in further detail in Sect. 5.

Algorithm 1 Dynamic KCA process

Input: f = sampling freq., m = time series subsequence length, l = min. size of S for sim. calc., σ = similarity threshold.
Output: Similarity "highlights".

```
S = Set of time series sub-sequences sofar (empty at start)
t_i = 1
loop
    if remainder t_i/f ≡ 0 then
        s = time series {p_i, ..., p_m}
        if usable(s) and |S| > l then
            simIndex = similarityIndex(s, S)
            if simIndex >= σ then
                highlight
            end if
            S = S ∪ s
        end if
    end if
    if end data stream then
        Exit loop
    else
        t_i = t_i + 1
    end if
end loop
```

5 Measuring the Similarity of Keystroke Time Series

The most significant part of the KCA process described above is the time series analysis element where a current keystroke time series sub-sequences s is compared with one or more previous series. Given two keystrokes time series sub-sequence s_1 and s_2, the simplest way to define their similarity *sim* is by measuring the corresponding distances between each point in s_1 and each point in s_2. In other words the Euclidean distance is measured between the points in the two series, summed and divided by the sub-sequence length to give an average distance. However, this approach requires both sub-sequences to be of the same length and, more significantly, tends to be over simplistic as it assumes a one-to-one correspondence. By returning to the time series sub-sequences given in Fig. 1a, b we can notice that the shape of the two series is similar but the "peaks" and "troughs" are offset to one another. Euclidean distance measurement will not capture this noticeable similarity. To overcome this limitation, Dynamic Time Warping (DTW) mechanism was adopted. The idea here being to measure the distance between every point in s_1 with every point in s_2, and recording these distances in a $|s_1| \times |s_2|$ matrix. This matrix can then be used to find the "best" path from the origin along to the opposite "corner". This path is referred to as the *Warping Path*, it length in turn can be used as a similarity measure for two time series.

DTW was first used as speech recognition technique to compare acoustic signals [14]. It has subsequently been adopted in the fields of data mining and machine learning [15]. Using DTW, the linearity of time series of different length is "warped" so that the sequences are aligned. Given two keystroke time series sub-sequences $s_1 = \{p_1, p_2, \dots, p_i, \dots, p_x\}$ and $s_2 = \{q_1, q_2, \dots, q_j, \dots, q_y\}$, where x and y are the lengths of the two series respectively, the two corresponding time series are used to constructed a matrix M of size $x \times y$. The value for each element $m_{ij} \in M$ is then computed by calculating the distance from each point $p_i \in s_1$ to each point $q_j \in s_2$:

$$m_{ij} = \sqrt{(p_i - q_j)^2} \tag{1}$$

A Warping Path ($WP = \{w_1, w_2, \dots\}$) is then a sequence of matrix elements (locations), m_{ij}, such that each location is immediately above, to the right of, or above and to the right of, the previous location (except at the location opposite to the origin, which is the warping path end point). For each location the following location is chosen so as to minimise the accumulated warping path length. The "best" warping path is the one that serves to minimise the distance from $m_{1,1}$ to $m_{|s_1|,|s_2|}$. The idea is thus to find the path with the shortest *Warping Distance WD* between the two-time series

$$WD = \sum_{i=1}^{i=|WP|} if \;\; m_i \in WP \tag{2}$$

WD is then an indicator of the similarity between the two keystroke time series under consideration. If $WD = 0$, the two time series in question are identical.

Figure 2 shows some results of the DTW process when applied to four of the time series sub-sequences used with respect to the evaluation of the proposed approach and reported on later in this paper. Figure 2(a) shows the *WP* for two keyboard time series sub-sequences from the same subject (user), while Fig. 2(b) shows the *WP* for two keyboard time series sub-sequences from two different subjects (users). The line included around the diagonal line in both figures indicates the WP that would have been obtained given two identical sub-sequences. The distinction between the generated WP in each can be observed from inspection of the figures.

6 Evaluation

In the foregoing a proposed KCA process was presented (Algorithm 1). Central to this KCA process was the ability to compare keyboard time series sub-sequences. The proposed mechanism for doing this was the DTW mechanism. To illustrate the

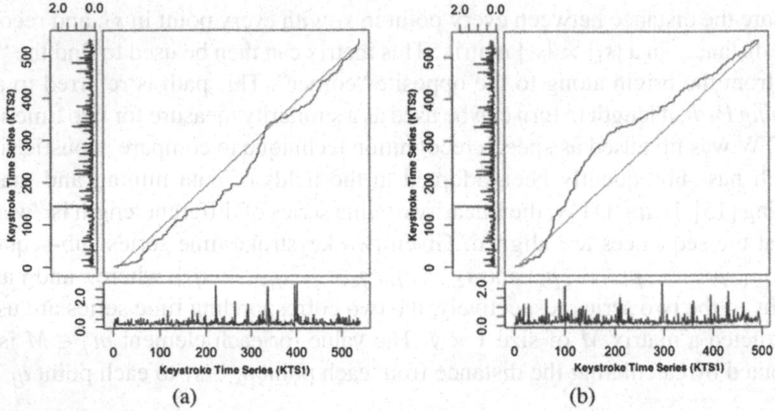

Fig. 2 Application of DTW: **a** Depicts WP for same subject where **b** Illustrates WP for different subjects

effectiveness of this mechanism this section presents the results obtained from a series of experiments conducted using DTW to compare keyboard time series. Two sets of experiments were conducted using a collection of 51 keyboard time series associated with 17 different subjects. The first set of experiments used DTW to compare time series in the collection; the objective was to determine how effectively DTW could be used to distinguish between time series. The second set of experiments compared the operation of the DTW technique with that when using a statistical feature vector based approach akin to that used in earlier work on KCA (see Sect. 2). The evaluation metrics used were: (i) accuracy, (ii) False Rejection Rate (FRR) (iii) False Acceptance Rate (FAR) and Mean Reciprocal Rank (MRR). The remainder of this evaluation section is organised as follows. We commence, Sect. 6.1, with a discussion of the data collection process. The outcomes from the experiments conducted to analyse the operation of the use of DTW within an overall KCA process are reported on in Sect. 6.2. A summary of the results obtained comparing the usage of DTW with the feature vector style approach, found in earlier work on KCA, is then presented in Sect. 6.3. Some discussion is present in Sect. 6.4.

6.1 Data Collection

Keystroke timing data was collected (in milliseconds) using a Web-Based Keystroke Timestamp Recorder (WBKTR) developed by the authors in JavaScript.[2] An HTML "front end" was used and subjects asked to provide answers to discussion questions. The idea was to mimic the situation where students are conducting online assessments. We therefore wished to avoid imposing constraints such as asking the subjects

[2]The interface can be found at: http://cgi.csc.liv.ac.uk/~hsaalshe/WBKTR3.html.

to use a specific keyboard or operating system. The idea was to allow subjects to type in the same manner as they would given an on line learning environment, in other words using different platforms, browsers and so on. JavaScript was used to facilitate the collection of data because of its robust, cross-platform, operating characteristics. This also offered the advantage that no third-party plug-ins were required to enable WBKTR to work. Another advantage of using JavaScript was that it avoided any adverse effect that might result from network delay when passing data to the "home" server, which might have affected the accuracy of recorded times, because the script function works at the end user station to record time stamp data within the current limitations of the accuracy end users's computer clock. The ability to paste text was disabled.

The subjects recruited were students and instructors working on online programmes (thus a mixture of ages). The data was collected anonymously. Additional information concerning the subjects was not recorded (such as gender and/or age). This was a deliberate decision so as minimise the resource required by subjects providing the data. Also because this data was not required, we are interested in comparing user typing patterns with themselves, not in drawing any conclusions about the nature of keyboard usage behaviour in the context of (say) age or gender. The subjects were also asked to type at least 100 words in response to each of three discussion questions (with no maximum limitation) so that adequate numbers of keystrokes could be collected. In [9] it was suggested that 100 keystrokes was sufficient for KCA (for convenience the WBKTR environment included a scripting function to count the number of words per question). Samples with a total number of keystrokes of less than 300 would have been discarded. During each session a JavaScript tool, with JQuery, transparently operated in the background to record the sequencing events KN and the flight time F^t between those events. A PHP script was used to store the identified attributes in the form of a plain text file for each subject. Once the keystroke data had been collected, time series were generated of the form described in Sect. 4 above. In this manner data from a total of 17 subjects were collected, three keyboard time series per subject, thus $17 \times 3 = 51$ time series in total. The data was used to create three data sets such that each data set corresponded to a discussion question. In the following the data sets are referred to using the letters (a), (b) and (c).

6.2 Effectiveness of DTW for Keyboard Usage Authentication

From the foregoing three keyboard time series data sets were collected. For the set of experiments used to determine the effectiveness of DTW each time series in each data set was compared with the time series in each other data set pair: (i) $a. \vee \{b, c\}$, (ii) $b. \vee \{a, c\}$ and (iii) $c. \vee \{a, b\}$.

For each experiment we refer to the first data set as data set 1, and the two comparator datasets as data sets 2 and 3. In each case we have 17 subjects numbered from 1 to 17. Consequently individual time series can be referenced using the notation

Table 1 Ranked average recorded *WD* values for DTW analysis ($a \vee b, c$), *correct matches* highlighted in bold font

	S1	S2	S3	S4	…	S13	S14	S15	S16	S17
All samples $(m-1)$ in the dataset a	0.042	0.084	0.057	0.064	…	**0.069**	**0.040**	0.035	**0.087**	**0.030**
	0.045	**0.085**	0.060	0.067	…	0.073	0.041	0.040	0.091	0.042
	0.048	0.090	0.062	**0.068**	…	0.074	0.042	0.041	0.093	0.042
	0.049	0.090	0.063	0.070	…	0.075	0.046	0.042	0.094	0.043
	0.050	0.094	0.064	0.071	…	0.076	0.048	0.042	0.095	0.044
	0.051	0.098	0.065	0.073	…	0.076	0.048	0.043	0.097	0.045
	0.053	0.099	0.065	0.075	…	0.078	0.049	0.043	0.099	0.045
	0.053	0.099	0.066	0.075	…	0.080	0.050	0.045	0.100	0.046
	0.055	0.100	0.067	0.078	…	0.081	0.054	0.045	0.101	0.046
	0.051	0.101	**0.068**	0.080	…	0.082	0.058	**0.047**	0.110	0.048
	…	…	…	…	…	…	…	…	…	…
Rank	**4**	**2**	**10**	**3**	…	**1**	**1**	**10**	**1**	**1**

Table 2 *WD* rankings for subjects when compared to themselves across datasets

Subjects	S1	S2	S3	S4	S5	S6	S7	S8	S9	S10	S11	S12	S13	S14	S15	S16	S17
Ranking list in Group *a*	4	2	10	3	4	5	5	2	4	2	3	10	1	1	10	1	1
Ranking list in Group *b*	1	4	10	15	3	3	2	2	2	2	6	3	1	1	4	1	1
Ranking list in Group *c*	1	10	10	5	1	5	1	3	10	1	3	6	1	2	7	5	3

s_{ij} where i is the data set identifier and j is the subject number. For each time series s_{1j} we compared with all time series in data sets 2 and 3 and a set of warping distance *WD* values obtained. For each pair of comparisons an average *WD* value was obtained and these values were then ranked in ascending order. Thus each comparison has a rank value r. The ranking outcome is shown in Table 1. Note that because of space limitation, the complete table is not shown in its entirely. In the table the rows and columns represent the subjects featured in the data sets. The values highlighted in bold font are the values where a subject is compared to itself. Ideally we would wish this comparison to be ranked first (recall that $WD = 0$ indicates an exact match). The last row in the table gives the ranking r' of each desired match (highlighted in bold font). All ranking values of corresponding samples of the same subject are listed in Table 2 with subjects represented by columns and data sets by the rows.

With respect to the above, the overall accuracy was computed as the ratio between the number of *incorrect matches* ℓ ranked prior to a *correct match* ($\ell = \sum_{i=1}^{i=m} \sum_{j=1}^{j=n} (r'_{ij} - 1)$ where: (i) m is the number of data sets, (ii) n is the number of subjects and (iii) r'_{ij} is the ranking of the desired match for subject s_{ij}). The total number of comparisons τ is given by $\tau = m \times n$. Thus, with respect to the results presented in Tables 1 and 2, $\ell = 52$ and $\tau = 867$, giving an accuracy of 94.00 % ($\frac{867-52}{867} \times 100 = 94.00$).

The False Rejection Rate (FRR) and False Acceptance Rate (FAR) were also calculated with respect to the outcomes of the time series analysis of typing patterns presented above. According to the European Standard for access control, the acceptable value for FRR is 1 %, and that for FAR is 0.001 % [16]. Thus, we used the FRR and FAR metrics to measure how far the operation of our proposed method, as a biometric authentication method, compared with this standard. For each average *WD* comparison, in each dataset grouping, we calculated FRR by computing the number of subjects n where the corresponding desired rank r' did not equal to 1, $\sum_{j=1}^{j=n} r'_j \neq 1$. If the equivalent sample's rank is not equal to 1, this means that the sample has been falsely rejected. In contrast, FAR is calculated by computing the number of subjects that are ranked higher than the desired sample. The subjects ranked before the desired subject can be considered to have been accepted as real users. Average FRR and FAR values, across the three data set groupings, of 5.99 % and 1.48 % were obtained.

6.3 *Effectiveness of Feature Vector Approach for Keyboard Usage Authentication*

To compare the operation of our proposed approach with the statistical feature vector style of operation found in earlier work on KCA (see Sect. 2), the keystroke data was used to define a feature vector representation. This was done by calculating the average flight time $\mu(f^t)$ for the most frequently occurring di-graphs found in the data

$$\mu(f^t) = \frac{1}{d} \sum_{i=1}^{i=d} Ft_i \tag{3}$$

where d is the number of identified frequent di-graphs and F^t is the flight time value between the identified di-graphs. Each identified di-graph was thus a feature (dimension) in a feature space with the range of average $\mu(f^t)$ values as the values for the dimension. In this manner feature vectors could be generated for each sample. The resulting representation was thus similar to that found in more traditional approaches to KCA [6, 9, 11, 13].

In the same manner as described in Sect. 6.2, we measured the similarity of each feature vector in the first data set with every other feature vector in the other two data sets using Cosine Similarity (CS) (note that this was done for all three data pairings as before). The CS values were calculated using Eq. 4, where $x \cdot y$ is the dot product

Table 3 Ranking lists of all samples in different groups when applying CS

Subjects	S1	S2	S3	S4	S5	S6	S7	S8	S9	S10	S11	S12	S13	S14	S15	S16	S17
Ranking list in Group *a*	5	20	8	13	20	2	20	10	25	20	14	15	1	3	8	1	1
Ranking list in Group *b*	5	18	15	13	15	2	20	11	25	18	20	10	2	3	8	8	5
Ranking list in Group *c*	4	19	7	7	20	8	18	3	15	20	20	13	1	8	10	5	1

between two feature vectors x and y, and $||x||$ ($||y||$) is the magnitude of the vector x (y). Note that in case of using CS, the feature vectors need to be of the same length (unlike in the case of DTW). As before averages for pairs of CS values were used, and as before the values were ranked but in this case in descending order ($CS = 1$ indicates a perfect match).

$$CS(x, y) = \frac{x \cdot y}{||x|| \times ||y||} \tag{4}$$

Table 3 presents an overview of the ranking results obtained using the feature vector approach in the same way as Table 2 presented an overview of the DTW rankings obtained. Detection accuracy was calculated in the same way as for the DTW experiments reported above. In this case $\ell = 163$ and, as before, $\tau = 867$. Thus an accuracy of 81.20 % ($\frac{867-163}{867} \times 100 = 81.20$). The average FRR and FAR values obtained using the feature vector representation, calculated as described above, 20.59 % and 1.69 % respectively.

6.4 Discussion

In the foregoing three sub-sections both the proposed DTW based and the established feature vector based approaches to keyboard authentication were analysed in terms of accuracy, FRR and FAR. A summary of the results obtained is presented in Table 4 with respect to the three data set combinations considered. Mean values are included at the bottom of the table together with their associated Standard Deviation (SD). The table also includes Mean Reciprocal Rank (MRR) values for each approach and data set combination. MRR is an alternative evaluation measure that can be used to indicate the effectiveness of authentication systems [17]. MRR is a standard evaluation measure used in Information Retrieval (IR). It is a measure of how close the position of a desired result (identification) is to the top of a ranked list. MRR is calculated as follows:

Table 4 Results obtained by representing keystroke features in the two approaches: (i) *Time series*, and (ii) *Feature vector*

Representation method	Time series				Feature vector			
	Metrics							
Dataset	FRR (%)	FAR (%)	MRR	Acc (%)	FRR (%)	FAR (%)	MRR	Acc (%)
Group(a) → $a. \vee \{b, c\}$	6.11	1.52	0.438	93.88	20.58	1.64	0.283	79.41
Group(b) → $b. \vee \{a, c\}$	5.17	1.41	0.520	94.82	21.64	1.76	0.155	78.35
Group(c) → $c. \vee \{a, b\}$	6.70	1.51	0.454	93.29	19.17	1.64	0.225	80.82
Mean	**5.99**	**1.48**	**0.471**	**94.00**	20.59	1.69	0.221	79.53
SD	**0.77**	**0.06**	**0.04**	**0.77**	1.24	0.07	0.06	1.24

$$MRR = \frac{1}{|Q|} \cdot \sum_{i=1}^{|Q|} \frac{1}{r_i} \tag{5}$$

where: (i) Q is a set of queries (in our case queries as to whether we have the correct subject or not), and (ii) r_i is the generated rank of the desired response to Q_i. An MRR of 1.00 would indicate that all the results considered are correct, thus we wish to maximise MRR. With reference to Table 4, the average MRR with respect to the proposed time series based approach to KCA proposed combination, is 0.471; while for the feature vector based approach it is 0.221. Returning to Table 4, inspection of the results indicates that the proposed DTW based approach to keyboard authentication outperforms the exciting feature vector based approach by a significant margin.

7 Conclusion

An approach to KCA using time series analysis has been presented that takes into consideration the ordering of keystrokes. The process operates by representing keystroke timing attributes as discrete points in a time series where each point has a timestamp of some kind and an attribute value. The proposed representation used a sequential keypress numbering system as the time stamp, and flight time as the attribute (distance between key presses). DTW was adopted as the time series comparison mechanism. For evaluation purposes data was collected anonymously using a bespoke web-based tool designed to mimic the process of conducting online assessments (responding to discussion questions). The evaluation was conducted by comparing every subject to every other subject to determine whether we could distinguish between the two using: (i) the proposed technique and (ii) a feature vector based approach akin to that used in established work on KCA. With respect to the first set of experiments, an overall accuracy of 94.00 % was obtained. This compared very favourably with an accuracy of 79.53 %, obtained with respect to the second set of experiments. The results demonstrated that the proposed time series based approach to KCA had significant potential benefit in the context of user authentication with respect to online assessments such as those used in online learning and MOOCS. The authors belief that further improvement can be realised by considering n-dimensional time series (time series that consider more than one keystroke attribute). Future work will also be directed at confirming the findings using larger datasets.

Acknowledgments We would like to express our thanks to those who participated in collecting the data and to Laureate Online Education b.v. for their support.

References

1. Gaines, R.S., Lisowski, W., Press, S.J., Shapiro, N.: Authentication by keystroke timing: some preliminary results, no. RAND-R-2526-NSF. RAND Corp., Santa Monica, CA (1980)
2. Bleha, S., Slivinsky, C., Hussien, B.: Computer-access security systems using keystroke dynamics. IEEE Trans. Pattern Anal. Mach. Intell. **12**(12), 1217–1222 (1990)
3. Joyce, R., Gupta, G.: Identity authentication based on keystroke latencies. Commun. ACM **33**(2), 168–176 (1990)
4. Ogihara, A., Matsumuar, H., Shiozaki, A.: Biometric verification using keystroke motion and key press timing for atm user authentication. In: Intelligent Signal Processing and Communications, 2006. ISPACS'06, pp. 223–226. IEEE
5. Syed, Z., Banerjee, S., Cukic, B.: Normalizing variations in feature vector structure in keystroke dynamics authentication systems. Softw. Qual. J. 1–21 (2014)
6. Ahmed, A.A., Traore, I.: Biometric recognition based on free-text keystroke dynamics. IEEE Trans. Cybern. **44**(4), 458–472 (2014)
7. Bours, P.: Continuous keystroke dynamics: a different perspective towards biometric evaluation. Inf. Secur. Tech. Rep. **17**(1), 36–43 (2012)
8. e Silva, S.R.D.L., Roisenberg, M.: Continuous authentication by keystroke dynamics using committee machines. In: Intelligence and Security Informatics, pp. 686–687. Springer, Berlin (2006)
9. Gunetti, D., Picardi, C.: Keystroke analysis of free text. ACM Trans. Inf. Syst. Secur. (TISSEC) **8**(3), 312–347 (2005)
10. Messerman, A., Mustafic, T., Camtepe, S.A., Albayrak, S.: Continuous and non-intrusive identity verification in real-time environments based on free-text keystroke dynamics. In: 2011 International Joint Conference on Biometrics (IJCB), pp. 1–8. IEEE (2011)
11. Shepherd, S.J.: Continuous authentication by analysis of keyboard typing characteristics. In: European Convention on Security and Detection, 1995, pp. 111–114. IET (1995)
12. Richardson, A., Kaminka, G.A., Kraus, S.: REEF: resolving length bias in frequent sequence mining using sampling. Int. J. Adv. Intell. Syst. **7**(1), 2 (2014)
13. Dowland, P.S., Furnell, S.M.: A long-term trial of keystroke profiling using digraph, trigraph and keyword latencies. In: Security and Protection in Information Processing Systems, pp. 275–289. Springer, US (2004)
14. Rabiner, L., Juang, B.H.: Fundamentals of speech recognition. Prentice Hall (1993)
15. Berndt, D.J., Clifford, J.: Using dynamic time warping to find patterns in time series. In: KDD Workshop, vol. 10, no. 16, pp. 359–370

16. Polemi, D.: Biometric techniques: review and evaluation of biometric techniques for identification and authentication, including an appraisal of the areas where they are most applicable. Reported prepared for the European Commision DG XIIIC, 4 (1997)
17. Craswell, N.: Mean reciprocal rank. In: Encyclopedia of Database Systems, pp. 1703–1703. Springer, US (2009)

Genetic Algorithms in Action

EEuGene: Employing Electroencephalograph Signals in the Rating Strategy of a Hardware-Based Interactive Genetic Algorithm

C. James-Reynolds and E. Currie

Abstract We describe a novel interface and development platform for an interactive Genetic Algorithm (iGA) that uses Electroencephalograph (EEG) signals as an indication of fitness for selection for successive generations. A gaming headset was used to generate EEG readings corresponding to attention and meditation states from a single electrode. These were communicated via Bluetooth to an embedded iGA implemented on the Arduino platform. The readings were taken to measure subjects' responses to predetermined short sequences of synthesised sound, although the technique could be applied any appropriate problem domain. The prototype provided sufficient evidence to indicate that use of the technology in this context is viable. However, the approach taken was limited by the technical characteristics of the equipment used and only provides proof of concept at this stage. We discuss some of the limitations of using biofeedback systems and suggest possible improvements that might be made with more sophisticated EEG sensors and other biofeedback mechanisms.

Keywords Interactive Genetic Algorithms · Brain Computer Interface · Evolutionary Audio Synthesis

1 Introduction

Genetic Algorithms (GAs) are a well-established evolutionary computing approach to problem-solving, whereby a set of candidate solutions is generated, individually evaluated for fitness (proximity to a desired outcome) by an evaluation function and then used to generate a new set of potential solutions using techniques analogous to natural genetics, this process being repeated until a candidate meets the desired criteria. Interactive Genetic Algorithms (IGAs) are a variation of this, in which the evaluation function is replaced by the conscious decisions of a human user. We have

C. James-Reynolds (✉) · E. Currie
School of Science and Technology, Middlesex University, London NW4 4BT, UK
e-mail: C.James-Reynolds@mdx.ac.uk

E. Currie
e-mail: E.Currie@mdx.ac.uk

M. Bramer and M. Petridis (eds.), *Research and Development in Intelligent Systems XXXIII*, DOI 10.1007/978-3-319-47175-4_25

developed this concept further, using biofeedback in the form of EEG signals to make these human decisions subconscious. The chosen domain for initial proof of concept was that of music synthesis, where the evaluation of short sequences of notes where the parameters of the sound used was generated by the iGA over successive generations, with rating of the sounds carried out by taking EEG measurements from subjects. The aim was to identify if the sounds generated converged on sounds that were "liked" by the subjects. Using an EEG approach to rate responses could be used in other domains such as image synthesis.

2 Background and Related Work

2.1 Genetic Algorithms

Genetic algorithms [1] are a category of evolutionary algorithm used in optimisation and search strategies, which loosely parallel Darwinian evolutionary theory. The GA process begins with an initial population of solutions, each of which is represented by the values of a given set of variables. In a traditional GA, a fitness function is then used to rate each individual on a scale according to the closeness of its characteristics to those of the desired outcome. In an iGA [2] the fitness function is the user, who similarly rates the individual solutions on the appropriate scale. In both the GA and the iGA, this rating is proportional to the probability of the individual being chosen as a parent for the next generation. Pairs of parents contribute variables (genes) to the new population, using a crossover. A mutation factor may be used to introduce some further random changes into the new population. This process is repeated until it converges to a solution that, in the opinion of the user, represents a desirable final state. Naturally, the number of individuals in a given generation for an iGA will be lower than that for a GA, as a human evaluator will be subject to such factors as the time it takes to observe or listen to the stimuli and also to fatigue, which are not an issue for the GA evaluation function.

It may be argued that, unless the solution space of an iGA is appropriately mapped to the cognitive space, users are not effectively supported by the system in terms of making good choices [3]. It is also difficult to measure the usefulness and usability of iGAs as tools without evaluating them on the basis of whether they reach a defined goal. It is possible to have fuzzy goals, where the user explores the solution space until they find a candidate that they find acceptable. This is an approach often taken in evaluating iGAs, but it is less rigorous than a goal-based approach [4] Furthermore, as Bauerly and Liu [5] points out, there is an assumption that users are consistent in their assignation of ratings across multiple generations, but we cannot be sure of the impact of user concentration changes and fatigue on the consistency of their judgement.

These possible inconsistencies in the behaviour of human evaluators seem to be a disadvantage of iGAs when compared to the traditional GA. However, there are some problems for which it is difficult to formalise the evaluation process. These might include evaluating the artistic merit of machine-generated images or sounds [6–9], the quality of music [10] or the stylishness of a machine-generated dress or suit design. Such evaluations require the judgement of a human being. Of course, these judgements are always going to be subjective to some degree, and to design a dress that will appeal to a large section of the population, it might be necessary to combine the opinions of a number of human individuals in generating each rating. This approach has been taken by Biles with 'GenJam Populi'. [11] The point is that, as long as the judgements are accurate enough to facilitate the required convergence, absolute precision is not required.

2.2 Biofeedback

The idea of using Electroencephalograph (EEG) data as a means of interfacing to a computer system is not a new one. Applications have included gaming, enabling technologies, and emotional response to valence and cognitive workload recognition. Brain Computer Interfaces (BCIs) are often problematic in practice. For example, in an application where a BCI is used by a subject to actively control a system, it can be difficult to identify when a signal is supposed to be associated with the system under control, and when it has been generated by some unrelated mental activity of the user. There is therefore a requirement for the users themselves to learn how to use such systems properly. Evidence suggests that not all users can be trained [12].

Bradley et al. [13] exposed subjects to images for 6 s while gathering data using a range of biofeedback readings such as skin resistance and Electromyographic measurement. Similar work by Franzidis et al. [14] only allowed a 1 s exposure, with 1.5 s breaks between images, while collecting EEG readings. Clearly, exposure times and choice of reading technique are parameters that must be optimised by experimentation. Franzidis's work [14] also explores the day-to-day variation in neurophysiological responses to the same stimulus and considers neurophysiological profiles.

There has been a substantial amount of work exploring more general bio-feedback approaches for managing stress and relaxation, although these will not be explored here. There has also been a significant amount of work on the use of neural networks to identify emotions from EEG signals, when giving subjects specific stimuli. For example, Murugappan et al. [15] took this approach and found a 10.01 % improvement with an audiovisual stimulus over a visual stimulus alone.

One of the difficulties with more complex EEG systems is setting them up, as they may have in excess of 100 electrodes. However, simpler EEG systems can still be effective. For example, Lee and Tan [16] claimed a 93.1 % accuracy in distinguishing between which of two tasks were being carried out, using a relatively low cost EEG setup.

There has been an increased use of EEG signals from gaming headsets [17] in the research community. These are often easy to configure and almost all of the signal processing is carried out onboard using proprietary Integrated Circuits, in the case of the Neurosky MindWave Mobile it is a Think Gear ASIC Module. This is responsible for noise filtering (in particular from muscular and skeletal movement signals and power cable interference). These headsets are often supplied with development kits that facilitate using the EEG data in custom applications or research [18]. The associated games are often limited to simple activities that rely on the user being able to focus or relax. One research application of such technologies is the measurement of cognitive workload and attention [19, 20]. Cernea et al. [21] and Moseley [22] used gaming headsets to measure emotional states/facial expressions and to induce specific user target states (i.e. meditative) respectively. We chose to use such a device to measure users' subconscious reactions in response to sound sequences, mapping these to quantified evaluations of the candidate solutions in an iGA. This represents a kind of 'halfway house' between a pure GA and an iGA; there is user interaction, but it is not consciously carried out.

3 Methodology

The Eugene system, described in [3] and shown in Fig. 1, is an Arduino based hardware controller with MIDI output, allowing the preview and rating of six individual candidate solutions in any given generation of an iGA using slider potentiometers set by the user.

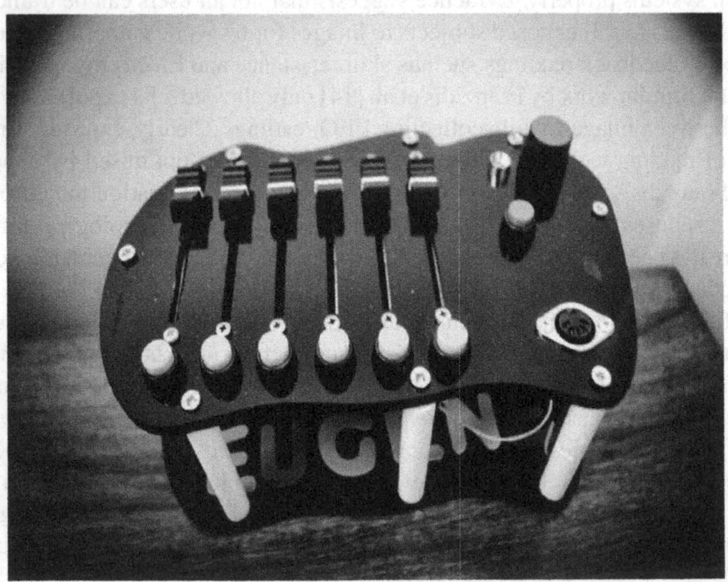

Fig. 1 The original Eugene interface with 6 sliders for rating [3]

Fig. 2 The equipment used, Arduino, BlueSmirf and Mindwave

The system has been adapted so that the data previously taken from the conversion of the potentiometers' voltage output is now taken from an EEG reading. The MIDI output (31250 baud) now utilises the Software Serial library and pin 11 of the Arduino, which is necessary because the Bluetooth board uses the transmit and receive pins. One advantage of this is that when updating code, the MIDI adaptor does not have to be disconnected. The six button switches in the original Eugene are also no longer required, as the sounds are played automatically after a successful EEG reading, rather than under user control. A design decision was made to keep the population to six individuals, as too many individuals might lead to user fatigue. The hardware used included a Neurosky MindWave mobile [17] EEG headset, with a single dry electrode. The on-board processing circuit outputs values over a Bluetooth data stream (57,600 baud). A BlueSmirf Gold [23] Bluetooth transceiver board was used to receive the signal from the Mindwave. The BlueSmirf uses the RX pin on the Arduino Uno (Fig. 2). The signals that can be obtained from the Mindwave include an attention and meditation signal, which are parsed by the Arduino.

The initial experiments were based on simple observation of the EEG signals captured when the users were listening to music that they "liked". A pilot study then explored how the development platform behaved and if the output of the iGA was converging. Finally a smaller experiment was carried out to identify if the output converged on sounds that were "liked" by the subjects in the context of the given musical sequence.

We recorded at the attention and meditation signals of 20 subjects listening to music they "liked".

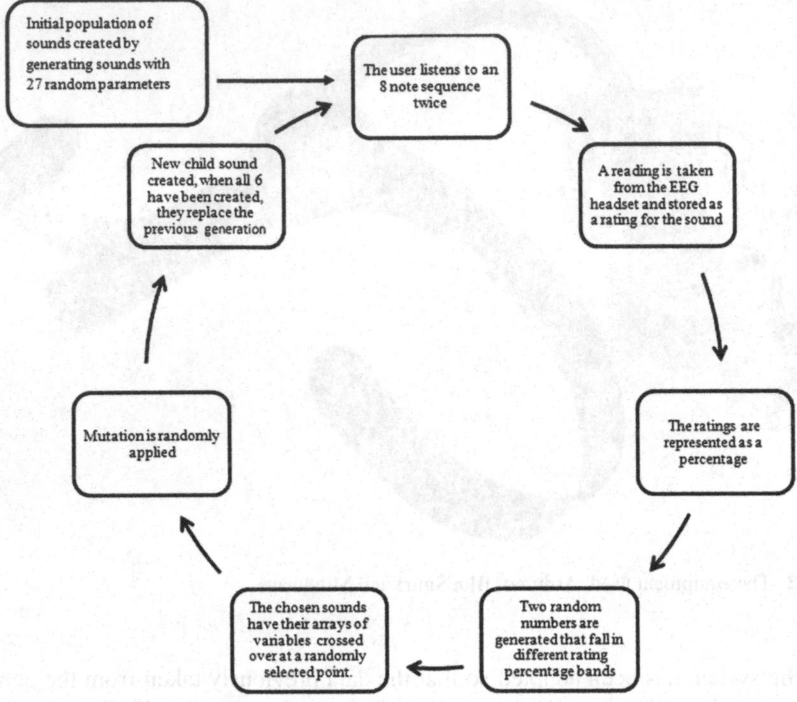

Fig. 3 The iGA process in the context of this work

We found that for 19 of them, the levels of both of these signals increased and the levels of both fell when the subjects listened to music they did not like. The sum of these two signals was then chosen to rate the sounds. Alternative possible strategies are considered in the discussion below, but these require further evaluation.

The initial population of sounds is generated randomly and selection of the parent sounds in each generation is by means of a roulette approach based on the sum of the two EEG ratings as stated above (Fig. 3). In the original Eugene controller elitism was used to keep the best sounds for the next generation, this has not been adopted here.

For each sound, 27 MIDI Continuous Controller (CC) values (genes) are assigned to the values of parameters that control a simple two-oscillator software synthesiser built in SynthEdit [24]. These parameters included attack, sustain, decay and release for two envelope generators, filter parameters of cut-off frequency and resonance, choice of oscillator waveforms and modulation parameters. The CC numbers used were from 5 to 28. SynthEdit is flexible in the mapping of CC values to the synthesiser parameters and also in the range of each parameter (Fig. 4).

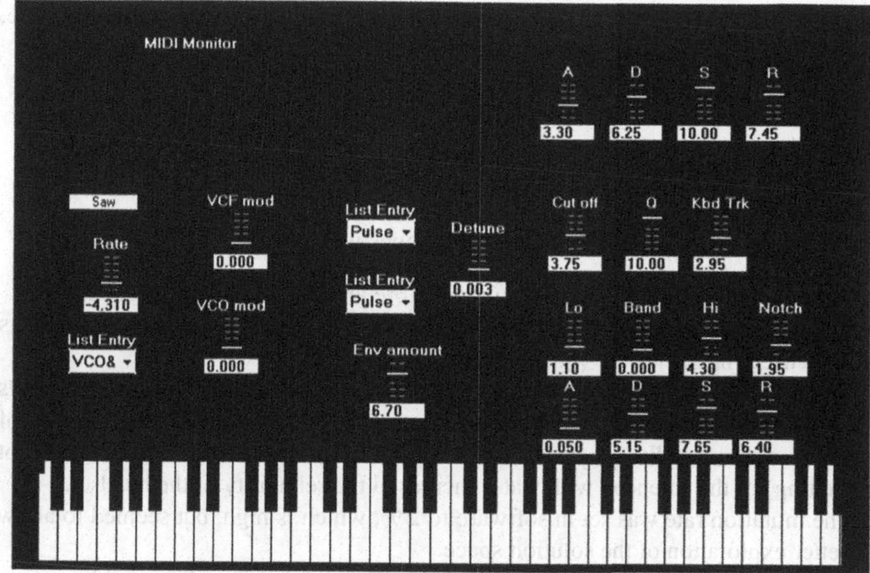

Fig. 4 The software synthesiser used

4 Pilot Study

A pilot study was conducted with two subjects, in which an 8 note sequence (C2, D#2, F2, G3, A#2, C3, D3) was repeated 4 times and the EEG data was read at the end of the sequence. It was possible to monitor the EEG data from the headset using the Arduino Serial Monitor. Although care had to be taken to ensure that the Serial Monitor was running at the start of the process as launching it caused the iGA process to start from the beginning.

A number of issues were identified:

- The sequence should only be repeated once or twice to reduce the time between generations and the effect of user fatigue, as the entire process of reviewing each generation was taking too long (over 30 s).
- The MindWave headset and headphones (for listening to the sounds) were not very comfortable to wear together, but in-ear headphones were not considered suitable to be shared by users.
- A warning needed to be generated when the signal quality was low, which happened occasionally (especially when headphones were adjusted) and interfered with the results.
- When a sound was very quiet or indeed inaudible, the EEG readings were no longer related to the sound, but were still used by the iGA in rating, this led to poor convergence.
- Some of the initial populations sounded similar and users did not think afterwards that there was much to choose between them.

The pilot study did not produce data about generating sound the user might like, but did identify a number of issues with the setup of the experiment.

5 Experiment

Following the pilot study, the following changes were made:

- The sequence was only repeated twice before an EEG reading was taken.
- Smaller headphones were used that provided more comfort when the two headsets were used together.
- An LED indicator was programmed to respond to any loss of signal quality. Tests ascertained that the Bluetooth connection was very reliable and that any loss of signal was normally the result of movement of the electrode. Dampening (not wetting) of the forehead with water increased the reliability of the headset.
- The mutation rate was set in software to 2 %, which is high, but seemed to allow better exploration of the solution space.
- The MIDI CC values could be restricted in their ranges, but this would increase the setup time and require uploading of the sketch (program) to the Arduino, every time there was a change. Using SynthEdit, it was possible to adjust the range of the parameters and mapping to CC values and this allowed many of those parameter interactions that produced no, or very low level audio output to be removed from the solution space. The remapped parameters were:
  - Amplifier Attack
  - Amplifier Sustain
  - Filter Attack
  - Filter Sustain
  - Filter Cutoff Frequency
  - Keyboard Filter Tracking

The experiment looked to identify if there was any convergence to a sound that was "liked" by the subject, in the context of the musical sequence of notes.

Three subjects took part in the test and each ran the controller three times. The subjects had no identified hearing difficulties. The results converged differently for each subject, with two of the subjects converging on variations of white noise based sounds in one of their trials. Monitoring indicated that this sound was producing a high meditation state reading from the subject, leading to a higher probability of selection. The subjects felt that the output after 5 to 7 generations had led to possibly viable solutions that they "liked", but not necessarily optimal ones. However, this was an improvement on the pilot study. Observation of the MIDI data and EEG readings showed that the iGA was working properly. Users also felt that this was a fairly time-consuming exercise, which was an interesting perception as the tests were only running for about 5 min for each.

6 Discussion

Interactive Genetic Algorithms traditionally ask the user to consciously rate each of the candidate solutions in each generation, and these ratings are used in the selection of parents for the next generation. Our approach involves the measurement of a subconscious response, which does not require any active decision making by the subjects. However, the results may be affected by the impact of the environment on the user; this was found to be the case when very low levels of sound, or no sound, were produced by the synthesiser. It might be reasoned that, because the data from the user is subconscious, it is not subject to distortion by factors such as the user second-guessing what those conducting the experiment might be expecting. However, the results can be manipulated by a user consciously choosing to relax or focus or by the user being distracted. A good example is counting backwards; this type of activity will distort the results obtained. A possible way to reduce distractions would be to conduct the experiments in a darkened room. A user's view on what may be a good output may change over time (even as the iGA is running) and possibly as they become aware of the solution space, we do not know the extent of this or its possible effect on the results.

The choice of notes in the sequence and the tempo would also have an impact on the results and users' choice of sounds. It is possible to make these into variables as part of the gene sequence or even as a second gene, but this would have significantly added to the complexity of the evaluation. Given that with some existing research, using very short exposure to stimuli before taking EEG readings, it might be possible to only use a 4-note sequence played once, which would speed up the process and produce results more quickly.

The strategy of adding together the meditation and attention signals was based on limited observation and the results, whilst improving the sound by removing sounds that did not suit the sequence or were unmusical, did not always converge to values that the users considered to be optimal, although they did generally converge to more desirable sounds. It seems that better algorithms for ascertaining rating values might be required. It is possible that some of the gaming headsets with more electrodes would perform better, as these have been used in other research successfully. When using gaming headsets, we are dependent on trusting the manufacturers' signal processing techniques and their software, but the loss of control over these things is offset by the gains in terms of portability, user acceptability and low project costs.

One of the issues identified by many researchers in the iGA area is that of user fatigue. The use of EEG and other biofeedback may provide some insights into this. More accurate rating systems will also allow us to explore the differences between conscious and subconscious rating, by capturing the EEG ranking and comparing it to conscious ratings using sliders.

The Arduino platform has limitations in terms of processing power, memory and speed, but these factors are not significant in this application. The Arduino also has the ability to connect to a wide range of low cost analogue and digital sensors, thus allowing other biofeedback rating strategies to be evaluated.

7 Conclusion

We have presented a novel approach using EEG signals for rating candidate solutions in an interactive Genetic Algorithm. This approach is highly appropriate for problem domains that are difficult to evaluate using a goal-based approach, due to the subjective nature of the evaluation. Our work so far has been with sound, but we have previously applied iGAs to visual media and the use of biofeedback data would be equally valid for this or indeed for any problem where the evaluation relies on the users' subjective satisfaction with the output.

The platform was developed from our original Eugene hardware-based iGA system [3], augmented with the necessary hardware and software necessary to capture and process the relevant EEG signals. The experiments conducted with this platform have established proof-of-concept for this system. This will also enable further research and we have suggested possible future directions for that research, including more sophisticated EEG capture and other biofeedback mechanisms. Our choice of using the sum of meditative and attention EEG signals to drive our evaluation was justified above, but nevertheless, experiments with more robust and tested signal processing algorithms and richer EEG data might produce a more optimal system, notwithstanding the extra costs in time and money associated with more sophisticated EEG capture. While the research has many limitations, it does enable us to explore the ways in which humans can interact with evolutionary systems and will hopefully lead to further useful insights in this area.

References

1. Holland, J.H.: Adaptation in Natural and Artificial Systems. University of Michigan Press, Ann Arbor (1975)
2. Todd, S., Latham, W.: Evolutionary Art and Computers. Academic Press (1992)
3. James-Reynolds, C., Currie, E.: Eugene: a generic interactive genetic algorithm controller. In: AI-2015: Thirty-Fifth SGAI International Conference on Artificial Intelligence, Cambridge, UK, 15–17 Dec 2015 (2015)
4. McDermott, J., O'Neill, M., Griffith, N.J.: Interactive EC control of synthesized timbre. Evol. Comput. **18**, 277–303 (2010)
5. Bauerly, M., Liu, Y.: Evaluation and improvement of interface aesthetics with an interactive genetic algorithm. Int. J. Hum. Comput. Interact. **25**, 155–166 (2009)
6. Caetano, M.F., Manzolli, J., Von Zuben, F.J.: Interactive control of evolution applied to sound synthesis. In: FLAIRS Conference, pp. 51–56 (2005)
7. Johnson, C.G.: Exploring the sound-space of synthesis algorithms using interactive genetic algorithms. In: Proceedings of the AISB'99 Symposium on Musical Creativity. Society for the Study of Artificial Intelligence and Simulation of Behaviour, pp. 20–27 (1999)
8. Dahlstedt, P.: Creating and exploring huge parameter spaces: Interactive evolution as a tool for sound generation. In: Proceedings of the 2001 International Computer Music Conference, pp. 235–242 (2001)
9. Yuksel, K.A., Bozkurt, B., Ketabdar, H.: A software platform for genetic algorithms based parameter estimation on digital sound synthesizers. In: Proceedings of the 2011 ACM Symposium on Applied Computing, pp. 1088–1089. ACM (2011)

10. Biles, J.: Genjam: a genetic algorithm for generating jazz solos. In: Proceedings of the International Computer Music Conference (1994)
11. Biles, J.A., Eign, W.G.: GenJam Populi: Training an IGA via Audience-Mediated Performance. San Francisco, USA (1995)
12. Allison, B.Z., Neuper, C.: Could anyone use a BCI? In: Tan, S.D., Nijholt, A. (eds.) Brain-Computer Interfaces: Applying Our Minds to Human-Computer Interaction, pp. 35–54. Springer, London (2010)
13. Bradley, M.M., Codispoti, M., Cuthbert, B.N., Lang, P.J.: Emotion and motivation I: defensive and appetitive reactions in picture processing. Emotion **1**, 276–298 (2001)
14. Frantzidis, C.A., Bratsas, C., Papadelis, C.L., Konstantinidis, E., Pappas, C., Bamidis, P.D.: Toward emotion aware computing: an integrated approach using multichannel neurophysiological recordings and affective visual stimuli. IEEE Trans. Inf. Technol. Biomed. **14**, 589–597 (2010)
15. Murugappan, M., Juhari, M.R.B.M., Nagarajan, R., Yaacob, S.: An Investigation on visual and audiovisual stimulus based emotion recognition using EEG. Int. J. Med. Eng. Inf. **1**, 342–356 (2009)
16. Lee, J.C., Tan, D.S.: Using a low-cost electroencephalograph for task classification in HCI research. In: Proceedings of the 19th Annual ACM Symposium on User Interface Software and Technology, pp. 81–90. ACM (2006)
17. NeuroSky, Biosensors. http://neurosky.com/biosensors/eeg-sensor/biosensors/ (2016). Accessed 7 June 2016
18. NeuroSky, Developer Program. http://developer.neurosky.com// (2016). Accessed 7 June 2016
19. Knoll, A., Wang, Y., Chen, F., Xu, J., Ruiz, N., Epps, J., Zarjam, P.: Measuring cognitive workload with low-cost electroencephalograph. In: Human-Computer Interaction–INTERACT 2011, pp. 568–571. Springer (2011)
20. Frey, J., Daniel, M., Castet, J., Hachet, M., Lotte, F.: Framework for Electroencephalography-based Evaluation of User Experience. arXiv preprint arXiv:1601.02768 (2016)
21. Cernea, D., Olech, P.-S., Ebert, A., Kerren, A.: EEG-based measurement of subjective parameters in evaluations. In: HCI International 2011–Posters' Extended Abstracts, pp. 279–283. Springer (2011)
22. Moseley, R.: Inducing targeted brain states utilizing merged reality systems. In: Science and Information Conference (SAI), 2015, pp. 657–663. IEEE (2015)
23. BlueSMiRF specification and datasheet. http://www.sparkfun.com/commerce/product_info.php?products_id=582 (2016). Accessed 6 Aug 2016
24. McClintock, Jeff SynthEdit. Current version available at http://www.synthedit.com (2002). Accessed 6 Aug 2016

Spice Model Generation from EM Simulation Data Using Integer Coded Genetic Algorithms

Jens Werner and Lars Nolle

Abstract In the electronics industry, circuits often use passive planar structures, such as coils and transmission line elements. In contrast to discrete components these structures are distributed and show a frequency response which is difficult to model. The typical way to characterize such structures is using matrices that describe the behaviour in the frequency domain. These matrices, also known as S-parameters, do not provide any insight into the actual physics of the planar structures. When simulations in the time domain are required, a network representation is more suitable than S-parameters. In this research, a network is generated that exhibits the same frequency response as the given S-parameters whilst allowing for a fast and exact simulation in the time domain. For this, it is necessary to find optimum component values for the network. This is achieved in this work by using an integer coded genetic algorithm with power mutation. It has been shown that the proposed method is capable of finding near optimal solutions within reasonable computation time. The advantage of this method is that it uses small networks with fewer passive components compared to traditional methods that produce much larger networks comprising of many active and passive devices.

Keywords Electronic circuit design · Integer cooled genetic algorithm · Planar coils

1 Introduction

In modern electronic circuits, planar coils are often used, for example on-chip inside integrated circuits or externally as geometric structures on printed circuit boards (PCBs). The design process is usually supported by simulation tools that operate in the frequency domain (e.g. Method of Moments [1, 2]). The derived simulation

J. Werner (✉) · L. Nolle
Jade University of Applied Sciences, Friedrich-Paffrath-Str. 101, 26389 Wilhelmshaven, Germany
e-mail: Jens.Werner@Jade-hs.de

L. Nolle
e-mail: Lars.Nolle@Jade-hs.de

M. Bramer and M. Petridis (eds.), *Research and Development in Intelligent Systems XXXIII*, DOI 10.1007/978-3-319-47175-4_26

results are given as frequency dependent scattering parameters (S-parameters) which provide a detailed insight into the electrical properties at the terminals (ports) of the coils. However, for certain investigations an analysis in the time domain is required. For this, an equivalent circuit model has to be designed which has to result in the same characteristics as the coil in both, the time and the frequency domain. A common method provided by commercial tools is known as broadband Spice (Simulation Program with Integrated Circuit Emphasis) model generation [3, 4]. In general, these methods generate networks of idealised components (lumped networks) that exhibit the required behaviour at their ports. In practice, these networks consist of a rather large number of active and passive elements, which do not resemble the physics of the actual coil. The proposed approach uses expert knowledge to generate a much smaller network that is closer to the actual physics. The challenge here is that the component values of this network have to be determined. Even for networks with few components the search space is too vast to find the optimum values manually. A computational optimisation algorithm is used in this work. Two practical applications of planar coils are introduced in the next section.

2 Practical Applications

Two application examples are described below to demonstrate that planar coils can be found in many popular electronic devices. The first example is about integrated wireless radio circuits; the second sketches an application of wired high-speed digital signal lines.

2.1 *On-chip Planar Coils in Oscillators*

In wireless receiver and transmitter circuits the development of fully integrated oscillators-without the need for external coils-has led to significant reduction of PCB area. This allows the integration of multiple radio frontends, like GSM, UMTS (3G), LTE (4G), WIFI (IEEE 802.11b/g/n/ac), Bluetooth, GPS et cetera, in a single portable device. All of these radio circuits require local oscillators that are based on a voltage controlled oscillator (VCO). These VCOs run at rather high frequencies in order to operate with small coils that are integrated on-chip in bipolar or CMOS semiconductor dies [5–7]. Figure 1 depicts a photograph of a digital cable tuner IC that is fabricated in 0.5 μm, 30 GHz BiCMOS technology. Four coils can be recognized that belong to four VCOs that are used for different frequency bands.

When a VCO is switched on, a critical situation can occur in which a proper oscillation might not be secured. Thus, a design engineer will perform various simulations in the time domain with numerous boundary conditions (e.g. supply voltage variation, rise time of supply voltage, spread of different circuit properties) before the costly manufacturing of the silicon wafers is started. For these kind of simulations a precise Spice model of the distributed and planar coil is beneficial.

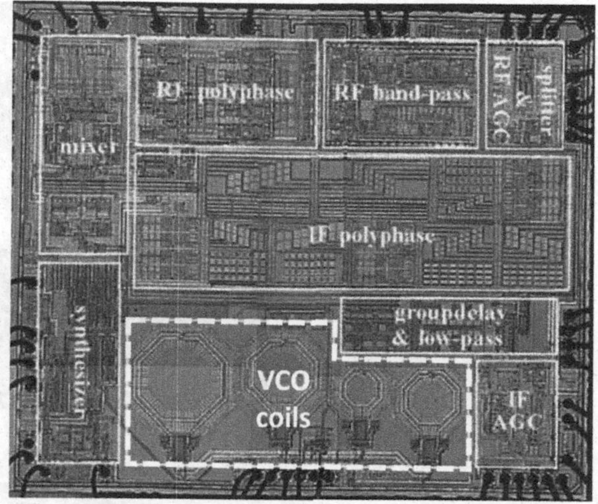

Fig. 1 Photograph of a single-chip wide-band tuner with fully integrated voltage controlled oscillators (VCO) [7]

2.2 *On-chip Planar Coils in Common Mode Filters*

In high-speed differential data lines (e.g. USB, SATA, PCIe) the spectrum used overlaps with wireless radio communication bands (e.g. GSM, LTE, WIFI IEEE 802.11ac). Electromagnetic interference (EMI) caused by USB 3 data lines can potentially disturb wireless applications in the 2.4 GHz band as reported in [8–11]. Likewise the spectrum of an USB 2 signal might interfere with the GSM 900 MHz downlink spectrum. This can cause a degradation of the receiver sensitivity in the mobile phone.

As wireless and wired signals utilise the same frequency band, a low pass filter can not be used to reduce unwanted interference from the data lines into the antenna and finally the receiver. Instead, a common mode filter (CMF) is applied. For a USB 2.0 transmission the potential interference into the GSM receiver of a mobile phone is described in [12]. The purpose of a CMF is to suppress unwanted common mode currents on the data lines, since those currents are responsible for the EMI. Ideally, the differential mode currents of the wanted high-speed signal are not affected. Typical CMF are built by two coupled coils. Figure 2 shows a view on a planar copper coil as part of an integrated on-chip CMF.

For all kind of digital high-speed transmission systems, a signal integrity (SI) analysis is performed in the time domain. This involves the simulation of a large number of transmitted data bits and the evaluation of their electrical characteristics at each sampling point in the time domain. The segmented and overlapped representation of this time domain data is known as eye diagram [13, 14]. Again, the simulation in the time domain benefits from an exact and compact Spice model of the involved planar structures.

In order to test the new approach a simplified case study is presented below.

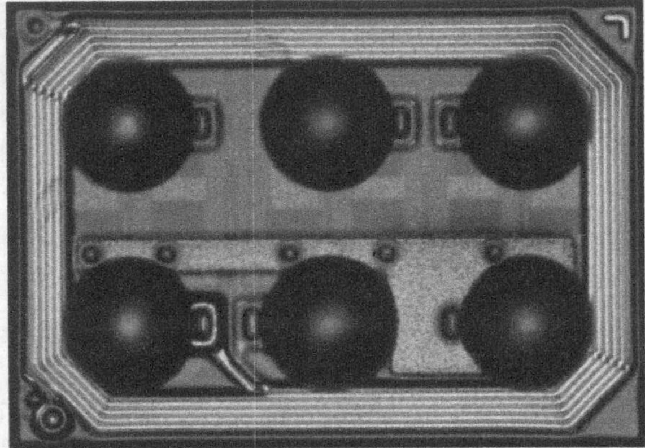

Fig. 2 Photograph of a common mode filter on a silicon substrate [12]. The planar copper coil in the upper layer is clearly visible. A second coil in the lower layer is hardly visible. *Dark* circles are solder balls for mounting the device on a printed circuit board. Dimensions are 1.3 mm by 0.9 mm

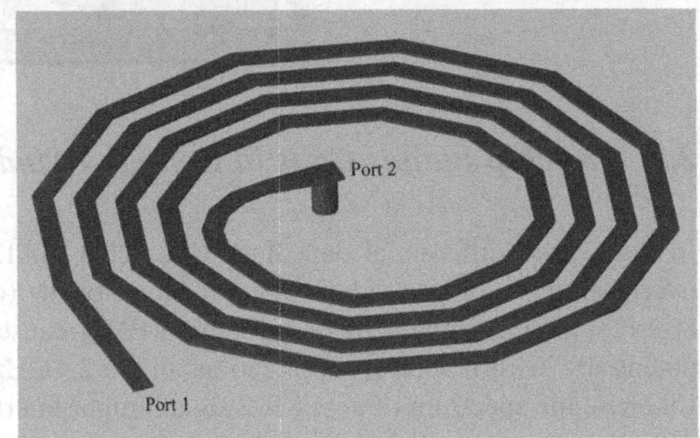

Fig. 3 Layout view of a single planar spiral inductor on a two-layered PCB. The complex impedance at port 1 is derived from EM simulation. Port 2 is connected to a GND plane on bottom layer. The dielectric substrate is hidden in this view

3 Test Bed

As shown in the application examples above, the specific geometry, number of coils, arrangement and other properties can be manyfold. Therefore a generic structure (see Fig. 3) has been chosen as a test bed for this study. It is a planar copper coil designed on a two-layer PCB where the bottom layer is completely filled with copper providing a distributed ground (GND) reference plane. The coil can be connected at two terminals (Port 1, Port 2) of which the second one is connected to GND by a through-hole via. This reduces the example to a one-port structure. In general, a two-terminal device can be characterised by scattering parameters $[\underline{S}]$ that describe the reflection of incident waves at each port ($\underline{S}_{11}$, $\underline{S}_{22}$) as well as the transmission from one port to another ($\underline{S}_{12}$, $\underline{S}_{21}$). All elements $\underline{S}_{ij}$ of this matrix are complex valued and frequency dependent.

In a Spice based network simulation [15, 16] the voltage $\underline{U}$ and the current $\underline{I}$ at port 1 can be calculated and their ratio yields the port impedance $\underline{Z}$ (Eq. 2). In order to express this impedance as reflection coefficient it can be converted by Eq. 3, [17].

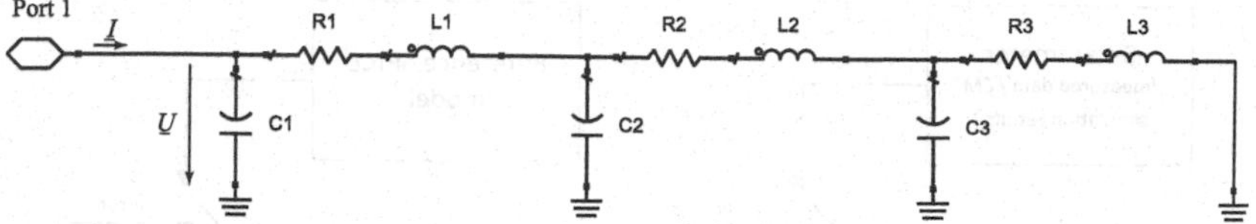

Fig. 4 Structure of tunable Spice circuit as used in optimisation loop

Here Z_0 is the characteristic impedance of the measurement system; it is 50 Ohms in many practical applications.

$$[\underline{S}] = \begin{bmatrix} \underline{S}_{11} & \underline{S}_{12} \\ \underline{S}_{21} & \underline{S}_{22} \end{bmatrix} \tag{1}$$

$$\underline{Z} = \frac{\underline{U}}{\underline{I}} \tag{2}$$

$$\underline{S}_{11} = \frac{\underline{Z} - \underline{Z}_0}{\underline{Z} + \underline{Z}_0} \tag{3}$$

Figure 4 depicts the structure of the tunable Spice network. It is built manually based on expert knowledge, but keeps the component values undefined since the determination is not trivial and is therefore performed by an optimisation algorithm.

In a real world application the planar structure is seen as a black-box where the S-parameter data characterises the behaviour at the terminals without revealing physics of the inner structure. The same is true for simulated distributed structures. The optimisation process has to tune the component values of the Spice network in order to minimize the error function against the given S-parameter data, Fig. 5 (left). However, in order to be able to evaluate the performance of a given optimisation algorithm the optimum component values for the lumped Spice model have to be known. Therefore, the Spice model with known component values in Fig. 6 is used as reference for the optimisation algorithm, Fig. 5 (right). This ensures that the error function has a know minimum of zero.

In order to analyse the complexity of the optimisation problem a sensitivity analysis of the component values has been performed. In the known optimum of the reference circuit each element of the optimum design vector $\mathbf{d}'$ is varied one at a time by a step of $(h \cdot d_i')$ with $h = 1 \times 10^{-6}$. This allows to approximate the derivative of the error function with respect to the ith component of $\mathbf{d}$ [18].

$$\left.\frac{\partial f(\mathbf{d})}{\partial d_i}\right|_{\mathbf{d}=\mathbf{d}'} \approx \frac{f(\mathbf{d}' + h \cdot d_i' \cdot \mathbf{e}_i) - f(\mathbf{d}')}{h \cdot d_i'} \tag{4}$$

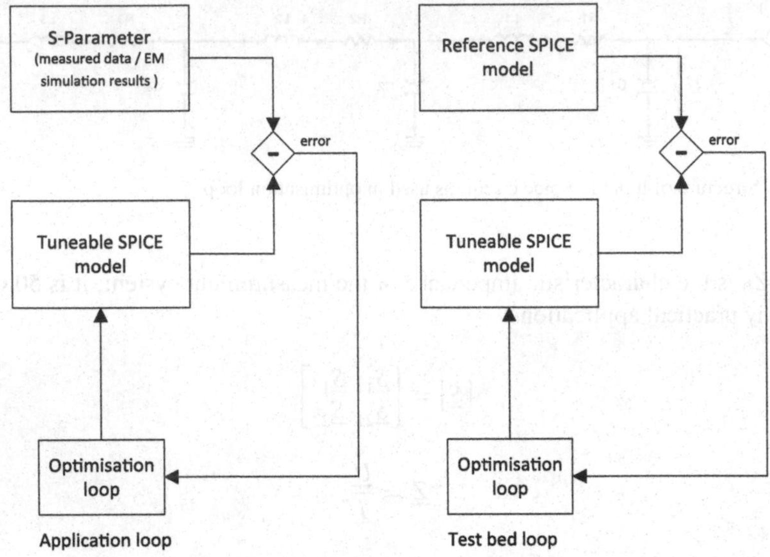

Fig. 5 Optimisation loop as used in real application to match the Spice model against given S-parameter data (*left*); optimisation loop during evaluation of the genetic algorithm within the test bed (*right*)

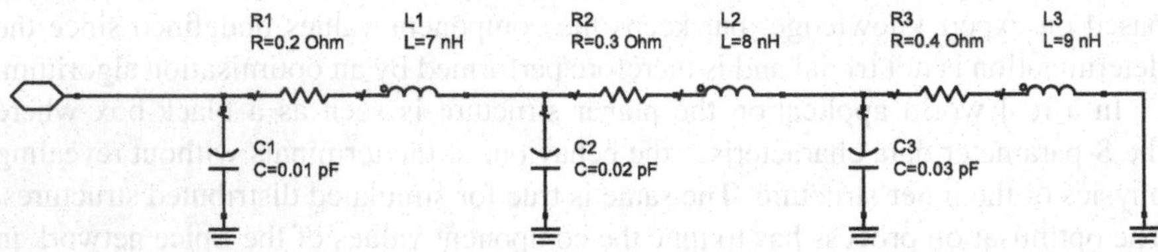

Fig. 6 Structure and component values of the reference Spice circuit

Here, f is the error function and e_i is the ith Cartesian basis vector. For all distinct frequency points at which the error function is evaluated the values of Eq. 4 are summed up and shown in Fig. 7 per each of the network components.

Although the network is only described by nine parameters the problem is rather complex; the component values are not linearly independent, which means they cannot be optimized independently from each other. Also, as it can be seen from Fig. 7 some of the parameters show a very high sensitivity, i.e. small variations in their values result in large changes of the error function. The nine-dimensional error landscape features a huge number of local optima. All this makes the problem at hand very demanding.

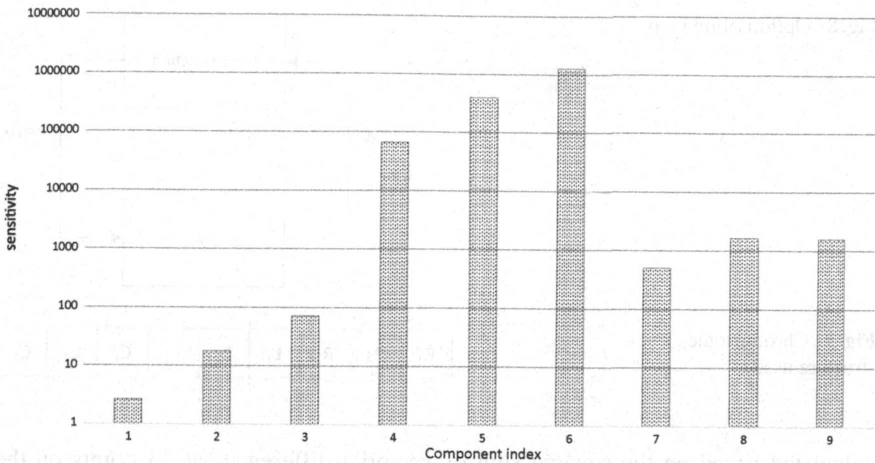

Fig. 7 Results of numerical sensitivity analysis of the reference Spice circuit at the optimum $\mathbf{d}'$

4 Optimisation of Component Values

A given lumped element network can be fully described by its design vector **d** (Eq. 5). The challenge in this research was to find an optimum design vector $\mathbf{d}'$ that produces the same frequency response as the original coil (Eq. 6).

$$\mathbf{d} = \begin{Bmatrix} R_1 \\ R_2 \\ R_3 \\ L_1 \\ L_2 \\ L_3 \\ C_1 \\ C_2 \\ C_3 \end{Bmatrix} \tag{5}$$

$$\mathbf{d}' = \arg\min f(\mathbf{d}) \tag{6}$$

For engineering design optimisation, it is recommended to use a fixed number of decimal places [19]. Thus, by using only two decimals, the optimisation problem was transformed into a discrete optimisation problem. For this, a genetic algorithm (GA) [20] was used to minimise the error function (Fig. 8).

Genetic algorithms are discrete optimisation algorithms which simulate the evolutionary mechanism found in nature by using heredity and mutation. They were first introduced in 1975 by Holland [21] who also provided a theoretical framework for Genetic Algorithms, the Schemata Theorem. In this work, the error is

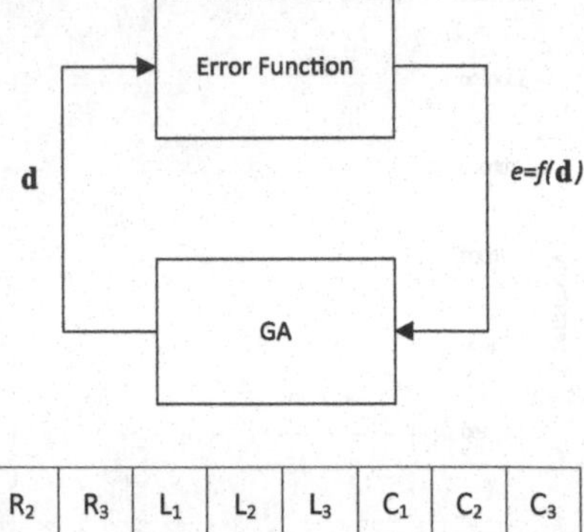

Fig. 8 Optimisation loop

Fig. 9 Chromosome structure used

R_1	R_2	R_3	L_1	L_2	L_3	C_1	C_2	C_3

calculated based on the squared sum of vectorial differences at 43 points on the frequency response curves (Eq. 7). The chosen frequency range ($f(1) = 100$ MHz, $f(43) = 10$ GHz) is stepped in a logarithmic scale with 21 points per decade (i.e. $f(n+1) = f(n) \times 10^{\frac{1}{21}}$). The conversion of a given S-parameter $\underline{S}_{11}$ into the complex Impedance $\underline{Z}$ is done according to Eq. 8, [17].

$$e = \sum_{n=1}^{43} \left| \underline{Z}_{ref}(f_n)) - \underline{Z}_n(f_n)) \right|^2 \tag{7}$$

$$\underline{Z} = \underline{Z}_0 \frac{1 + \underline{S}_{11}}{1 - \underline{S}_{11}} \tag{8}$$

For the optimisation problem under consideration, an integer coded genetic algorithm [22] was used. Figure 9 shows the chromosome structure employed.

Each gene holds integer numbers from the range 1...1000 and is decoded into its phenotype by dividing it by 100 in order to represent the fractional component values of the network.

Figure 10 shows a flowchart of the basic algorithm. After choosing the mutation probability p_m, the crossover probability p_c, the number n of individuals in the gene pool, the number r of tournaments used for the selection of one parent, and the maximum number of iterations i_{max}, the gene pool is randomly initialised. In each of the n generations, parents from the current generation are selected for the mating pool using tournament selection with r tournaments each [23]. Then, uniform crossover [24] is used with probability p_c to produce offspring from the mating pool. After power mutation [25] is applied to the offspring with the probability p_m, parents that perform worse than their offspring are replaced by their offspring. The algorithm terminates after n generations.

The next section provides the results of the experiments conducted.

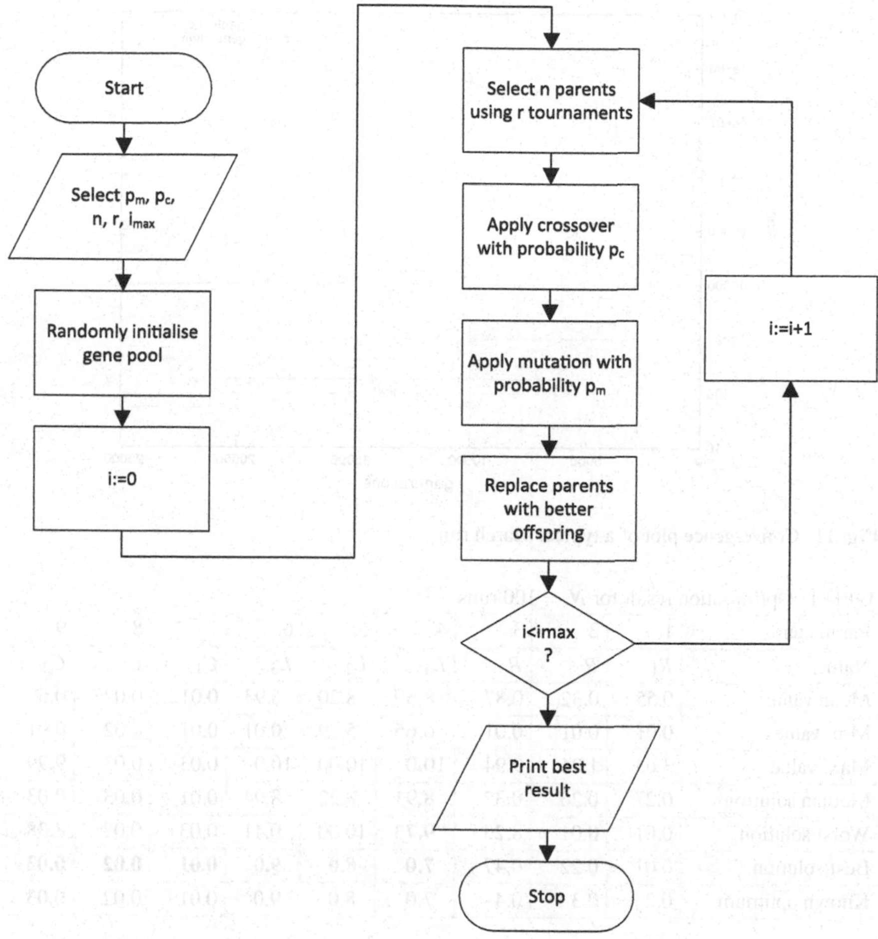

Fig. 10 Flowchart of the genetic algorithm used

5 Experiments

For the experiments, the population size was set to 1000. The crossover probability was chosen to be $p_c = 0.6$ and the mutation probability was $p_m = 0.001$. Five individuals competed in each tournament and the number of generations was set to $n = 25{,}000$. All these parameters were determined empirically.

Figure 11 shows a convergence plot of a typical search run where the solid line represents the best individual of each generation whereas the dashed line shows the average error of the whole population. It can be seen that the error was successfully reduced from 5.9×10^7 down to 54 (arbitrary units).

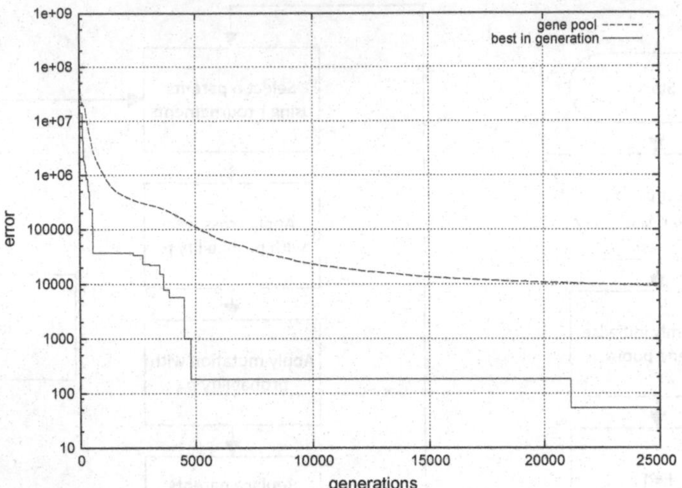

Fig. 11 Convergence plot of a typical search run

Table 1 Optimisation result for $N = 100$ runs

Parameter	1	2	3	4	5	6	7	8	9
Name	R_1	R_2	R_3	L_1	L_2	L_3	C_1	C_2	C_3
Mean value	0.55	0.32	0.87	8.59	8.20	6.93	0.012	0.027	0.69
Min. value	0.01	0.01	0.01	6.65	5.29	0.01	0.01	0.02	0.01
Max. value	3.63	1.26	9.94	10.0	10.00	10.0	0.03	0.03	9.99
Median solution	0.27	0.26	0.32	8.91	8.22	8.94	0.01	0.03	0.03
Worst solution	0.01	0.01	8.25	9.73	10.00	0.41	0.03	0.02	8.38
Best solution	0.01	0.22	0.47	**7.0**	**8.0**	**9.0**	**0.01**	**0.02**	**0.03**
Known optimum	0.2	0.3	0.4	7.0	8.0	9.0	0.01	0.02	0.03

Table 1 summarises the solutions found in the 100 experiments, each taking approximately eight minutes on a standard PC. It shows the worst, the best and the median solution found in the experiments, as well as the theoretical optimum. It also presents the component values for the network found by the optimisation algorithm. As it can be seen, in the best solution found during the 100 runs six of the nine component values (in bold) match the optimum values exactly. The remaining three, namely R_1, R_2, R_3, are also very close to the optimum. It is noticeable that those six parameters with the highest sensitivity (i.e. $L_1, L_2, L_3, C_1, C_2, C_3$) have been matched to the theoretical optimum, which indicates that they affected the contribution of the less sensitive parameters. This might be worthy further investigations.

In Fig. 12 the frequency response of the input reflection coefficient $\underline{S}_{11}$ is shown for the best solution and the median solution found during 100 runs as well as the optimum solution. This frequency response is characterised by its magnitude and

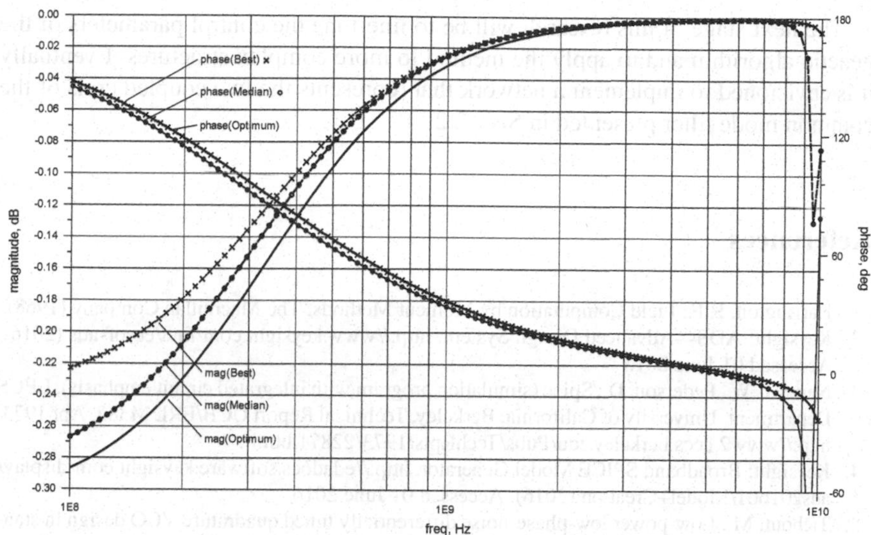

Fig. 12 Frequency response of $\underline{S}_{11}$ comparing best and median result over 100 runs versus the optimum solution

phase. As it can be seen, in both cases, the phase matches the optimum over the whole frequency range very well. For the magnitude, the median solution is close to the optimum at low frequencies (1×10^8–4×10^8 Hz) but shows a significant deviation from the optimum at frequencies above 8×10^9 Hz. On the other hand, the best solution models the frequency response above 8×10^9 Hz very accurately, whilst showing a worse matching in the lower frequency range.

From the hundred runs many converged into local optima. This indicates that the control parameters of the genetic algorithm are not tuned optimally.

6 Discussion and Future Work

Compared to the network list generated by a commercial broadband spice generator, which consisted of 16 voltage controlled current sources (VCCS) and five passive elements, the network tuned by the genetic algorithm consisted of only nine passive elements. Both networks modelled the reference perfectly in the phase frequency response, whilst showing similar differences in the amplitude response.

A huge advantage of the presented approach is the reduced number of network elements. Also, in principle, for more complex structures the new method will deliver more compact network lists compared to traditional methods. On the other hand, it is necessary to run the optimisation algorithm a sufficient number of times in order to be sure to approach the global optimum.

The next stage of this research will be to fine-tune the control parameters of the genetic algorithm and to apply the method to more complex structures. Eventually it is envisioned to implement a network that represents the two coupled coils of the common mode filter presented in Sect. 2.2.

References

1. Harrington, R.F.: Field Computation by Moment Methods. The Macmillan Company (1968)
2. Keysight: ADS—Advanced Design System. http://www.keysight.com/find/eesof-ads (2016). Accessed 01 June 2016
3. Nagel, L.W., Pederson, D.: Spice (simulation program with integrated circuit emphasis), EECS Department, University of California, Berkeley, Technical Report UCB/ERL M382, Apr 1973. http://www2.eecs.berkeley.edu/Pubs/TechRpts/1973/22871.html
4. Keysight: Broadband SPICE Model Generator. http://edadocs.software.keysight.com/display/ads201601/Model+Creation (2016). Accessed 01 June 2016
5. Tiebout, M.: Low-power low-phase-noise differentially tuned quadrature VCO design in standard CMOS. IEEE J. Solid-State Circuits **36**(7), 1018–1024 (2001)
6. Fanori, L., Mattsson, T., Andreani, P.: A 2.4-to-5.3 GHz dual-core CMOS VCO with concentric 8-shaped coils. In: 2014 IEEE International Solid-State Circuits Conference Digest of Technical Papers (ISSCC), pp. 370–371, Feb 2014
7. van Sinderen, J., Seneschal, F., Stikvoort, E., Mounaim, F., Notten, M., Brekelmans, H., Crand, O., Singh, F., Bernard, M., Fillatre, V., Tombeur, A.: A 48-860 MHz digital cable tuner IC with integrated RF and IF selectivity. In: 2003 IEEE International Solid-State Circuits Conference, 2003. Digest of Technical Papers. ISSCC, vol. 1, pp. 444–506, Feb 2003
8. Intel: USB 3.0* Radio Frequency Interference Impact on 2.4 GHz Wireless Devices. http://www.intel.com/content/www/us/en/io/universal-serial-bus/usb3-frequency-interference-paper.html (2012). Accessed 28 May 2015
9. Chen, C.H., Davuluri, P., Han, D.H.: A novel measurement fixture for characterizing USB 3.0 radio frequency interference. In: 2013 IEEE International Symposium on Electromagnetic Compatibility (EMC), pp. 768–772, Aug 2013
10. Franke, M.: USB Wireless E M I Training. http://www.testusb.com/EMI%20Training%20edit.pdf (2012). Accessed 14 Mar 2016
11. Bluetooth & USB 3.0—A guide to resolving your Bluetooth woes. http://www.bluetoothandusb3.com/ (2016). Accessed 14 Mar 2016
12. Werner, J., Schütt, J., Notermans, G.: Sub-miniature common mode filter with integrated ESD protection. In: 2015 IEEE International Symposium on Electromagnetic Compatibility (EMC), pp. 386–390, Aug 2015
13. Gao, W., Wan, L., Liu, S., Cao, L., Guidotti, D., Li, J., Li, Z., Li, B., Zhou, Y., Liu, F., Wang, Q., Song, J., Xiang, H., Zhou, J., Zhang, X., Chen, F.: Signal integrity design and validation for multi-GHz differential channels in SiP packaging system with eye diagram parameters. In: 2010 11th International Conference on Electronic Packaging Technology High Density Packaging (ICEPT-HDP), pp. 607–611, Aug 2010
14. Ahmadyan, S.N., Gu, C., Natarajan, S., Chiprout, E., Vasudevan, S.: Fast eye diagram analysis for high-speed CMOS circuits. In: 2015 Design, Automation Test in Europe Conference Exhibition (DATE'15), pp. 1377–1382, March 2015
15. Nagel, L.W.: What's in a name? IEEE Solid-State Circuits Mag. **3**(2), 8–13 (2011)
16. Ngspice, a mixed-level/mixed-signal circuit simulator. http://ngspice.sourceforge.net. Accessed 01 June 2016
17. Pozar, D.M.: Microwave Engineering, 4th edn. John Wiley & Sons, Inc. (2012)

18. Bischof, C., Carle, A.: Automatic differentiation principles, tools, and applications in sensitivity analysis. In: Second International Symposium on Sensitivity Analysis of Model Output, pp. 33–36, April 1998
19. Nolle, L., Krause, R., Cant, R.J.: On practical automated engineering design. In: Al-Begain, K., Bargiela, A. (eds.) Seminal Contributions to Modelling and Simulation. Springer (2016)
20. Goldberg, D.E.: Genetic Algorithms in Search, Optimization and Machine Learning, 1st edn. Addison-Wesley Longman Publishing Co., Inc., Boston, MA, USA (1989)
21. Holland, J.: Adaptation in natural and artificial systems (1975)
22. Abbas, H.M.: Accurate resolution of signals using integer-coded genetic algorithms. In: 2006 IEEE International Conference on Evolutionary Computation, pp. 2888–2895 (2006)
23. Miller, B.L., Goldberg, D.E.: Genetic algorithms, tournament selection, and the effect of noise. Complex Syst. **9**, 193–212 (1996)
24. Syswerda, G.: Uniform crossover in genetic algorithms. In: Schaffer, J.D. (ed.) ICGA, pp. 2–9. Morgan Kaufmann (1989)
25. Deep, K., Thakur, M.: A new mutation operator for real coded genetic algorithms. Appl. Math. Comput. **193**(1), 211–230 (2007)

Short Papers

Dendritic Cells for Behaviour Detection in Immersive Virtual Reality Training

N.M.Y. Lee, H.Y.K. Lau, R.H.K. Wong, W.W.L. Tam and L.K.Y. Chan

Abstract This paper presents a cross-disciplinary research of artificial intelligence (AI) and virtual reality (VR) that presents a real application of an aircraft door operation training conducted in an immersive virtual environment. In the context of the study, AI takes an imperative role in distinguishing misbehaviour of trainees such as inappropriate steps and positions in the virtual training environment that mimics a real training scenario. Trainee's behaviours are detected by the classical Dendritic Cell Algorithm (DCA) which is a signal-based classification approach that is inspired from the segmented detection and interaction with the signal molecules mechanisms of the human dendritic cells. The resulted approach demonstrated accurate detection and classification processes that are evidence in the experimental studies. This position paper also reveals the ability of the DCA method in human behaviour detection/classification in a dynamic environment.

Keywords Artificial Immune Systems · Dendritic Cell Algorithm · Human behaviour detection · Virtual reality · Field operations training

1 Introduction

Recent development of artificial intelligence (AI) is moving rapidly in tackling real world problems, such as autonomous robot control and anomaly detection. In recent studies, AI has also been applied to classification and recognition of human action

N.M.Y. Lee (✉) · H.Y.K. Lau · R.H.K. Wong · W.W.L. Tam · L.K.Y. Chan
University of Hong Kong, Pokfulam, Hong Kong
e-mail: myleenicole@graduate.hku.hk

H.Y.K. Lau
e-mail: hyklau@hku.hk

R.H.K. Wong
e-mail: rockywonghk@gmail.com

W.W.L. Tam
e-mail: tamwlhku@hku.hk

L.K.Y. Chan
e-mail: lkychan@hku.hk

M. Bramer and M. Petridis (eds.), *Research and Development in Intelligent Systems XXXIII*, DOI 10.1007/978-3-319-47175-4_27

in virtual training [1], for rehabilitation [2–4], risk presentation [5–7], and language/scripting/reading [8]. In the aforementioned applications, intelligent agents recognise speech, behaviour and posture of the trainee in order to provide automatic and instantaneous feedback to the trainee instead of solely relying on direct supervision of trainers. In the forgoing studies, evaluation of misbehaviours and decision-making are usually involved with hundreds and thousands of data collected from a highly dynamic and complex virtual environment. In additions, high computational complexity and false positive are often encountered when the knowledge is incorporated with rule-based systems (RBS) such as Perti net [9, 10], or artificial neural network (ANN) [11, 12], particularly for real-time reaction. In view of that, Dendritic Cell Algorithm (DCA) [13] is proposed to produce accurately classification results with the aids of signal detection agents (inspired by DCs) that are not commonly used in the aforementioned approaches. This represents one of the latest development in the family of Artificial Immune Systems (AIS).

In the context of the study, the DCA is adapted in a case of operational training in virtual reality training that is conducted in an immersive environment—the imseCAVE virtual reality system [14]. In the training process, the misbehaviour and abnormalities such as performing incorrect procedures, and moving in a wrong location of the trainee are automatically detected and classified as "dangers" in the metaphor of DCA. The experimental details and the capabilities of the algorithm are elaborated in Sects. 3 and 4 respectively.

2 Dendritic Cells and the DCA

DCA [13] which is inspired by the Danger Theory, is proposed by Matzinger [15]. The theory asserts that the immune responses are stimulated by the danger signals released from the distressed cells (e.g. injury and stress in the host), instead of the presence of the non-self cells [16]. These danger signals are typically detected by a specialised antigen presenting cells (APCs), namely, Dendritic cells (DCs) in the innate immune system. The maturity of DCs and the magnitude of the immune responses depend on the sensed antigens, input signals from the host and their severity. The captured attributes (of the segmented surface receptors of antigens) are interpreted by an intracellular signal-based classifiers or detectors [17, 18] (with pseudo code is presented in Fig. 1) that embedded in individual DCs. Primarily, the attributes are represented by the input signals, namely, PAMPs, DAMPs and safe according to their nature. In a computational context, the larger deviation generally indicates the hazardous to the host, for instances of the safety error in the operation training. The converted signals are further used for the assessment of a set of output signals (denoted by C in the signal processing function as presented in Eq. (1)) of co-stimulatory molecules (CSM), mature cytokines and semi-mature cytokines (are denoted by CSM, mDC and $smDC$ respectively).

$$C_{[CSM,mDC,smDC]} = \frac{(W_p * C_p) + (W_d * C_d) + (W_s * C_s)}{(W_p + W_d + W_s)} \quad (1)$$

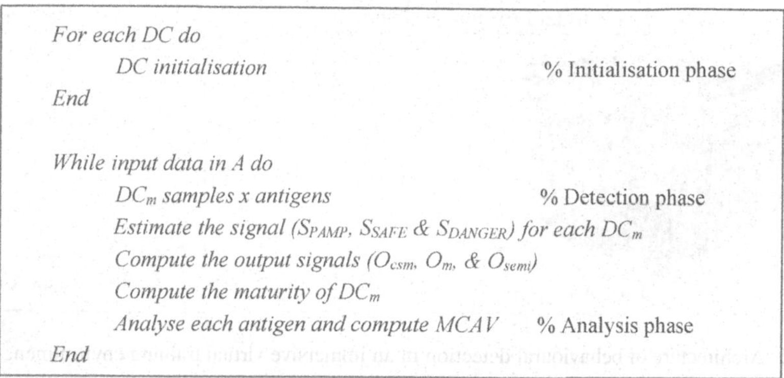

```
For each DC do
        DC initialisation                                   % Initialisation phase
End

While input data in A do
        DC_m samples x antigens                             % Detection phase
        Estimate the signal (S_PAMP, S_SAFE & S_DANGER) for each DC_m
        Compute the output signals (O_csm, O_m, & O_semi)
        Compute the maturity of DC_m
        Analyse each antigen and compute MCAV               % Analysis phase
End
```

Fig. 1 Pseudo code for the classical DCA. The input data (denoted by A_n where $n = 1 \ldots \delta_n$) are collected continuously in a given environment, a population of DCs (denoted by DC_m where $m = 1 \ldots \delta_m$) sample and acquire the information from the input data for signal transformation and maturation computation

Table 1 The weighting value for the signal-based detector adopted in Eq. (1), that is derived empirically from the immunology data [17]

W	CSM	$smDC$	mDC
p	2	0	2
d	1	0	1
s	2	3	−3

Resulted concentration obtained from Eq. 1 is regulated by the stimulative and suppressive control of the production of cytokines, which are now modeled by a set of weighting values as suggested in [17] (Table 1). The concentration here implies the potency of the maturity of DCs. That means, a high concentration of CSM indicates a higher potency of the trafficking of the DCs to the lymph nodes and more mature of the DCs. According to the obtained concentration of all kinds of output cytokines, the degree of anomaly of a given antigen (or objects/activities) is estimated by the mature context antigen value (MCAV). This measures the context of the antigens presenting in matured DCs. The antigens with a MCAV that exceeds the threshold of anomaly (approx. 0.65), are classified as misbehavior and vice versa.

3 Behaviour Detection in an Immersive Virtual Environment

Attributes of trainee's activities and behaviour are analogous to the antigens and signals respectively (Table 2), which are captured and streamed from the immersive training environment powered by the Unity Game Engine (https://unity3d.com) to the Matlab-based processor where the attributes of the sampled segment data are evaluated by the DCs. The architecture of the behaviour detection-enable training system is depicted in Fig. 2.

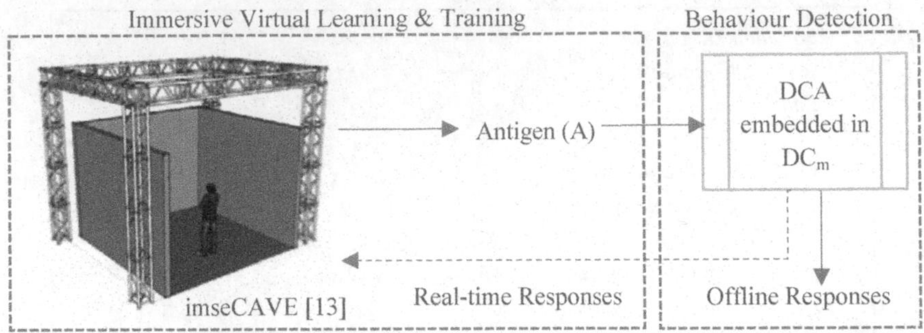

Fig. 2 Architecture of behavioural detection in an immersive virtual training environment

Table 2 The analogy for behaviour detection

Human immune system	Notation	DCA for behaviour detection
DCs	DC_m	Detectors
Antigens	A_n	Activities and behaviour of the given training (captured x frames)
Signals	S_i	Attributes that are used to differentiate the correctness of the activities and human behaviours (refer to Table 3)

Table 3 Categorization of signals for the implication of an immersive virtual training

Signals	Denotations	Implication for operational training
PAMP (p)	A signature of abnormal behaviour	Irrelevant tasks (or steps)
DAMP (d)	Significantly increase (or a large deviation) in response to abnormal behaviour	(a) Action error (b) Safety related error (c) Target object error (d) Condition error
SAFE (s)	A confident indicator of normal behaviour	Interactions with the target objects

According to the definitions given to human misbehaviour in [6, 19], which are categorised into, for instances, the number of irrelevant steps and various types of errors (e.g., the distance between the safety gate and the trainee in the aircraft door operation training). In the context of DCA, they are analogous to the input signals of PAMPs and DAMPs respectively as according to the analogy proposed in Table 3.

4 Results and Observations

In the context of the experiments, ten runs are conducted for each scene comprising forty DCs, which involve (a) 120 and (b) 140 antigen presentations. That means, a segment of 30–35 erratically captured antigens are presented four times in each

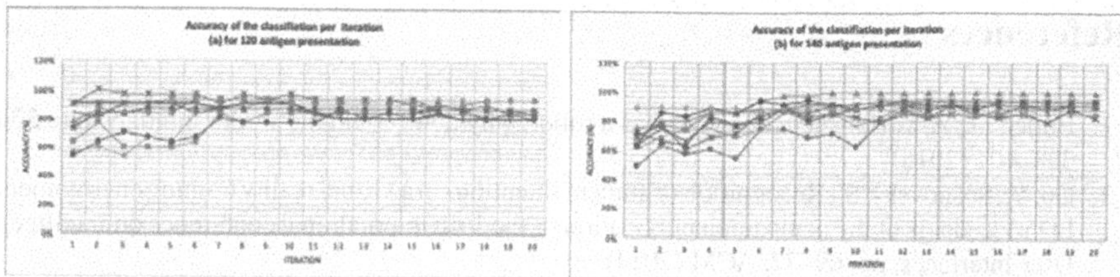

Fig. 3 Classification accuracy for the scenes of **a** 120 and **b** 140 antigen presentation

iteration. The detection and classification of misbehaviour process are implemented with Matlab on an iMac i7 3.40 GHz.

To recapitulate the objective of the study that is intended to demonstrate the capability of DCA in producing precise and accurate results in human behaviour detection with primarily concerns of both true positive and true negative rate. In brief, the study is evaluated based on the accuracy obtained in each of (i) individual iteration and (ii) captured antigens.

Based on the experiments with twenty iterations as depicted in Fig. 3, the results show that segmentation potentially generates finer gained results. In both scenes (a) and (b), the classification accuracy reaches 95 and 100 % (from 50 %+ at the early stage of detection) respectively. Though the deviation (approx. 4–5 %) of the performed runs can definitely influence the consistency and stability of the overall performance, the classification accuracy can be improved by adopting additional antigen presentation as demonstrated in (b).

As for each sampled antigen, 60–70 % of the activities are correctly classified in each iteration through the experimental studies. Though the obtained results are not significant, the capabilities of the DCA and other findings presented in this paper are indeed meaningful as pointers to further study in future. The encountered deviations and the impact due to the scene settings should be further investigated.

5 Conclusion and Future Work

This study presented in this paper demonstrates the capabilities of DCA in differentiating abnormal human behaviours in a virtual training environment. The experimental results also revealed the potential of the framework and provide pointers for future research. The authors are currently working on improving the consistency of the algorithms and the impacts due to the scene settings as aforementioned.

References

1. Butler, D.: A world where everyone has a robot: why 2040 could blow your mind. Nature **530**, 398–401 (2016)
2. Gu, Y., Sosnovsky, S.: Recognition of student intentions in a virtual reality training environment. In Proceedings of the companion publication of the 19th international conference on Intelligent User Interfaces, pp. 69–72. ACM (2014)
3. Zhang, Z., Liu, Y., Li, A., Wang, M.: A novel method for user-defined human posture recognition using Kinect. In: 7th International Congress on IEEE Image and Signal Processing (CISP), pp. 736–740 (2014)
4. Parsons, T.D., Trost, Z.: Virtual reality graded exposure therapy as treatment for pain-related fear and disability in chronic pain. In: Virtual, Augmented Reality and Serious Games for Healthcare, vol. 1, pp. 523–546 (2014)
5. Amokrane, K., Lourdeaux, D.: Virtual reality contribution to training and risk prevention. In: Proceedings of the International Conference of Artificial Intelligence (ICAI'09). Las Vegas, Nevada, USA. (2009)
6. Amokrane, K., Lourdeaux, D., Michel, G.: Serious game based on virtual reality and artificial intelligence. ICAART **1**, 679–684 (2014)
7. Barot, C., Lourdeaux, D., Burkhardt, J.M., Amokrane, K., Lenne, D.: v3s: a virtual environment for risk-management training based on human-activity models. Presence Teleoperators Virtual Env. **22**(1), 1–19 (2013)
8. Kamath, R. S.: Development of intelligent virtual environment by natural language processing. Special issue of International Journal of Latest Trends in Engineering and Technology (2013)
9. Lin, F., Ye, L., Duffy, V.G., Su, C.J.: Developing virtual environments for industrial training. Inf. Sci. **140**(1), 153–170 (2002)
10. Fillatreau, P., Fourquet, J.Y., Le Bolloc'H, R., Cailhol, S., Datas, A., Puel, B.: Using virtual reality and 3D industrial numerical models for immersive interactive checklists. Comput. Ind. **64**(9), 1253–1262 (2013)
11. Chaudhary, A., Raheja, J.L., Das, K., Raheja, S.: Intelligent approaches to interact with machines using hand gesture recognition in natural way: a survey (2013)
12. Barzilay, O., Wolf, A.: Adaptive rehabilitation games. J. Electromyogr. Kinesiol. **23**(1), 182–189 (2013)
13. Greensmith, J.: The dendritic cell algorithm. Nottingham Trent University (2007)
14. Chan, L.K., Lau, H.Y.K.: A tangible user interface using spatial Augmented Reality. In: IEEE Symposium on 3D User Interfaces (3DUI), pp. 137–138 (2010)
15. Matzinger, P.: Tolerance, danger, and the extended family. Annu. Rev. Immunol. **12**, 991–1045 (1994)
16. Burnet, F.M., Fenner, F.: The Production of Antibodies, 2nd edn. Macmillan, Melbourne (1949)
17. Greensmith, J., Aickelin, U., Cayzer, S.: Introducing dendritic cells as a novel immune-inspired algorithm for anomaly detection. Artif. Immune Syst. 153–167 (2005)
18. Kim, J., Bentley, P.J., Aickelin, U., Greensmith, J., Tedesco, G., Twycross, J.: Immune system approaches to intrusion detection-a review. Nat. Comput. **6**(4), 413–466 (2007)
19. Amokrane, K., Lourdeaux, D., Burkhardt, J.: HERA: learner tracking in a virtual environment. Int. J. Virtual Real. **7**(3), 23–30 (2008)

Interactive Evolutionary Generative Art

L. Hernandez Mengesha and C.J. James-Reynolds

Abstract Interactive Genetic Algorithms may be used as an interface for exploring the solution spaces of computer art generative systems. In this paper, we describe the application of "Eugene" a hardware interactive Genetic Algorithm controller to the production of fractal and algorithmic images. Processing was used to generate the images. In tests with users, it was found that there was adequate exploration of the solution space, although the results were not always as expected by the user. Some additional testing explored this and the results indicate that mapping strategies used in encoding the solution are important if the interface allows for goal oriented user tasks.

Keywords Interactive Genetic Algorithms · Human Computer Interaction · Generative Art

1 Introduction

This paper summarises work completed as part of a final year undergraduate project exploring the production of generative art controlled by a hardware based interactive Genetic Algorithm (iGA) Controller [1]. This was developed around an Arduino Nano and had a population size of 8. The data produced was sent to Processing [2] and was mapped to variables used in the creation of generative art. In part this work was carried out to effectively demonstrate that this approach was viable, but more importantly it allowed testing with users to understand issues with the controller, its evaluation and this approach. The methodology was similar to a cognitive walk-through, where users were asked about what they were doing and their thoughts on the process as they carried out two tasks. Firstly, exploring the solution space and secondly a goal based task where they were asked to generate a specific image.

L. Hernandez Mengesha · C.J. James-Reynolds (✉)
Middlesex University, London NW4 4BT, UK
e-mail: carl9@mdx.ac.uk; C.James-Reynolds@mdx.ac.uk

L. Hernandez Mengesha
e-mail: LM1107@live.mdx.ac.uk

M. Bramer and M. Petridis (eds.), *Research and Development in Intelligent Systems XXXIII*, DOI 10.1007/978-3-319-47175-4_28

2 Background and Context

Generative art or in this case computer art, is where the artist or creator uses an autonomous system to generate artworks. It is the practice of any art in which "the artist uses a system, such as the set of natural language rules, a computer program, a machine or a procedural invention which results in an artistic artwork" [3] Algorithmic art can be seen as more than just an algorithm that creates something. It implies that the algorithm is the step by step "developed artistic procedure" created by the artist, which is then translated and interpreted as a mathematical algorithm by the autonomous system or computer program [4]. Geometry created with iterative functions rather than formulae include fractals such as the Mandelbrot set [5] and Julia set [6], which are closely related mathematically. These work on the basis of self-repetition and may be self-similar. They are often coloured using rules such as the escape time algorithm [7], which is based on the number of iterations used to find if the pixel lies in the appropriate set. A unique aspect of fractals is that they are infinite. Creating such artwork often requires software where the user has to interact with formulae and parameters. However, iGAs offer an alternative approach.

Turing [8] proposed the creation of a machine which could learn and evolve by applying nature's principle of natural selection, however there was little uptake of this idea until Holland [9] popularised it.

A Genetic Algorithm applies the idea of natural selection by treating potential solutions as individuals which are chromosomes. Each chromosome or solution consists of the variables (gene sequence) that make up the solution. The full range of possible solutions is known as the solution space. We can have many solutions, but a test is required in order to identify how well any given solutions can perform. This test in nature is survival; in a GA it is performance against a fitness function that will provide a quantitative performance rating. Where we are evaluating a piece of artwork, the rating is a subjective one and in an iGA the rating is given by the user, who replaces the fitness function.

The complete iGA process is as follows:

We randomly generate an initial population of potential solutions (chromosomes or gene sequences).

The user rates all the individual solutions that make up the population.

The rating given is used to calculate the probability of the solution being chosen as a parent, using a roulette wheel approach. This allocates a number of slots on a roulette wheel proportional to the rating. Random numbers are then generated and where they fall on the roulette wheel selects parents, after a check that the same parent has been not chosen twice.

The genes that make up the parents are strings of variables and in a process similar to that of reproduction, the genes of each parent are cut and a child is created from the crossover of the two genes. This gives the child variables from each parent.

We continue this process until we have enough child members of the population to replace the parents.

In nature we also have the idea of mutation and this is also used in the Genetic Algorithm, where we introduce small random changes to the gene. In practice this serves to ensure that we explore as much of the solution space as possible. It may be that the process is left to run for a given number of generations otherwise the process will continue until the best (fittest) solution is achieved [10].

Much of work behind these processes was carried out by Bremmerman [11] who had allowed the field to expand by exploring and adapting solutions to optimization problems, mutation and selection [12, 13].

3 Interactive Generative Art Controller (Eugene)

The user interface (shown in Fig. 1) is composed of 8 sliders, an Arduino Nano 328 microcontroller, and a play push button switch for each slider. The iGA and Arduino combined make the interactive physical interface. Its workings are as follows:

- **MIDI Socket**: The socket allows a MIDI to USB lead to be plugged into the controller and allows communication between the controller and Processing sketches. MIDI [14] is a communication protocol that was designed for musical equipment; it is used to send values between 0 and 127 from the controller.
- **Sliders**: These are used to provide the algorithm with the user rating which goes from 0 to 1023 (50–950 to be precise). The algorithm then employs the rating to adjust the probability of being selected. Users have the option of keeping or discarding individual offspring via the sliders, where a rating of 0 discards the individual, replacing it with a new individual created with random values. A rating of 1023 allows the individual to be kept where it becomes part of the gene pool for the next generation. This is an example of elitism [15].
- **Evolve Button**: This is used to create the next generation by starting the selection and crossover process.
- **Mutate**: A mutate value is stored as a variable in the code.

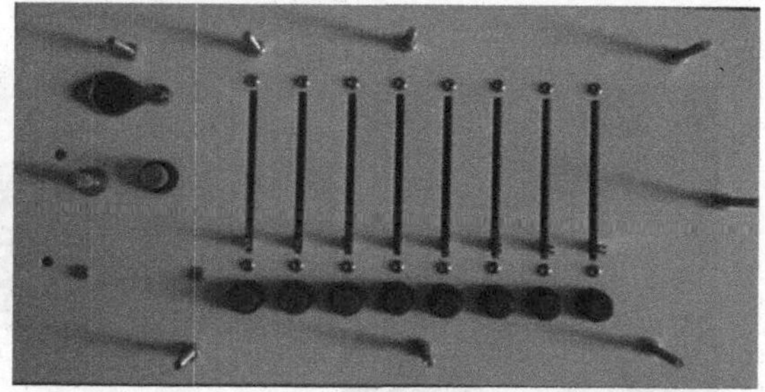

Fig. 1 The controller

4 Evaluation

Functionality and colour mapping were initially tested with simple sketches, in order to determine the ordering of the variables and mapping of values. Some of the results are shown below in Fig. 2. Further testing allowed users to play with the interface in order to gather a user perspective. The user evaluation consisted of observation and short questions to ascertain if users liked the output and to identify if they felt in control. The algorithms used in testing included standard implementations of the Mandelbrot set, Julia set, a Fibonacci spiral and a Circles algorithm.

When mapping MIDI CC values to the values used to create and colour the artwork, it was important to understand the implications of linear or logarithmic mapping, and bounds that could result in the graphic output being at too large or too small a size for the display area. It is possible to have values that produce no results; these were avoided by remapping to reduce user confusion.

4.1 User Evaluation

Tests were targeted and non-targeted with a group of 14 users; these were students from a range of backgrounds. Users tested the controller with both approaches. Targeted tests gave the users a specific image goal to work towards, and the non-targeted tests allowed users to play with the interface. This was to retrieve feedback with regards to the interface design and performance. From observation and discussion, some users found the usage of the interface more complex than others and had an expectation of the sliders visibly changing a parameter. Some did not seem to understand how the rating system operated and because the internal algorithm was hidden, found it difficult to believe that the concept might work. Half the test subjects got very close to the target in the target based tests. All the subjects could explore the solution space, with two of the subjects not really interested in playing with the controller. Students with more technical backgrounds seemed to be more comfortable with the overall concept and performed better in the targeted tests. Figure 3 shows some of

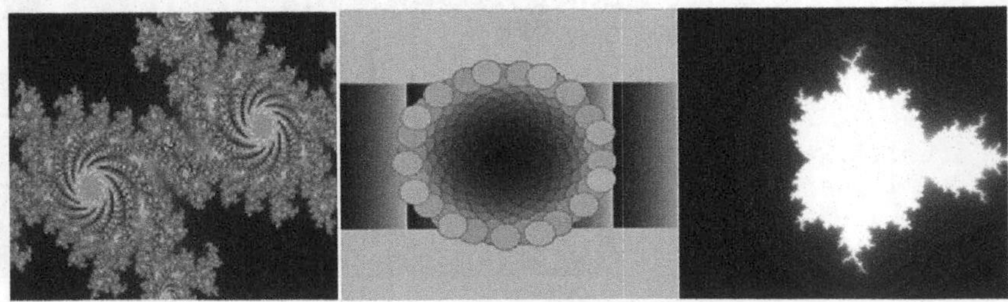

Fig. 2 Julia and Fibonacci spiral (implementation with *circles*)

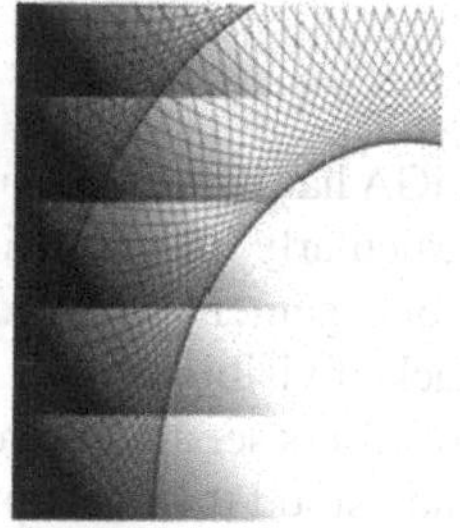
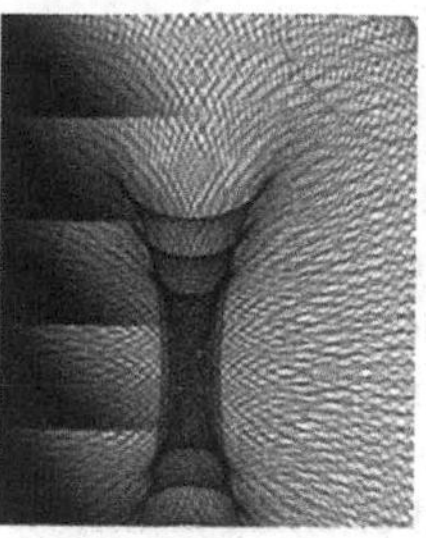

Fig. 3 Example output solutions of first test with circles sketch

the outputs generated with 6 variables mapped as Red, Green, Blue, Radius, Ellipse centre height, Ellipse centre width. There are 1080 circles drawn for each image.

5 Discussion

Parameter mapping is the key to viable output solutions. Eugene outputs MIDI CC values from 0 to 127 and these should be mapped to appropriate ranges within the target application for the specific implementation. When the mapping is incorrect, images can be distorted or hidden. This differed for each of the different image generation algorithms. It can be argued that these are still viable solutions within the solution space, but it was decided to restrict this. As each algorithm requires a different number of variables (genes), the Eugene code needed to be re-uploaded each time. A future improvement would be the ability to select the number of genes from the front panel. Allowing some variables to be averaged would be similar to co-dominance in traditional genetics, this approach would seem particularly effective for attributes such as colour, although it would need to be mapped to more intuitive dimensions such as HSL rather than RGB values.

Grouping and ordering of the variables was also important in the performance of the controller and although this was tweaked to give better results, further testing might reveal optimization of this. In the testing of the interface, the tests with the Julia set were limited, as rendering was slow; however the other tests with the Mandelbrot set and simpler geometric patterns ran efficiently. Although the number of test subjects was limited, the responses were very similar. During user testing, it was determined that users did not feel in control when manipulating images with the interface, although those with a technical background appreciated the strategy and most users liked the results. When users were given a target to reach, at times they did not feel there was appropriate convergence of solutions, this was because they felt that sometimes images returned to previous states, (this was not true as values were being observed and there were minimal changes). Half of the users felt that they were able to get close to the target image, despite these observations.

6 Conclusions

This work explored the use of an iGA hardware controller as the interface for producing evolutionary art. We were particularly interested in a user perspective and issues in adapting the controller. User engagement varied considerably and depended on their interest, but the perceived lack of visible change when sliders were adjusted led to a feeling of not being in control. Users seemed more able to successfully complete goal oriented tasks when they understood the concept behind the interface.

References

1. James-Reynolds, C., Currie, E.: Eugene: A Generic Interactive Genetic Algorithm. Research and Development in Intelligent Systems XXXII, pp. 361–366. Springer International Publishing, London (2015)
2. Fry, B., Reas, C.: The Processing Language. https://processing.org/. Accessed 20 Aug 2016
3. Galanter, P.: What is generative art? Complexity theory as a context for art theory. In: GA2003—6th Generative Art Conference (2003)
4. Verostko, R.: Notes on algorithmic art in the ISEA'94 Catalogue. In: The Fifth International Symposium on Electronic Art, p. 61. University of Art & Design, Helsinki, Finland (1994)
5. Mandelbrot, B.: Fractal aspects of the iteration of $z \rightarrow \lambda z(1\text{-}z)$ for complex λ and z. Non-Linear Dynamics. Ann. N. Y. Acad. Sci. **357**, 249–259 (1979)
6. Julia, G.: Mémoire sur l'iteration des fonctions rationnelles. Journal de Mathématiques Pures et Appliquées **8**, 47–245 (1918)
7. Barnsley, M.: Fractals Everywhere. Academic Press, Boston (1988)
8. Turing, A.M.: Computing Machinery and Intelligence. Mind **49**, 433–460 (1950)
9. Holland, J.H.: Adaptation in Natural and Artificial Systems. University of Michigan Press, Ann Arbor (1975)
10. Mitchell, M.: An Introduction To Genetic Algorithms. Massachusetts London, England, Cambridge (2002). (Fifth printing)
11. Bremermann, H.J.: Optimization through evolution and recombination. Self-organizing Syst. **93**, 106 (1962)
12. Dianati, M., Song, I., Treiber, M.: An introduction to genetic algorithms and evolution strategies. Technical report. University of Waterloo, Ontario, Canada (2002)
13. Taylor, T., Dorin, A., Korb, K.: Digital genesis: computers, evolution and artificial life. In: Presented at the 7th Munich-Sydney-Tilburg Philosophy of Science Conference: Evolutionary Thinking, University of Sydney, 20–22 (2014)
14. The MIDI 1.0 Specification. https://www.midi.org/specifications. Accessed 20 Aug 2016
15. De Jong. K.: Analysis of the Behavior of a Class of Genetic Adaptive Systems Technical Report no 185 University of Michigan (1975)

Incorporating Emotion and Personality-Based Analysis in User-Centered Modelling

Mohamed Mostafa, Tom Crick, Ana C. Calderon and Giles Oatley

Abstract Understanding complex user behaviour under various conditions, scenarios and journeys is fundamental to improving the user-experience for a given system. Predictive models of user reactions, responses—and in particular, emotions—can aid in the design of more intuitive and usable systems. Building on this theme, the preliminary research presented in this paper correlates events and interactions in an online social network against user behaviour, focusing on personality traits. Emotional context and tone is analysed and modelled based on varying types of sentiments that users express in their language using the IBM Watson Developer Cloud tools. The data collected in this study thus provides further evidence towards supporting the hypothesis that analysing and modelling emotions, sentiments and personality traits provides valuable insight into improving the user experience of complex social computer systems.

Keywords Emotions · Personality · Sentiment analysis · User experience · Social networking · Affective computing

---

The original extended version of this paper is available here: http://arxiv.org/abs/1608.03061.

M. Mostafa · T. Crick (✉) · A.C. Calderon
Department of Computing & Information Systems, Cardiff Metropolitan University, Cardiff, UK
e-mail: tcrick@cardiffmet.ac.uk

M. Mostafa
e-mail: momostafa@cardiffmet.ac.uk

A.C. Calderon
e-mail: acalderon@cardiffmet.ac.uk

G. Oatley
School of Engineering & Information Technology, Murdoch University, Murdoch, Australia
e-mail: g.oatley@murdoch.edu.au

M. Bramer and M. Petridis (eds.), *Research and Development in Intelligent Systems XXXIII*, DOI 10.1007/978-3-319-47175-4_29

1 Introduction

As computer systems and applications have become more widespread and complex, with increasing demands and expectations of ever-more intuitive human-computer interactions, research in modelling, understanding and predicting user behaviour demands has become a priority across a number of domains. In these application domains, it is useful to obtain knowledge about user profiles or models of software applications, including intelligent agents, adaptive systems, intelligent tutoring systems, recommender systems, e-commerce applications and knowledge management systems [12]. Furthermore, understanding user behaviour during system events leads to a better informed predictive model capability, allowing the construction of more intuitive interfaces and an improved user experience. This work can be applied across a range of socio-technical systems, impacting upon both personal and business computing.

We are particularly interested in the relationship between digital footprint and behaviour and personality [6, 8]. A wide range of pervasive and often publicly available datasets encompassing digital footprints, such as social media activity, can be used to infer personality [4, 9] and development of robust models capable of describing individuals and societies [5]. Social media has been used in varying computer system approaches; in the past this has mainly been the textual information contained in blogs, status posts and photo comments [1, 2], but there is also a wealth of information in the other ways of interacting with online artefacts. From sharing and gathering of information and data, to catering for marketing and business needs; it is now widely used as technical support for computer system platforms.

The work presented in this paper is builds upon previous work in psycholinguistic science and aims to provide further insight into how the words and constructs we use in our daily life and online interactions reflect our personalities and our underlying emotions. As part of this active research field, it is widely accepted that written text reflects more than the words and syntactic constructs, but also conveys emotion and personality traits [10]. As part of our work, the IBM Watson Tone Analyzer (part of the IBM Watson Developer Cloud toolchain) has been used to identify emotion tones in the textual interactions in an online system, building on previous work in this area that shows a strong correlation between the word choice and personality, emotions, attitude and cognitive processes, providing further evidence that it is possible to profile and potentially predict users identity [3]. The *Linguistic Inquiry and Word Count* (LIWC) psycholinguistics dictionary [11, 13] is used to find psychologically meaningful word categories from word usage in writing; the work presented here provides a modelling and analysis framework, as well as associated toolchain, for further application to larger datasets to support the research goal of improving user-centered modelling.

The rest of the paper is structured as follows: in Sects. 2 and 3 we present our data, the statistical analysis and identify the key elements of our model; in Sect. 4 we summarise the main contributions of this paper, as well as making clear recommendations for future research.

2 Data Analysis and Feature Extraction

Our dataset comes from an online portal for a European Union international scholarship mobility hosted at a UK university. The dataset was generated from interactions between users and a complex online information system, namely the online portal for submitting applications. The whole dataset consists of users (N = 391), interactions and comments (N = 1390) as responses to system status and reporting their experience with using the system. Google Analytics has been used to track user behaviour and web statistics (such as impressions); this data from has been used to identify the server's status and categorised the status as two stages: *Idle*, where the system had a higher number of active sessions; and marked as *Failure*, where the system had a lower number of sessions engaged. Interactions were first grouped by server status, then sent to the IBM Watson Tone Analyzer to generate emotion social tone scores. In what follows *Failure* status shows a significant difference in overall *Anger* in different status; furthermore, the *Joy* parameter shows a significant difference with the system in *Idle* and *Failure* status. However *Fear* and *Sadness* parameters is almost the same, even with the system in *Idle* status.

We identified the user's personality based on analysis of their Facebook interactions, namely by collecting all comments from the users, again using the IBM Watson Personality Insights tool. However, a number of users in the dataset had completed the Big Five questionnaire (N = 44); for these users, their Big Five scores have been used instead. The second stage involved grouping the comments based on server status and segmenting these interactions by user; this allowed us to investigate the impact of server status in the emotion of the user and investigate the Big Five dimension as a constant parameter. By investigating the relationship between personality trait dimensions and the social emotion tones, we are able to find the highest correlation to identify the key elements of the potential model by applying linear regression and Pearson correlation. This allowed building of a neural network multilayer perception using the potential key elements with higher correlations.

The data collected from the social media interactions was grouped by users and via IBM Watson Personality Insights, we were able to identify the Big Five personality traits for each user. Using the IBM Watson Tone Analyzer, the data was grouped by user's comments and server status (*Failure*, *Idle*) to identify the social emotion tone for each user.

3 Statistical Analysis and Key Elements of Model

As part of modelling the users' responses and behaviour, in order to build a conceptual framework model, we applied linear regression to investigate the relationship between the Big Five personality dimensions and the emotion tones.

Linear regressions (presented in Table 1 and Fig. 1) do not show significant correlations between the Big Five dimensions and the social emotion tones. There are,

Table 1 Linear regression coefficients

	Openness			Extraversion			Conscientiousness			Agreeableness			Neuroticism		
	B	t	Sig	B	t	Sig	B	t	Sig	B	t	Sig	B	t	Sig
(Constant)	0.356	3.282	0.001	0.162	1.642	0.101	0.16	1.623	0.105	0.297	2.831	0.005	0.828	9.934	0
Anger	−0.063	−0.735	0.463	0.064	0.831	0.406	0.124	1.592	0.112	0.024	0.293	0.769	0.116	1.767	0.078
Disgust	0.478	4.354	0	0.114	1.142	0.253	0.255	2.551	0.011	−0.061	−0.574	0.566	−0.363	−4.303	0
Fear	0.065	0.534	0.594	0.172	1.549	0.122	0.04	0.356	0.722	0.093	0.783	0.434	−0.023	−0.241	0.81
Joy	0.066	0.561	0.575	0.446	4.179	0	0.436	4.058	0	0.188	1.652	0.099	−0.487	−5.39	0
Sadness	−0.226	−2.118	0.035	−0.185	−1.906	0.057	−0.03	−0.313	0.754	0.014	0.132	0.895	0.233	2.841	0.005

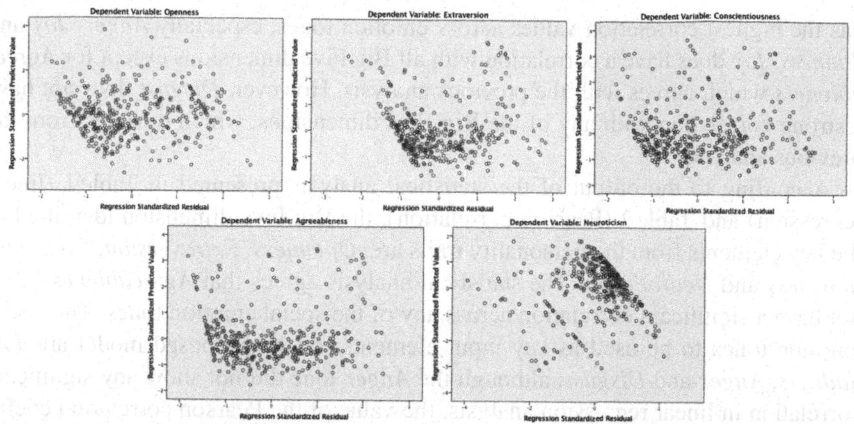

Fig. 1 Scatterplots of the Big Five dimension (dependent variables) and social emotion tones (independent variables)

however correlations that can be used as key elements for the model; namely the correlation of *Openness* and *Disgust* (0.479), *Extraversion* and *Joy* (0.446 with p-value of 0), *Conscientiousness* and *Joy* (0.436),and *Disgust* with 0.255.*Agreeableness*, does not appear to have a high impact in the social emotion parameters, with the highest correlation being 0.188 with *Joy*, which can be overlooked as a useful factor in the model. *Neuroticism* and *Disgust* is −0.363, *Joy* is −0.487 and p-value is zero is both cases; and *Sadness* with 0.233. All correlation values are <0.5. However, *Agreeableness* does not have a linear relationship with any of the social emotion tones. Furthermore, the social emotion tones that have a potential linear relationship are *Disgust*, *Joy* and *Sadness*, since the three tones have a correlation between >0.3 and <0.5.

Previous linear regression analysis suggested that the following Big Five dimensions: *Openness*, *Extraversion*, *Conscientiousness* and *Neuroticism* have the highest correlation with the social emotion tones: *Joy*, *Sadness* and *Disgust*. For further analysis, the Pearson correlation for the same dataset has been performed to compare the output with the linear regression correlations. As noted in Table 2, there is no significant correlation in both; however, in the Pearson correlation, *Neuroticism*

Table 2 Pearson correlations

	Anger	Disgust	Fear	Joy	Sadness
Openness	−0.098	0.231	0.043	0.035	−0.151
Conscientiousness	−0.111	−0.001	−0.113	0.267	−0.19
Extraversion	−0.175	−0.077	−0.071	0.349	−0.291
Agreeableness	−0.068	−0.089	−0.027	0.14	−0.069
Neuroticism	0.375	−0.037	0.153	−0.488	0.379

has the highest correlation values across emotion tones, especially *Anger*, *Joy* and *Sadness*. *Joy* does have a correlation with all Big Five dimensions except for *Agreeableness* which agrees with the previous analysis. However, *Disgust* does not have a strong correlation with any of the Big Five dimensions, which deviates from the previous analysis.

According to the output of the statistical analysis presented in Table 1 (linear regression) and Table 2 (Pearson correlation), the Big Five dimension identified as the key elements from the personality traits are: *Openness*, *Extraversion*, *Conscientiousness* and *Neuroticism*. The statistical analysis agrees that *Agreeableness* does not have a significant correlation across any of the social emotion tones. The social emotion tones to be used as key input elements for the proposed model are *Joy*, *Sadness*, *Anger* and *Disgust*; although the *Anger* tone did not show any significant correlation in linear regression analysis, the value of the Pearson correlation coefficient is between 0.3 and 0.5 which can be used as input for the model.

The dataset used to build this model is based upon a number of users (N = 391), eight inputs (*Openness*, *Extraversion*, *Conscientiousness*, *Neuroticism*, *Joy*, *Sadness*, *Anger* and *Disgust*) and the class/output variable as the server status (where No: System *Failure* and Yes: System *Idle*). The total number of the instances for the testing set is 57; the output of the model shows a 75.44 % corrected predicted instances and 24.56 % incorrectly classified instances (kappa statistic: 0.5295; mean absolute error: 0.3432; RMS error: 0.4246). As this has been performed on a small subset of the overall larger project dataset, the output data is encouraging and provides the infrastructure for further analysis and research to exploit the full dataset.

4 Conclusions and Future Work

In this paper, preliminary results from a larger ongoing theme of research to profile online/digital behaviour have been presented[1]; the objective is to build a conceptual framework to improve user experience and computer system architecture design. The research also concerns applicability in interested in profiling complex behaviours and psychopathies using social network analysis, particularly for crime informatics [7]. Previous work in this space analysed the document uploading behaviour (such as motivation letters, and social media interactions) of the applicants of the international scholarship mobility portal; by examining the upload footprint for the users we were able to determine several classes of behaviour [9].

Social media is used, not only as a content and sharing platform, but also as a platform for technical support for various of online applications and services. This paper demonstrates the analysis of one such online application that has used Facebook as technical support platform for the users. We have produced a model that can predict server status based on personality traits and social emotion tones, by investigating the linear regression and Pearson correlation to identify the key

[1] See original extended paper: http://arxiv.org/abs/1608.03061.

elements to be used as input for the neural network to build this model (*Openness*, *Extraversion*, *Conscientiousness*, *Neuroticism*, *Joy*, *Sadness*, *Anger* and *Disgust*). The produced model shows a good potential start for further data analysis, with 75 % accuracy in predication based on 57 test cases. Furthermore the available of high-quality, low-cost and adaptable tools provided by the IBM Watson Developer Cloud, provide significant further opportunities to integrate linguistic analysis into this research domain. The model produced from this work provides a number of recommendations for future research to further incorporate emotion and personality-based analysis in user-centered modelling. In particular, expanding the dataset by gathering more data from similar types of interactions, as well as technical queries; annotating and categorising the dataset by gender to investigate the relationship between gender and emotion raised by the user in different computer system statuses; as well as exploring different computer events not only limited to *Idle* and *Failure*, but including more complex events e.g. account hacked, system speed, unexpected error and unsaved data.

References

1. Blamey, B., Crick, T., Oatley, G.: R U :-) or :-(? Character- vs. word-gram feature selection for sentiment classification of OSN Corpora. In: Research & Development in Intelligent Systems XXIX (2012)
2. Blamey, B., Crick, T., Oatley, G.: 'The First Day of Summer': parsing temporal expressions with distributed semantics. In: Research & Development in Intelligent Systems XXX (2013)
3. Fast, L.A., Funder, D.C.: Personality as manifest in word use: correlations with self-report, acquaintance report, and behavior J. Pers. Soc. Psychol. **94**(2), 334–346 (2008)
4. Lambiotte, R., Kosinski, M.: Tracking the digital footprints of personality. Proc. IEEE **102**(12), 1934–1939 (2014)
5. Lazer, D., et al.: Computational social science. Science **323**(5915), 721–723 (2009)
6. Oatley, G., Crick, T.: Changing faces: identifying complex behavioural profiles. In: Proceedings of HAS 2014, LNCS, vol. 8533, pp. 282–293
7. Oatley, G., Crick, T.: Measuring UK crime gangs: a social network problem. Soc. Netw. Anal. Min. **5**(1), 33 (2015)
8. Oatley, G., Crick, T., Bolt, D.: CCTV as a smart sensor network. In: Proceedings of DASC 2015
9. Oatley, G., Crick, T., Mostafa, M.: Digital footprints: envisaging and analysing online behaviour. In: Proceedings of AISB Symposium 2015
10. Pennebaker, J., King, L.: Linguistic styles: language use as an individual difference. J. Pers. Soc. Psychol. **77**(6), 1296–1312 (1999)
11. Pennebaker, J.W., Francis, M.E., Booth, R.J.: Linguistic Inquiry and Word Count. Erlbaum Publishers (2001)
12. Schiaffino, S., Amandi, A.: Intelligent user profiling. In: Artificial Intelligence: An International Perspective, LNCS, vol. 5640, pp. 193–216 (2009)
13. Tausczik, Y.R., Pennebaker, J.W.: The Psychological meaning of words: LIWC and computerized text analysis methods. J. Lang. Soc. Psychol. **29**(1), 24–54 (2010)

An Industrial Application of Data Mining Techniques to Enhance the Effectiveness of On-Line Advertising

Maria Diapouli, Miltos Petridis, Roger Evans and Stelios Kapetanakis

Abstract Nowadays, online behavioural targeting is one of the most popular and profitable business strategies on the display advertising. It is based on data analysis of web user behaviours with the usage of machine learning aiming to optimise web advertising. The objective of this paper is to identify consumers who have no previously observed an advert but are "possible prospects" more likely to purchase an advertisement's product. By identifying prospect customers, online advertisers may be able to optimise campaign performance, maximise their revenue as well as deliver advertisements tailored to a variety of user interests. Our work presents various benchmark machine-learning algorithms and attribute pre-processing techniques in the context of behavioural targeting. The performance of the experiments is evaluated using the key performance metric which is the predicted conversion rate. Our experimental results indicate that the presented data mining framework can significantly identify prospect customers in the vast majority of cases. Our results seem promising, indicating that there is a need for further studies in the area of data mining in online display advertising.

Keywords Knowledge Discovery and Data Mining · Machine Learning · Intelligent Decision Support Systems

M. Diapouli (✉) · M. Petridis · R. Evans · S. Kapetanakis
University of Brighton, Brighton BN2 4GJ, UK
e-mail: M.Diapouli@brighton.ac.uk

M. Petridis
e-mail: M.Petridis@brighton.ac.uk

R. Evans
e-mail: R.P.Evans@brighton.ac.uk
S. Kapetanakis
e-mail: S.Kapetanakis@brighton.ac.uk

M. Bramer and M. Petridis (eds.), *Research and Development in Intelligent Systems XXXIII*, DOI 10.1007/978-3-319-47175-4_30

1 Introduction

According to statistics published by "eMarketer" online advertisers in US have spent more than 1.1 billion US-dollars in 2010 on behavioural targeted advertising a figure which exceeded 2.0 billion in 2014. The estimate represented steady growth rates of about 20 % from 2010 to 2014 [1]. Behavioural targeting and customer prospecting are very promising and challenging aspects in display advertising. Promising since the more information of user behavioural activity exist the better targeted advertisements could be delivered to end users. It is also challenging since display advertising is a rather complex ecosystem which involves multiple interested parties such as end users, advertisers, publishers, and ad platforms. Additionally, the size of data generated and collected from any involved parties is significantly large. Billions of website requests every day trigger millions of advertisements that are displayed on millions of users.

For evaluating the performance of an adverting campaign two of the most well-known key performance indicators are Click-Through Rate (CTR) and conversion rate. According to Lewis and Reiley (2014) [2] the effect of online advertising on sales is not fully associated with CTR. Their collaboration with Yahoo! Research and an eminent retailer had reported that 78 % of the lift in retailer sales was originated from users who had viewed ads but had not clicked them, while only 22 % was attributed to those who had clicked. In our investigation we found that the online advertising campaign had substantial impact on the users who simply viewed the ads. In particular based on thorough analysis we identified that impressions are more strongly correlated to conversions than clicks. Most interestingly, clicks had a very trivial correlation (correlation = 0.00000115) with conversion. These findings suggest that the most meaningful metric for evaluating campaign performance is conversions instead of clicks.

The work we present in this paper handles a challenging area in the online display advertising marketplace, this of customer prospecting. Customer prospecting identifies web users who are likely to purchase a product after an advertisement view. We developed a data mining methodology based on an advertising campaign implemented by an ad network provider. We collected and analysed campaign data that comprised audience demographic information and audience behavioural segments to predict whether a user who had not previously seen an advert is likely to convert. The ultimate goal of this research was to increase an individual advertising campaign performance by augmenting its cost per action (CPA) ratio.

2 Methodology

Customer prospecting is related with predictive modelling in data mining terms. Predictive modelling also referred as supervised data mining, aims to predict the possible future event on the basis of previous historical knowledge [3]. The appropriate selection of data samples is important for effective analysis and prediction based on underlying patterns [4]. In this research decision trees were preferred over the commonly used logistic regression and collaborative filtering classification methods. Decision trees were used since they are more effective in building profiles for the web users who have converted in the past and then predict whether a new web-user is likely to convert. For predictive modelling decision trees seem more powerful and prevalent tools [5] compared to logistic regression and collaborative filtering since:

- Logistic regression could be used to examine the model's exponentials of the coefficients to explore which user attributes affect the likelihood of conversion. In such way we would be able to explore the necessary coefficients but be unable to explore the underlying ruleset which could indicate and predict online "target" users willing to convert.
- Collaborative Filtering was also considered since it has been proven effective in finding prospect customers based on past customer behaviours (training samples) [6]. In such way a successful model would be retrained on a regular basis in order to include recent user activity information. However, in our investigated dataset such information was not available and this research was not able to benefit from "live ad-feeds" including: Ad ids, ad-width, ad-height, visibility time viewed, format, etc.

Therefore, based on the limitations of the dataset and the model target audience, our methodology for selection was based on data-mined patterns for ruleset generation in order to understand and predict successful (or not) online user-conversions.

2.1 The Data

The dataset used in this experiment was retrieved from advertising campaigns deployed by an advertising company. The company was using a no-sql distributed database management system (DDMS) based on Apache Hadoop. Any marketing data was extracted from DDMS and stored in tab separated files (tsv) textual format for further processing and analysis. During a one-week campaign 20 million impressions were displayed to web users, a figure which was increased exponentially over more campaigns and longer campaign times or series of campaigns.

The used dataset comprised three different types of ad-logs, described as: impressions, clicks, and conversions. Any available log data were organised and aggregated based on the user id feature on a row-per-record basis. Each record contained three information types: (i) Behavioural data (columns 1,2,6,7,8,9) (ii) Interest profiles (columns 1,2,4,6,7,8,9,10,11,12) and (iii) Intent profiles (columns 1,2,3,6,7,8,9).

The dataset used in this experiment was gathered as a day-campaign and it consisted of 3,425,119 impressions that were displayed on 3,407,293 users. Among them 8,082 users clicked on the displayed advert (click response rate 0.24 %) and 913 converted (convert response rate 0.03 %). Due to the very low number of response rate both for click and conversion the data were highly skewed. To overcome this limitation of imbalanced data a sampling technique was adopted. In the under-sampling approach data from the majority class were re-moved in order to balance the training set [7]. For our experiments the under-sampling approach was used where cases from the dependent feature conversion rate were randomly eliminated. The size of the data set had 3,407,293 observations. Each observation characterised a web user and was described by 46 independent features and 1 dependent feature. The independent features were both categorical and nominal. These features were related to a user's browsing history (URLs that user has visited in the past). The dependent feature was a nominal one-named conversion rate which illustrated whether a user had made a purchase in the past.

3 The Experiments

3.1 *Modelling*

For this work IBM SPSS Modeler 15.0 was used for all experiments. Different algorithms were assessed in order to benchmark the most appropriate and accurate dataset for classification and prediction. The data set was separated into training and testing set in order to build and evaluate our decision tree model. We assigned 70 % of data for the training set and 30 % for the test set. In the training phase the model was processed by using the training set and then tested in order to evaluate our model's accuracy.

3.2 *Comparing the Performance of the Different Algorithms*

Different decision tree algorithms have been selected for searching patterns in data. This process included deciding which algorithm provided the lowest average classification error. The selected algorithms were: classification and regression tree (C&RT), Chi-squared automatic interaction detector (CHAID) and C5.0.

C5.0 shows a higher likelihood (90.63 %) to predict the event for someone converting on an advertisement compared to the other baselines (88.25 % for C&RT, and 89.34 % for CHAID). In particular, C5.0 is showing 99.02 % sensitivity (the portion of users that were correctly predicted to convert) and 90.62 % specificity (the portion of users that did not convert and were successfully predicted) which accounts for the overall accuracy of 90.63 %. The sensitivity and specificity measures are used to ascertain the model validity and accuracy [8].

Table 1 Model accuracy

Partition	Training set	Percentage training set (%)	Testing set	Percentage testing set (%)
Correct	1,482	93.09	3,088,032	90.67
Wrong	110	6.91	317,669	9.33
Total	1,592		3,405,701	

Table 2 Training set

	Predicted value: 0	Predicted value: 1
Actual value: 0	1,032	103
Actual value: 1	7	450

Table 3 Testing set

	Predicted value: 0	Predicted value: 1
Actual value: 0	3,087,578	317,667
Actual value: 1	2	454

3.3 Comparing the Performance of Different Data Sets

Typically, the performance of machine learning algorithms is evaluated using predictive accuracy. The evaluation and interpretation of the mined patterns in terms of reliability and accuracy of the derived rules have taken place in the evaluation phase.

We performed our experiments using 10-fold cross validation. The original dataset was randomly divided into ten (10) subsets. Each time, one of the 10 subsets was used as the test set and the other 9 subsets were combined to form the training set. For the conversion field, there were two classes, the positive class, assigned as 1, that comprised "converted" users and the negative class consisted of "no converters", assigned as 0.

Table 1 illustrates the correct and wrong prediction of our model. Tables 2 and 3 are coincidence matrices that show the pattern of matches between each predicted field and its conversion/target field. The rows defined by actual values and columns defined by predicted values, with the number of records having that pattern in each cell.

4 Conclusions

In this paper we demonstrated data mining as an effective tool for direct marketing which can improve online marketing campaigns. The majority of the existing research in this area so far focuses on computational and theoretical aspects of direct marketing though little efforts have been put on technological aspects of applying data mining in the process of direct marketing. The complexity of data mining models makes it difficult for marketers to use it, hence we outlined a simplified framework to guide marketers and managers in making use of data mining methods and focus their advertising and promotion on those categories of people in order to reduce time and costs. We explained the steps and tasks that are carried out at each stage of the data mining framework and showed some examples of the type of predictive efficiency that can be achieved through the use of the proposed approach. This has shown that substantial gains can be achieved by adopting this pragmatic and exploratory approach to predict user behaviour in on-line advertising.

The results obtained so far, are very promising and encourage us to continue experimentation with more sophisticated models or other algorithms to further improve the performance of the system. In particular, it seems sensible to experiment with the following settings in future work:

- introducing the temporal dimension to our model, in particular, to apply time series analysis techniques to build the model
- combining the model with content-based approach
- additional category-based attributes specifying the times spent on each of the categories, with possible division into work-days and week-days, for example a different choice of categories.

As future work, we also would like to incorporate more user and publisher information obtained from third party media providers into data hierarchies to improve model prediction.

Acknowledgments This work was funded by Innovate UK (formerly known as the Technology Strategy Board).

References

1. Silverman, D.: Interactive Advertising Board (IAB) internet advertising revenue commissioned report (Updated June 2016). http://www.iab.com/insights/iab-internet-advertising-revenue-report-conducted-by-pricewaterhousecoopers-pwc-2/. Accessed May 2016
2. Lewis, R.A., Reiley, D.H.: Online ads and offline sales: measuring the effect of retail advertising via a controlled experiment on Yahoo!. Quant. Mark. Econ. **12**(3), 235–266 (2014)
3. Chapman, P., Clinton, J., Kerber, R., Khabaza T., Reinartz, T., Shearer, C., Wirth, R.: CRISP-DM 1.0 Step-by-step data mining guide. Crisp DM Consortium (Updated 2010) (1999). https://www.the-modeling-agency.com/crisp-dm.pdf. Accessed Apr 2016

4. Kim, Y.: Toward a successful CRM: variable selection. Sampl. Ensemble Decis. Support Syst. **41**(2), 542–553 (2006)
5. Reddy, C.S., Vasu, V., Kumara Swamy Achari, B.: Effective decision tree learning. Int. J. Comput. Appl. **82**(9) (2013)
6. Anastasakos, T., Hillard, D., Kshetramade, S., Raghavan, H.: A collaborative filtering approach to ad recommendation using the query-ad click graph. In: ACM Conference on Information and Knowledge Management, 1927–1930 (2009)
7. Kubat, M., Matwin, S.: Addressing the curse of imbalanced training sets: one-sided selection. In: International Conference on Machine Learning, pp. 179–186 (1997)
8. Han, J., Kamber, M., Kaufmann, M.: Data Mining: Concepts and Techniques, 2nd edn. (2006)

Printed in the United States
By Bookmasters